# Nanotechnology and Global Sustainability

# PERSPECTIVES IN NANOTECHNOLOGY

## Series Editor
Gabor L. Hornyak

# Nanotechnology and Global Sustainability

Edited by
## Donald Maclurcan
## Natalia Radywyl

CRC Press
Taylor & Francis Group
Boca Raton   London   New York

CRC Press is an imprint of the
Taylor & Francis Group, an **informa** business

The photos across the top of the cover were courtesy of Dr. Richard Wuhrer from the Microstructural Analysis Unit of the University of Technology, Sydney.

CRC Press
Taylor & Francis Group
6000 Broken Sound Parkway NW, Suite 300
Boca Raton, FL 33487-2742

© 2012 by Taylor & Francis Group, LLC
CRC Press is an imprint of Taylor & Francis Group, an Informa business

No claim to original U.S. Government works

Printed in the United States of America on acid-free paper
Version Date: 20120229

International Standard Book Number: 978-1-4398-5576-8 (Hardback)

### Library of Congress Cataloging-in-Publication Data

Nanotechnology and global sustainability / editors, Donald Maclurcan, Natalia Radywyl.
    p. cm. -- (Perspectives in nanotechnology)
    Includes bibliographical references and index.
    ISBN 978-1-4398-5576-8 (hardback)
    1. Nanotechnology. 2. Sustainable engineering. I. Maclurcan, Donald. II. Radywyl, Natalia.

T174.7.N3723 2011
620'.5--dc23
                         2011043469

**Visit the Taylor & Francis Web site at**
**http://www.taylorandfrancis.com**

**and the CRC Press Web site at**
**http://www.crcpress.com**

# Contents

## Section I   Limits

## Section II   Capacity

## Section III   Appropriateness

## Section IV   Governance

# List of Figures

# List of Tables

# Foreword

**Vijoleta Braach-Maksvytis**

When Donnie Maclurcan approached me in 2004 to help guide some of his groundbreaking PhD research on the societal implications of nanotechnology, I was delighted to discover a like-minded colleague who shared such a consideration. As his PhD conclusions sharpened, Donnie was already beginning to collate the work of others into a volume that would take his dissertation findings about nanotechnology and global inequity one step further. With a steadfastness of vision, unswerving integrity, and belief in the better characteristics of us as global peoples, this book was created.

Yet this work has much deeper foundations. In the late 1950s, the field of nanotechnology was foreshowed with Nobel Laureate physicist Richard Feynman's dream of taking advantage of a "new world" available at the nanoscale—the level of atoms and small molecules.

What is it about nanoscience that has created so much attention? It has opened a world of new materials and properties simply by the reduced dimensions of familiar materials on the nanoscale. This is because of three main characteristics: The nanoscale is the scale of nature's building blocks, such as DNA and proteins; at this scale, materials have more surface than volume, increasing the importance of surface-interaction properties; and, at nanoscale, the effects of quantum physics begin to dominate over classical physics. Take, for example, the simple interaction of light with gold metal. Light on a golden wedding ring tells us that gold is gold colored. Light interacting with a 20 nanometer-sized nanoparticle of gold tells us that gold has a deep red color—not a trace of gold in sight!

It was not until the early 1990s that Feynman's dream became a frontier science and, even then, it needed the advent of analytical tools that were capable of shedding "light" on the nanoscale before it could really take off.

The field began to grow in the late 1990s with an aspiration and approach that, in part, contrasted sharply with previous emerging technology areas such as nuclear power, stem cell research, and genetic engineering. The usual path of scientific research is driven by curiosity or funding, and typically develops in isolation from societal frameworks. Scientists are trained and respected for their abilities within a particular field, and they dedicate their lives to the pursuit of knowledge. Most of science methodology or insight remains a mystery to the general population. Over the past decade or so, attempts have been made to connect science to other areas such as business and policy, but scientists are still being trained without appropriate consideration for the social context or ramifications of their work. When new breakthroughs are made, such as the examples given above, they descend upon the population as existing technology—and only then is consideration given to the debates about human and environmental safety, ethics, legalities, long-term effects, unexpected up- and downsides, and commercial versus global needs.

Yet, at the turn of the millennium, there was a different mood among those who took a more public lead in the area of nanotechnology, marking the first of three distinguishing characteristics for the field with respect to science and society. Resulting from preceding experiences, some recognition was given to issues relating new science to societal impacts. Many of us also thought that perhaps this time we could learn from history and avoid repeating the same mistakes. Hence, safety and societal implications were fostered at the same time as the fledgling scientific research expanded. In Australia, for example, the philosophy school at Charles Sturt University set up a nanotechnology investigative program in the early 2000s. Similarly, as Cornell University established its major nanotechnology research center, it included projects looking at the interaction of nanomaterials with biological materials to foreshadow safety and toxicity issues. This provision of thought and resources to the possible future ramifications of technologies simultaneous to their emergence had never been done before.

However, connecting science and society more broadly was extremely difficult to put into practice among the members of the nanotechnology community itself. In the early 2000s, I convened a meeting of nanoscientists to discuss the societal implications of nanotechnology. "What has that got to do with us?" came the response because they felt it was their job to only create the knowledge; what happened after that was not their business and it was not that for which they had been trained. I followed up with a nanotechnology lecture to the Australian Academy of Sciences, ending my presentation by raising the possibility that perhaps this new field of scientific application could consider a change in the usual approach to research. I suggested that by collaborating with disciplines other than science, we could seize a unique and historic opportunity to directly address the urgent issues currently facing the world. However, for some this was an unwelcome view and there was a sense of indignation at the suggestion that people outside the field of science might determine areas scientists could or could not research.

The second characteristic that marked this new field in the early 2000s was a deliberately forged, immediate international connectivity. The usual path for new areas of research is usually decided by individual scientists, schools, departments, research organizations, or funding agencies. At the Australian Nanotechnology Network in Australia, however, we linked our nanotechnology activities with key policy groups around the world and created cross-country collaborations (such as with the EU). This was both possible and particularly beneficial because of the enormous ramifications that nanotechnology holds across almost all applications. We were far from alone in recognizing such possibilities.

The third unusual aspect influencing nanotechnology's development was the funding of collaborations between developing and developed countries. This funding arrangement created entry points for new researchers at the start of an emerging field, undermining long-held conservative views that developing countries were not sufficiently educated to be at the forefront of science, let alone capable of making early and significant contributions. This challenge to traditional perspectives about scientific development subsequently began to drive a realization that the huge issues facing the world are the responsibility of us all as global citizens. This new awareness was spurred on by a range of international developments, such as the UN Millennium Development Goals and the "living" philanthropy exemplified by people such as Bill and Melinda Gates. Irrespective of one's views on their outcomes, these happenings helped substantiate, for many, a sense of interconnectivity across the planet and a renewed desire to ensure, for all, the provision of basic needs in areas such as health care, food, and water. Nanotechnology has weighed in on these debates, offering new avenues for consideration in our approaches to vast and long-standing problems such as those relating to water purification, malaria detection, and targeted drug delivery.

Have these approaches worked and are they truly beneficial? Has the field succeeded in opening up the closed world of science research to other disciplines and players, enough to actually deeply inform the work being carried out? Although some progress has undoubtedly been made when compared to previous frontier fields, the historical approaches to science research still hold fast and, for me, the progress is not nearly ambitious or rapid enough in the "how" of nanotechnology.

As the authors in this book explain, despite the best of intentions, technology remains fundamentally constrained by the market-defined framework in which it presently operates.

But the ground is indeed shifting. The cracks have appeared, and the world has turned with the tiniest shift in its axis of rotation—a shift that is irreversible. This book is a product of that shift.

There are some truly unique connections made in this book—and timely ones, too. Donnie Maclurcan and Natalia Radywyl highlight and exemplify the kind of systems approach required to shape positive and sustainable paths we can jointly pursue. They have not held back with their boldness of vision. In light of the truly engaging collection of questions and viewpoints they have assembled in this book, their boldness is well justified. In piecing together such a range of views, this book differs from previous multi-authored efforts by collectively proffering a glimpse, however small, of how that vision could play out. Its fundamental significance and contribution to the field of nanotechnology and, more broadly, our shared futures are undoubtedly important and unique.

# Editors

**Donnie Maclurcan** is an honorary research fellow with the Institute for Nanoscale Technology at the University of Technology, Sydney, and cofounder of the Post Growth Institute, an international network inspiring and equipping people to explore paths to global prosperity that do not rely on economic growth. Incorporating fieldwork in Thailand and Australia, his PhD from the University of Technology, Sydney, was an original and comprehensive assessment of nanotechnology's possible consequences for global inequity. This work developed into the book *Nanotechnology and Global Equality* (2011, Pan Stanford) and research that has been translated into more than twenty languages. Dr. Maclurcan is a past member of the Global Nanotechnology Taskforce on Implications and Policy, was interim chair of the International Free and Open Source Software Foundation, and was appointed a fellow of the Royal Society of the Arts in 2010. In 2006, he founded Project Australia, a community organization helping people launch not-for-profit initiatives, and later launched Australia's first professional speakers' bureau for social innovators: uThinc.

**Natalia Radywyl** is a social researcher and honorary research fellow in the School of Culture and Communication at the University of Melbourne. As someone working across multiple disciplines, sectors, and projects, she advocates the role of critically informed design, seeking a common language to design for positive change. With interests in user-centered and spatial design, Dr. Radywyl specializes in ethnographic approaches to understanding user experience and facilitating public engagement. Recent research includes her PhD dissertation, *Moving Images: The Museum and a Politics of Movement* (2008), which employed ethnography, social theory, and cultural policy analysis to appraise visitor experience and codify the spatial ecology in the Australian Centre for the Moving Image, Melbourne; visitor engagement with urban screens and public space (2009–2010); and viewer engagement with social media in the context of Australian public broadcasting (2011). As a social researcher, she also has a strong grounding in policy and is a government consultant on public behavior change and agenda setting.

Dr. Radywyl's practice and research background reflect her interdisciplinary inclinations. Having completed projects spanning languages, creative arts, music, and interactive media, and having lectured in new media, media policy, theories of consumption, and urban culture, she continues to publish information about the role of public institutions such as broadcasters, museums, and common urban spaces in mediating meaningful experiences in everyday life.

# Contributors

**Sunandan Baruah** completed his master's degree in microelectronics and his PhD in nanotechnology at the Asian Institute of Technology, Thailand. Dr. Baruah's research investigates the use of inexpensive wet-chemical methods to fabricate innovative materials and device components. His research areas include the synthesis of nanoparticles (such as ZnO, ZnS, $Zn_2SnO_4$, Au, Ag) as building blocks for new materials (nanowires, nanorods, microwires) and using nanoparticles for practical applications like photocatalysis, solar cells, air and water filters, and gas sensors. He is working with biotechnologists to develop fluorescent quantum dots into biolabels and with agriculturists for degrading pesticides through photocatalysis using metal oxide nanoparticles. Another area of interest is the toxic effect of engineered nanoparticles on the ecosystem. His research publications include nineteen papers in peer-reviewed international journals, eighteen conference papers, and four book chapters.

**Deb Bennett-Woods** is a professor and chair of the Department of Health Care Ethics in the Rueckert-Hartman College for Health Professions, and the director of the Center for Ethics and Leadership in the Health Professions at Regis University in Denver, Colorado. In addition, she is a board-certified fellow of the American College of Health Care Executives.

Dr. Bennett-Woods currently teaches courses with an applied focus in health care ethics across all graduate and undergraduate programs in the Rueckert-Hartman College for Health Professions. Her book, *Nanotechnology: Ethics and Society* (CRC Press, 2008), reflects her interest in emerging technologies. She is published on a variety of topics in ethics and leadership and is a frequent public speaker, facilitator, and trainer. In her role with the Center for Ethics and Leadership in the Health Professions she serves as an ethics consultant to local health care facilities.

**Diana M. Bowman** is an assistant professor in the Risk Science Center and the Department of Health Management and Policy in the School of Public Health at the University of Michigan, and a visiting scholar in the Department of International and European Law, KU Leuven (Belgium). With a background in both science and law (BSc in physiology, LLB and PhD in law), Dr. Bowman's research has primarily focused on legal, regulatory, and public health policy dimensions relating to new technologies. In 2007 she coauthored, with Karinne Ludlow and Graeme Hodge, an Australian government-commissioned report analyzing the readiness of Australia's regulatory frameworks for the introduction of nanoproducts. She has also co-edited three books looking at regulatory issues relating to nanotechnologies: *New*

*Global Frontiers in Regulation: The Age of Nanotechnology* (2007, with Hodge and Ludlow), *Nanotechnology Risk Management: Perspectives and Progress* (2010, with Matthew Hull), and the *International Handbook on Regulating Nanotechnologies* (2010, with Hodge and Andrew Maynard). In 2010, Dr. Bowman accepted a position as a member of the Australian government's National Enabling Technology Strategy's Expert Forum.

**Vijoleta Braach-Maksvytis** is a pioneer in establishing nanotechnology as a frontier field in Australia and internationally. She was on the editorial board for the first comprehensive volume on nanotechnology (*Dekker Encyclopedia of Nanoscience and Nanotechnology*, 2004) and initiated Australia's first National Nanotechnology Network. This network drew together science, industry, government, investors, and social sectors as well as international linkages and led to the establishment of the Australian Federal Government Office of Nanotechnology and interdisciplinary funding initiatives. Dr. Braach-Maksvytis was the inaugural director of Commonwealth Scientific Industrial Research Organisation (CSIRO) Nanotechnology. She received her PhD in biophysics from the University of Sydney in 1992 and holds more than 25 patents in nanotechnology. Her previous roles include CSIRO executive for Science Strategy, director of CSIRO Global Development, principal for Global Research Alliance, head of office of the Chief Scientist of Australia, deputy vice chancellor of Innovation and Development at the University of Melbourne, member of the UNESCO Committee, and advisor to the United Nations Development Programme (UNDP) and developing countries. Dr. Braach-Maksvytis is currently the director of her own innovation strategy company and a non-executive company director. Her career in innovation spans research, business, science, policy, commercialization, global development, community sectors, indigenous partnerships, the arts, philosophy, leadership development, and mentoring. She also sits on a number of not-for-profit boards such as those of the Institute for the Economics of Peace, the Intellectual Property Research Institute of Australia, and UNICEF.

**Nupur Chowdhury** has an LLM in public international law from the Amsterdam Law School at the University of Amsterdam (2007), an LLB from the University of Delhi (2004), and a BA in political science from Lady Shri Ram College (2001). She specializes in environmental law and has researched extensively on the issues of natural resources law, intellectual property rights, and biodiversity, biotechnology, and nanotechnology regulation.

She has lectured at TERI University, New Delhi, on biodiversity, wildlife, forestry law, and international environmental law, and after an associate fellowship with the Science and Technology Area of the Resources and Global Security Division at TERI, she joined the Department of Legal and Economic Governance Studies as a PhD candidate in 2009. She is a member of the Centre for International Sustainable Development Law Research Group within the McGill Law Faculty, the Bar Council of Delhi, the Netherlands Institute of Government, and the International Union for Conservation of Nature Commission on Environmental Law. She has been a member of the editorial team of International Law in Domestic Courts (ILDC) and is currently coauthoring the American Society of International Law (ASIL) Reports on International Organization Report on the World Intellectual Property Organization (WIPO).

**Joydeep Dutta** is vice president (Academic Affairs) of the Asian Institute of Technology (AIT), director of the Center of Excellence in Nanotechnology, and professor of microelectronics at AIT, Bangkok, Thailand. His broad research interests include nanomaterials in nanotechnology and the self-organization and application of nanoparticles. Dr. Dutta is a fellow of the Institute of Nanotechnology and the Society of Nanoscience and Nanotechnology; a senior member of the Institute of Electrical and Electronics Engineers, United States; a founding member of the Thailand Nanotechnology Society; and a member of several other significant professional bodies. His research accomplishments include the award-winning *Fundamentals of Nanotechnology* (American Library Association's Choice Award for Outstanding Academic Title, 2010), two additional academic textbooks, more than 170 research publications, eight chapters in the Science and Technology reference series, more than sixty invited and keynote lectures, and three patents.

**Ron Eglash** has a BS in cybernetics, an MS in systems engineering, and a PhD in the history of consciousness, all from the University of California. Supported by a Fulbright postdoctoral fellowship, he undertook field research on African ethnomathematics and subsequently authored *African Fractals: Modern Computing and Indigenous Design* (Rutgers University Press, 1999), which also recently appeared in his TED (Technology, Entertainment, Design) talk. Dr. Eglash is a professor of science and technology studies at Rensselaer Polytechnic Institute, where he teaches design of educational technologies and graduate seminars in social studies of science and technology. His "Culturally Situated Design Tools" software, offering math and computing education from indigenous and vernacular arts, is available for free at www.csdt.rpi.edu. Recently funded work includes his National Science Foundation "Triple Helix" project, which brings together graduate fellows in science and engineering with local community activists and K–12 educators to seek new approaches for putting science and innovation in the service of underserved populations.

**David J. Grimshaw** is the head of International Programme: New Technologies at the Schumacher Centre for Technology and Development, Practical Action, and a senior research fellow with the U.K. Department for International Development. He is a visiting professor in Information and Communication Technologies for Development (ICT4D) at Royal Holloway (University of

London) and Development and Technology (Coventry University), and has held visiting fellowships across Australia, the United States, Russia, and Malaysia. His earlier academic career was with Warwick Business School, University of Warwick; University of Leeds; and Cranfield School of Management. Dr. Grimshaw was also a member of the expert advisory panel on ICT for Rural Livelihoods Programme, Overseas Development Institute, and the World Bank (2007), and chair of demos at the ICTD 2010, London.

Dr. Grimshaw's research interests explore the use of geographical knowledge by business, knowledge exploitation, and e-business. Recently completed projects include *Connecting the First Mile*; *Podcasting in the Andes*; *Nano-Dialogues in Zimbabwe, Peru, and Nepal*; and *Delivering Public Value from New Technologies* in collaboration with the Universities of Sussex, Lancaster, and Durham, supported by the Economic and Social Research Council. He is the author of *Bringing Geographical Information Systems into Business* (2nd ed., 1999, John Wiley), co-editor of *IT in Business: A Manager's Casebook* (1999) and *Strengthening Rural Livelihoods: The Impact of Information and Communication Technologies in Asia* (2011). Dr. Grimshaw has also been a consultant for many companies on strategic information systems and geographic information systems, has helped start the charity Science for Humanity, has worked with SciDev.Net on its new technologies gateway, and is on the steering group of MATTER, an "action-tank" devoted to making new technologies work for us all.

**David J. Hess** is a professor of science and technology studies at Rensselaer Polytechnic Institute. His research investigates the social science dimensions of environmental issues and policy with a focus on the role of social movements in science and technology. His most recent books are *Alternative Pathways in Science and Industry* (2007, MIT Press) and *Localist Movements in a Global Economy* (2009, MIT Press). Dr. Hess has also published on the environmental, health, and safety implications of nanotechnology in *Science as Culture*.

**Graeme A. Hodge** is a professor of law and director of the Monash Centre for Regulatory Studies, Monash University, Melbourne, Australia. He is a leading policy analyst on regulation, privatization, and public–private partnerships. Dr. Hodge is an internationally recognized scholar, having published eleven books and 100 papers on management, social and economic policy, public administration, and regulation. His most recent book publications have been the *International Handbook on Regulating Nanotechnologies* (2010) with Diana Bowman and Andrew Maynard, and the *International Handbook on Public–Private Partnerships* (2010) with Carsten Greve and Anthony Boardman. Dr. Hodge has worked with the Organisation for Economic Cooperation and Development (OECD), the European Commission, the United Nations, and the Asian Development Bank, as well as serving as a special adviser to several Australian parliamentary committees and inquiries. He has acted as a consultant on governance matters in Australasia, Europe, Indonesia, the Philippines, and China and is a regular media commentator.

**Louis Hornyak** has been involved with nanotechnology for more than twenty years. From 1978 to 1990, he worked in the materials and process engineering group for an aerospace company in San Diego, California. His responsibilities included research and development of anodized and polymer coatings and advanced aerospace materials for adhesives and insulation. His work contributed to the Atlas launch vehicle, MD-11, and the space shuttle programs. From 1997 to 2002, Dr. Hornyak worked at the National Renewable Energy Laboratory in Golden, Colorado, where he investigated ways to store hydrogen gas on carbon nanotubes. Later, Dr. Hornyak founded and directed the Colorado Nanotechnology Initiative, a group dedicated to bringing awareness of the potential and promise of nanotechnology to Colorado from 2003 to 2005. He is currently interim director of the Center for Learning Innovation and Quality, which oversees academic and research quality at the Asian Institute of Technology, Bangkok, Thailand, and has developed curricula and mentoring graduate students at the Center of Excellence in Nanotechnology at AIT. He has also coauthored *Introduction to Nanoscience* (2008, CRC Press) and *Fundamentals of Nanotechnology* (2008, CRC Press), and is editor of the *Perspectives in Nanotechnology Series* (CRC Press), which examines the societal issues of nanotechnology.

**Stephanie Howard** has worked with civil society organizations in Europe, North America, and New Zealand on new technologies. She is currently projects director of the Sustainability Council of New Zealand.

**Anna Lamprou** is a PhD student in science and technology studies at Rensselaer. With a background in chemistry, she is conducting dissertation research on comparative policies for nanotechnology in the United States and European Union.

**Kristen Lyons** has been engaged in social research, advocacy, and education on topics related to food, agriculture, and the environment, as well as new technologies and social justice, for the past fifteen years. She currently teaches sociology of the environment and development at the University of Queensland in Brisbane, Australia. In her current research, Dr. Lyons is examining the social and environmental implications of emerging agri-food nanotechnologies, as well as questions related to governance and public engagement related to new technologies. Dr. Lyons is also undertaking research related to new technologies, agricultural change, and land grabs in Africa. She is a supporter of the Friends of the Earth Australia Nanotechnology Campaign and has participated in and critiqued a number of attempts at public engagement related to nanotechnologies in Australia.

**Usman Mushtaq** is undertaking a master's in civil engineering at Queen's University, Canada. His current research investigates the development of socially just design cycles, processes, and methodologies. Broadly, he is interested in applying critical, Marxist, and anarchist social theory to engineering practice.

Mr. Mushtaq worked a researcher at the NanoElectronics and Devices Laboratory at the University of Pittsburgh for two years. During his time there, he was involved in a nanowire sensor project as well as a nanomaterials fuel cell project. He is also an editor for both the *International Journal for Service Learning in Engineering*, which seeks to nurture service learning in engineering as a distinct body of knowledge, and the *International Journal of Engineering, Social Justice, and Peace*, which seeks to expand on engineering practices and education that enhance gender, racial, class, and cultural equity and are democratic, nonoppressive, and nonviolent. In his free time, he is involved with organizing migrant justice.

**Joshua M. Pearce** is currently a cross-appointed associate professor in the Department of Materials Science and Engineering and the Department of Electrical and Computer Engineering at Michigan Technological University. He received his PhD in materials engineering from Pennsylvania State University, developing low-cost mixed-phase nanocrystalline and amorphous silicon solar photovoltaic materials and devices, before being appointed assistant professor of physics by Clarion University of Pennsylvania to develop both a nanotechnology program and the first sustainability program in Pennsylvania's state system of higher education. Dr. Pearce formerly held a post at the Department of Mechanical and Materials Engineering and the School of Environmental Studies at Queen's University, Canada, where he coordinated the Queen's Applied Sustainability Research Group (research available, open source, at http://www.appropedia.org/QAS) and designed the new Queen's graduate program of applied sustainability.

Dr. Pearce's research concentrates on the use of open source appropriate technology to find collaborative approaches to challenges in sustainability, poverty reduction, and international development. His work spans the areas of electronic device physics, nanotechnology, and materials engineering of solar photovoltaic cells, and also includes applied sustainability, energy policy, and engineering service learning. Dr. Pearce has more than fifty peer-reviewed publications and has won the Golden Apple Award for engineering education excellence. He is also an editor for the *International Journal for Service Learning in Engineering*, which seeks to nurture service learning in engineering as a distinct body of knowledge.

**Gyorgy Scrinis** teaches in the graduate environment program at the University of Melbourne. His research focuses on the science, technology and politics of food production and consumption, including genetically-modified foods, nanofoods, nutrition science, and functional foods. He is currently completing a book project on the topic of nutritionism, or nutritional reductionism.

**Nidhi Srivastava** is a lawyer by training and works as an associate fellow at the Centre for Resources and Environmental Governance at The Energy and Resources Institute (TERI), New Delhi, India. She holds a bachelors degree from the University of Delhi and a masters degree from Katholieke Universiteit, Leuven. She has worked on the regulation of new and emerging technologies, especially in the context of nanotechnology. Her work has been a part of larger multi-disciplinary research on capabilities and governance in nanotechnology. With her background in environmental law, she has focused on regulatory aspects of nanotechnology, including risk regulation. In addition, she has worked on issues of environmental protection and resource development, including energy resources.

**Kathy Jo Wetter** works as a researcher for ETC Group, a Canadian civil society organization that monitors corporate concentration and emerging technologies. Dr. Wetter holds a PhD in art history from the University of North Carolina at Chapel Hill.

**James Whelan** is the research director for the Public Service Research Program at the Centre for Policy Development in Sydney. Dr. James' research history includes lecturing and leading social science research programs with several universities and research institutions on topics that include deliberative governance, environmental politics, and social movements. He has published widely and contributed to national and international conferences. Dr. James has also been a campaigner and strategist with social and environmental justice organizations and networks. He is founder and director of the Change Agency, a not-for-profit education and action research initiative that supports progressive social movements in the Asia–Pacific region.

**Fern Wickson** is a cross-disciplinary researcher at the GenØk Centre for Biosafety in Tromsø, Norway. Her primary research interests include environmental philosophy, ecotoxicology, the politics of risk and uncertainty, and the governance of emerging technologies. With a long-standing interest in the theory and practice of cross-disciplinary research, Dr. Wickson undertook a bachelor of arts/bachelor of science double degree at Australian National University and attained a first-class honors degree in environmental politics from the University of Tasmania. She completed a cross-disciplinary PhD with the Schools of Biological Sciences and Science, Technology, and Society at the University of Wollongong, with a thesis on Australia's environmental regulation of genetically modified crops. She then conducted interdisciplinary investigations into the social and ethical aspects of nanotechnology as a postdoctoral researcher at the Centre for the Study of the Sciences and the Humanities, University of Bergen. This work included examination of the different narratives told about the relationship between nanotechnology and nature, as well as critical perspectives on the discourse and practice of public engagement as a governance tool. Dr. Wickson has published in both social and natural science journals about the environmental governance of emerging technologies. She is also co-editor of the recent book *Nano Meets Macro: Social Perspectives on Nanoscale Sciences and Technologies* (2010, Pan Stanford).

# Acronyms

| | |
|---|---|
| AAO | Anodic aluminum oxide |
| AFM | Atomic force microscope |
| AGRA | Alliance for a Green Revolution in Africa |
| AIT | Asian Institute of Technology |
| BIOTEC | National Centre for Genetic Engineering and Biotechnology (Thailand) |
| BSD | Berkeley Software Distribution |
| $C_{60}$ | Carbon 60 |
| CBD | Convention on Biological Diversity (UN) |
| CDRI | Central Drug Research Institute (India) |
| CDSCO | Central Drugs Standards Control Organization |
| CFCs | Chlorofluorocarbons |
| CFTRI | Central Food Technology Research Institute (India) |
| CIS | Copper indium (di)selenide |
| CLARA | Latin American Cooperation of Advanced Networks |
| CNT | Carbon nanotube |
| CoEN | Center of Excellence in Nanotechnology (Thailand) |
| CPCB | Central Pollution Control Board (India) |
| CSDT | Culturally Situated Design Tool |
| CSIR | Council for Scientific and Industrial Research (India) |
| CdTe | Cadmium telluride |
| CTA | Constructive technology assessment |
| DBT | Department of Biotechnology (India) |
| DCGI | Drug Controller General of India |
| DOD | Department of Defense (U.S.) |
| DIWO | Do-It-With-Others |
| DIY | Do-It-Yourself |
| DSIR | Department of Scientific and Industrial Research (India) |
| DSSC | Dye-sensitized solar cells |
| DST | Department of Science and Technology (India) |
| EHS | Environment, Health, and Safety |
| EPA | Environment Protection Act (India) |
| EPFL | Swiss Federal Institute of Technology |
| ETC Group | Action Group on Erosion, Technology, and Concentration |
| EULA | End User License Agreement |
| FAO | Food and Agriculture Organization of the United Nations |
| FoE | Friends of the Earth |
| FoEA | Friends of the Earth Australia |
| FONAI | Focus Nanotechnology Africa Inc. |
| FSSA | Food Safety and Standards Authority (India) |

| GE | Genetic engineering/genetically engineered |
|---|---|
| GEAC | Genetic Engineering Approval Committee (India) |
| GHG | Greenhouse gas |
| GM | Genetically modified |
| GMO | Genetically modified organism |
| GRAS | Generally recognized as safe |
| IAASTD | International Assessment of Agricultural Knowledge, Science and Technology for Development |
| IBSA | India–Brazil–South Africa Dialogue Forum |
| ICMR | Indian Council of Medical Research |
| ICPC | International Cooperation Partner Countries |
| ICT | Information and Communications Technology |
| IFCS | International Forum on Chemical Safety |
| IITR | Indian Institute of Toxicology Research |
| IP | Intellectual property |
| ISO | International Organization for Standardization |
| IT | Internet technology |
| L-B-L | Layer-by-layer |
| MCM | Multicriteria mapping |
| MIT | Massachusetts Institute of Technology |
| MoHFW | Ministry of Health and Family Welfare (India) |
| MTEC | National Metal and Materials Technology Centre (Thailand) |
| NanoAfNet | Nanosciences Africa Network |
| NANOTEC | National Nanotechnology Center (Thailand) |
| NECTEC | National Electronics and Computer Technology Centre (Thailand) |
| NEPA | National Environmental Protection Agency (India) |
| NGO | Nongovernment organization |
| NIPER | National Institute of Pharmaceutical Education and Research (India) |
| NNI | National Nanotechnology Initiative (U.S.) |
| NREL | National Renewable Energy Lab (U.S.) |
| NSDTA | National Science and Technology Development Agency (Thailand) |
| NSTI | Nano Science and Technology Initiative (India) |
| ODI | Overseas Development Institute (U.K.) |
| OECD | Organisation for Economic Cooperation and Development |
| OSAT | Open source appropriate technology |
| OS Nano | Open source nanotechnology |
| P2P | Peer-to-peer |
| PCAST | President's Council of Advisors on Science and Technology (U.S.) |
| PEN | Project on Emerging Nanotechnologies (U.S.) |
| PMNT | Poor man's nanotechnology |
| R&D | Research and development |

| | |
|---|---|
| REACH | Registration, Evaluation, Authorisation and Restriction of Chemicals (EU) |
| RS-RAE | Royal Society and Royal Academy of Engineering (U.K.) |
| RTTA | Real-time technology assessment |
| SAICM | Strategic Approach to International Chemicals Management |
| SEM | Scanning electron microscope |
| SPCB | State Pollution Control Board (India) |
| STM | Scanning tunneling microscope |
| SWCNT | Single-wall carbon nanotube |
| TB | Tuberculosis |
| TEM | Transmission electron microscope |
| TFPP | Thin Film Partnership Program |
| UN | United Nations |
| UNICEF | United Nations Children's Fund |
| USPTO | U.S. Patent and Trademark Office |
| UV | Ultraviolet |
| WHO | World Health Organization |
| WIPO | World Intellectual Property Organization |
| WPMN | Working Party on Manufactured Nanomaterials (OECD) |
| WPN | Working Party on Nanotechnologies (OECD) |
| WTO | World Trade Organization |
| ZnO | Zinc oxide |

# Section I

# Limits

# 1

## Nanotechnology and Limits to Growth

Donnie Maclurcan and Natalia Radywyl

**CONTENTS**

### 1.1 Introduction

A new era of human collectivity is emerging. Born of desire, fear, sensibility, and ingenuity, it rides on the back of past struggles and our increasing capacity to share information across the globe.[1] The individualistic milieu that epitomized neoliberal structures of the previous four decades is beginning to change form. A common, underlying goal is progressively connecting us to one another: a desire to better the present as a means to secure viable futures on this planet. Rising climate awareness, for example, is being translated into practical community action via movements such as transition towns, collaborative consumption, and slow food. This marks a palpable shift in commitment toward what we view as sustainability: addressing the needs of the present without compromising the ability of future generations to also meet their own needs (Brundtland 1987). Nanotechnology is one such significant site where scientific innovation, new forms of collaboration, and new thinking about sustainability could indeed coalesce to pave the way to more equitable futures for individuals and communities around the world.

However, while the hope of a new collectivity is inspiring, we believe that to understand this book's aims it is crucial to recognize that current big-banner issues—such as climate change—are not the root causes of our present global predicament. Indeed, these issues are a symptom of a much more vicious, seemingly untouchable malady: our addiction to economic, consumption, and population growth in a world of finite resources—or, as Serge Latouche (2010) explains, the colonization of our imaginations by the growth paradigm.

Arguments placing questions about growth and its limits at the heart of sustainability debates are not new (see, for example, Barnett and Morse 1965; Meadows et al. 1972; Ehrlich and Holdren 1972). However, they have been revitalized in recent years (see Turner 2008; Hall and Day 2009; Latouche 2010; Martínez-Alier et al. 2010; Simpson, Toman, and Ayres 2005), in part because of growing support for the notion of socially equitable and environmentally sustainable "degrowth" in the Global North (Reichel and Seeberg 2009).[2] Furthermore, such arguments now exist atop strong evidence of "ecological overshoot" since the 1980s, whereby humanity's demand on the environment—in terms of the amount it takes to produce all the living resources we consume as well as to absorb our carbon dioxide emissions—has continually exceeded the biosphere's regenerative capacity (Wackernagel et al. 2002).

---

## 1.2 Rethinking Sustainable Innovation

Understanding these issues is fundamental to any investigation of emerging technology and its interaction with sustainability. The time has come to think closely about what sustainability means and the forms it could take. For example, the modern scientific model—still the dominant paradigm for scientific innovation—flourished in the context of the Industrial Revolution and has therefore been customarily driven by a process of modernization inextricably linked with economic growth. Gains in efficiency, productivity, and utility have constituted the often unspoken drivers of assessment for technological innovations such as the steam train, antibiotics, and even the "green revolution" of the 1940s—an agricultural approach that focused on increasing crop yields via the application of new plant varieties and modern agricultural techniques.[3] As Ellul (1964) noted in the 1960s, "modern technology has become a total phenomenon for civilization, the defining force of a new social order in which efficiency is no longer an option but a necessity imposed on all human activity" (17). In this respect, little has changed in recent times (Scrinis and Lyons 2007). From a contemporary sustainability perspective, mainstream critiques of emerging innovation rarely venture beyond questions of which environmental efficiency gains can be made, which effects productivity impacts might have on financial sustainability, and which abilities each technology holds to sustain and improve levels of perceived human comfort. For sustainable development, ecological modernization is proffered as *the* answer, where it is believed that economic growth can be decoupled from environmental degradation via changes in production processes and institutional adaptation (Blowers 1997). It is this kind of thinking that leads popular economic commentators to suggest that "innovation is what allows you to use a finite amount of resources more efficiently, yielding the kind of growth that is sustainable" (*Economist* 2009).

The limitations of such approaches become immediately apparent when held to more serious scrutiny. As Tim Jackson (2009, 488) shows, based on realistic demographic and lifestyle expectations, carbon intensities "would have to fall, on average, by more than 11% per year to stabilize the climate, 16 times faster than they have fallen since 1990."[4] Additionally, in what is generally known as the rebound effect, environmental benefits from efficiency gains are typically offset by corresponding increases in overall consumption (see Polimeni et al. 2008).[5] There are clearly limits to growth.

This is not to deny the many fruits of modernization that may be—and have been—of great importance for meaningfully adapting to limits to growth.[6] Developments in cradle-to-cradle manufacturing, open source software, the transnational movements outlined above, and critical theory itself, spring to mind. However, what we seek to highlight is that a whole *other* set of benchmarks hold increasing levels of legitimacy when it comes to exploring the implications of emerging technologies, such as the extent to which technological innovation can be decentralized, locally appropriate, and democratically controlled. Benchmarks such as these call into question the current range of dominant assumptions: How narrowly has sustainability been defined, and how might this definition be pragmatically interrogated and expanded? What is the broader potential here for science and sustainability? Are there alternative forms of scientific knowledge and production that can be looked to for grappling with questions of sustainability in a deeper and also more global sense—especially forms that are able to incorporate the voices and needs of communities who are not yet represented within mainstream science?

Given that present and future ecosystems' stability is tied directly to the ways in which we collectively act (Pachauri and Reisinger 2007), addressing these questions can afford no further wait. The problems of unfettered growth and the role of emerging technology in growth's perpetuation are now so great that we must either adapt more holistic approaches to the processes of technological creation, or risk their continuation as a medium through which we move emphatically toward self-ruin. On a more positive note, we believe that a timely opportunity has now arisen in light of the increasing attention by scholars and the broader community to the prospect of futures without growth (see, for example, notions of prosperity without growth [Jackson 2010]; property beyond growth [Alexander 2011]; and managing without growth [Victor 2008]). Apart from commentary by Peter Victor (2008) about technology assessment, technological innovation is given little consideration in these works. Thus, in this book we add to such alternatives the notion of innovation without growth—a scenario in which further scientific innovation does not equate with further increases in national gross domestic product (GDP).[7]

## 1.3 Lines of Inquiry

Building on previous work (Maclurcan 2011), we believe that holistic approaches to creating futures founded upon innovation without growth require the following four fundamental attributes:

1. *Recognition of limits, ecological overshoot, and unsustainable trajectories*—an alternative starting point for innovation that acknowledges the need for a new scientific approach—one that is impregnable to co-option by the shortsightedness often underscoring capitalist ventures, such as "greenwashing."

2. *Decentralized capacity*—an alternative global infrastructure for innovation that responds to the various detrimental divides between and within the Global North and Global South (the terms we prefer to use to describe what are more commonly referred to as the developed and developing countries, respectively) through greater decentralization and autonomy.[8]

3. *Local appropriateness*—alternative approaches to technological designs that ensure sensitivity to human needs, cultural norms, and environmental effects.

4. *Democratic governance*—alternative methods for overseeing innovation that are participatory, enable the empowerment of people, and influence innovation trajectories.

By reflecting upon these four attributes, it is possible to consider how new technologies can actually offer so much more than scientific breakthroughs as they are traditionally conceived. In short, we believe new technologies can also become an advanced platform for defining social values—values that are constituted "in both direct and not-so-direct ways ... [highlighting] the reciprocal relationship between the role of artefacts in *reflecting* social priorities and their role in *reinforcing* or stabilizing those priorities" (Nieusma 2010, 223). In this sense, along with our authors, we value the various forms of equity that lie at the heart of each of the four attributes we have mentioned: environmental equity (limits), equity of power (capacity), equity of needs (appropriateness), and participatory equity (governance). We foreground equity in this way in part to acknowledge that such attributes are critical for technology to be considered truly appropriate (Schumacher 1973). But our larger interest is in acknowledging the need to address the strong correlation between inequity and unsustainability and the converse, long-held belief that social and environmental sustainability go hand in hand (Wilkinson and Pickett 2009).[9]

## 1.4 Introducing Nanotechnology

Thus, with a steadfast belief in our ability to cocreate equitable, sustainable futures, we turn our attention to one of the most controversial of all recently emerging technologies—nanotechnology: "the application of scientific knowledge to control and utilize matter in the nanoscale [the scale of atoms and small molecules], where properties and phenomena related to size or structure can emerge" (ISO Technical Committee 2008). Here, we use a definition agreed upon by the International Organization for Standardization (ISO) to highlight our consideration for nanotechnology in its accepted form (compared to its speculative form as molecular manufacturing, which brought with it limited and populist notions such as "gray goo" and "robots in the bloodstream").[10]

Although what nanotechnology is remains contested (and, therefore, any reading must keep this limitation in mind), at a fundamental level, it does have the following six defining features: it is based upon a size or length scale (the nanoscale—generally agreed as 1–100 nanometers, with 1 nanometer equal to 1 billionth of a meter); it involves the ability to control, manipulate, or engineer on that scale; it involves exploiting properties unique to the nanoscale; it is the practical application resulting from this exploitation; it is often the product of conducting "old science" in a new way; and it is the natural (but sometimes unconscious) progression for those working in cutting-edge areas of science and is therefore a new field rather than a new discipline (Maclurcan 2009).

A range of scientific developments have been fundamental to nanotechnology's emergence. The materialization of "tools to see, measure, and manipulate matter at the nanoscale" (Ratner and Ratner 2002, 39) has included the discovery of the scanning tunneling microscope (STM) in 1981 and the atomic force microscope (AFM) in 1986.[11] Utilizing various forms of surface interaction, these instruments have enabled imaging of a sample's topography, composition, and scientific properties at the nanoscale. Furthermore, the ability of the STM to move single atoms on surfaces has provided humans with a means by which to engineer with atomic precision (Harper 2003).[12] New techniques have also driven nanotechnology's emergence, including quantum mechanical computer simulation, soft x-ray lithography, and new synthesis methods such as chemical vapor deposition, all spurring an ever-accelerating understanding of scientific endeavor at the level of atoms and small molecules.[13] The final significant piece in nanotechnology's scientific evolution has been the discovery of materials such as quantum dots, circa 1983 (see Brus 1984); fullerenes—including the spherical forms known as buckyballs—in 1985 (see Kroto et al. 1985); and nanotubes—particularly carbon-based—in 1991 (see Iijima 1991).[14]

Playing a significant role in nanotechnology's development is the luxury it has enjoyed as a financially well-supported field (to be explored in Chapter 4). Nanotechnology has received a wide uptake in research and development (R&D), particularly in the Global North, with a range of applications across many sectors, already available for purchasing. This relative ease of development in the Global North has been aided by favorable policy conditions and agreements with industry, which have given innovators relative freedom to develop and exploit the technology (as will also be explored in Chapter 4).

## 1.5 Perspectives on Nanotechnology

More broadly, for some, nanotechnology also provides new hope for global equity (Court et al. 2004; Juma and Yee-Cheong 2005; Salamanca-Buentello and Daar 2005) and national economic growth decoupled from environmental degradation (Court et al. 2004; Barker et al. 2005; Juma and Yee-Cheong 2005; El Naschie 2006; Esteban et al. 2008). In light of present trajectories and historical precedence, others, including ourselves, are more skeptical about these two claims (ETC Group 2008; Invernizzi and Foladori 2005; Maclurcan 2011).[15] As Nieusma (2010, 212) notes, the market is presently failing to "direct nanotechnology in many of the directions that are both possible and broadly desirable." Moreover, Foladori and Invernizzi (2008) believe that nanotechnology has yet to offer evidence of an ability to technologically fix detrimental consumer addiction to the use of ever diminishing nonrenewable resources, and the market's compliance in supporting such an addiction. As Slade (2010) notes, the dominant values driving technological innovation are inherently economic, and that "economic values rarely serve as preferred end-state values for public policies, especially those that relate to health and wellbeing" (70).

However, with the mass of political, business, and financial support for nanotechnology seemingly guaranteeing its near-term continuation in some form or another, in this book we seek to cut through the polarizing discourse outlined above by advancing with critically-informed hope. In short, we seek to engage deeply with the challenges posed by nanotechnology and, where possible, from asset-based positions—that is, by taking existing social, physical, economic, and political assets into account rather than working within a needs-based, problem-solving paradigm (see Mathie and Cunningham 2003). Hence, the writing in this book is as much concerned with how current, emerging, and future efforts in nanotechnology can play constructive roles in shifts toward holistic approaches to sustainability as it is with questions of whether nanotechnology offers hope for greater equity or, indeed, sustainability. This does not necessarily spell a compromised position nor, one would hope, a platform for co-option by market forces. Rather, it is about understanding the importance of cocreating optimistic alternatives that both resonate with and are understandable to the mainstream and, as such, can enter popular discourse for debate and scrutiny.

In this light, we regard two present, era-defining trends as holding great importance for our particular engagement with nanotechnology and sustainability: increasing information openness and greater global interconnectivity. For researchers, open-access publishing is a watershed for broadening horizons. But perhaps more importantly, interconnectivity offers us, as editors, the ability to rapidly pull together disparate ideas from these vast horizons. The result, in cases such as this book, is a cross-fertilization of insights and ideas from authors who collectively bridge historically impassible boundaries of gender, discipline, sector, and geography. The propensity for holistic systems-thinking and shifts in approaches becomes possible in such circumstances.

As we shall reflect upon in our conclusions (Chapter 13), we believe the collective meeting of our authors within this text allows the drawing of new and valuable associations. Our contributors write from a range of perspectives, experiences, professions, and locations around the world and raise the issues they regard to be most pressing and urgent. Inevitably, when drawn together, they form a chorus of multifarious perspectives, some dissonances, but also unlikely congruity—and we believe that this is where the richness of this text lies: It allows us to think about scientific innovation and sustainability in new ways. This co-presence may, in fact, assist in transcending the identified competition between instrumentalist and contextualist perspectives on nanotechnology and global development (see Liao 2009; Maclurcan 2011).[16]

In this regard, our work in this book marks a departure from other contributions examining the relationship between nanotechnology and sustainability. There have been numerous investigations of nanotechnology's potential environmental impacts, each presenting its own valuable perspective.[17] Some have written comprehensively on nano-applications to protect and enhance our natural environment and humanity's place in it (Garcia-Martinez and Moniz 2010; Smith and Granqvist 2010). Others have detailed matters relating to nanotechnology's environmental risks (Bottero 2007; Karn 2004; Sellers 2009). Yet, as an understandable result of their scientific focus, these efforts have all stuck to a narrow remit with respect to sustainability rather than tackle the philosophical heart of the issues at hand. In this book we adopt a much broader remit, regarding sustainability as less about sustaining the ways we live and more about cocreating sustainable ways of living.[18] Along with increasingly shared risks around issues such as climate change and emerging infectious disease (see Chapter 7), this is why the title of our book refers to *global* sustainability, flagging that we will extend our discussion beyond the investigative realms of technological efficiency, productivity, and utility.

In a book with global implications, we are compelled to explain our decision to collate writing that focuses on nanotechnology and the Global South.[19]

This is primarily because, irrespective of outcomes, many of the North's most sustainable aspirations would seem to be reflected in approaches proposed for the Global South and, from a sustainability perspective, the South offers different wisdoms and knowledge that presently have an indefensible lack of influence over contemporary and dominant scientific paradigms.[20] Furthermore, with an equitable response to ecological overshoot requiring degrowth of the economies in the Global North toward a steady-state global economy, the philosophies and approaches of appropriate technology (which place local and ecological contexts at the heart of innovation), as often envisaged for the Global South, become increasingly relevant to the Global North (Wicklein and Kachmar 2001).

## 1.6 Chapters and Themes

The chapters we have collected have been organized around the four values we regard as fundamental to a sustainability approach to science and technology (although, as would be expected, there are some chapters that could be seen to straddle a number of these values). The book is therefore divided into four sections to reflect each of these themes. Each section includes a broad thematic critique as well as more specific criticisms, case studies, and proposed alternatives for many of the challenges raised.

In this first section, *Limits*, we have collated work that provides a starting point for considering the limits to growth. In "Nanotechnology and the Environment" (Chapter 2), David Hess and Anna Lamprou review significant ways in which nanotechnology could assist in "greening" the economy. Placing their discussion in a broad framework of limits and the need for a steady-state economy, they investigate nanosolar as an avenue for increasing energy efficiency, improving energy storage, enabling renewable energy technologies, and shifting to more flexible solar designs. This framework results in a critique of nanosolar, raising the issue of exposure to toxic nanoparticles and thereby highlighting the care that must be given to new technologies, especially when arriving on the back of hype and claims of a "single-bullet" solution. Yet Hess and Lamprou's analysis sees the convergence of solar energy and nanotechnology not only as an exemplar of the complex mix of benefits and risks that nanotechnology poses, but also as an opportunity to develop an analysis of how environmental social theory and environmental policy might be brought together. They thereby navigate a new path for environmental sociology that can respond pragmatically to the ecological impacts of overaccumulation.

In "Nanotechnology and Traditional Knowledge Systems" (Chapter 3), Ron Eglash takes a very different approach to investigating nanotechnology within the context of limits. He uses case studies of nanomaterials to explore relationships between nanotechnology and the traditional knowledge of ancient state and nonstate indigenous societies. Eglash's detailed discussion helps us expand our understanding of how indigenous practices can contribute to contemporary science and technology while opening up new perspectives on the ability for cutting-edge innovation to exist without growth. By also connecting nanotechnology with traditional knowledge at the macroscale as a "boundary object," he expands the horizons of science education, highlighting the value (especially among indigenous descendants) of cultural connection to nanotechnology. Upon returning to traditional knowledge in concluding discussion, he explores the implications for patenting of nanoproperties, including those of purely inorganic origin or organic–inorganic hybrids.

Section 2, *Capacity*, draws together contributions ranging from geopolitical pressures of governance, barriers and promises in the agri-food sector, to unlikely, bespoke-style innovation that is exceeding expectations in Thailand. In "Nanotechnology and Geopolitics: There's Plenty of Room at the Top" (Chapter 4), Stephanie Howard and Kathy Jo Wetter provide a thorough survey of the emerging geopolitical landscape in which nanotechnology is situated. Their review is crucial to understanding the various capacities of features shaping this landscape, including funding distribution, research orientation, and policy engagement. The authors map how the technologies and their ownership, control, and governance are evolving, before engaging with rising areas of debate such as intellectual property. Using nano "clean tech" as an example, they also question the rigor of economic analysis underpinning the case for a nano-economy, especially in relation to governmental support. Moving to discussion of the responsible-development governance culture that many governments have adopted, they conclude by reviewing public participation in nanotechnology.

In "Nanotechnology, Agriculture, and Food" (Chapter 5), Kristin Lyons, Gyorgy Scrinis, and James Whelan draw attention to the competing visions for technological innovation shaping the future of agriculture and food in which nanotechnologies are being developed and applied. Their aim is to examine the extent to which the agricultural and food industries have embraced nanotechnologies and the contribution such technologies could make to address the agri-food crisis. In doing so, their work provides a comprehensive review of the wide variety of applications (and the range of processing functionalities) being researched and commercialized in relation to agriculture and food. The authors raise questions around the claims associated with each, especially with "smart" applications, and their likely impact on the future of agriculture and food. They subsequently move to discussion of environmental and human side effects, particularly given the greater potency, reactivity, and bioavailability associated with nanoparticles (as compared to conventional counterparts). Throughout their contribution, Lyons, Scrinis, and Whelan explore the potential for nanotechnology to be co-opted by corporate interests in the food and agriculture sector and to also drive technological solutions to systemic problems in dietary patterns, food quality, poverty, and socioeconomic structures.

Moving from a macro to a micro analysis, in "Poor Man's Nanotechnology—From the Bottom Up (Thailand)" (Chapter 6), Sunandan Baruah, Louis Hornyak, and Joydeep Dutta present an inspiring and at times humorous case for different thinking when it comes to the requirements for conducting nanotechnology research. Drawing on their work at Thailand's Centre of Excellence in Nanotechnology, they write from a self-identified "perspective of the poor man's laboratory." After sharing the story of their research institute's humble beginnings, they detail the processes involved in developing a number of technologies through bottom-up nanomaterial synthesis inspired by the natural world. Their work explores the extent to which cost and other resource barriers to high-quality research are indeed surmountable and whether nanotechnology research can realistically be oriented toward human needs.

In Section 3, *Appropriateness*, our authors consider a range of contexts, including innovations across the global health sector, facilitating discussions between diverse stakeholders in the Global South, and sharing knowledge through open source access to scientific innovation. In "Nanotechnology and Global Health" (Chapter 7), Deb Bennett-Woods examines the relationship between the emerging potential of nanotechnology and existing needs in global health, particularly with respect to the Global South. Viewing health as a useful lens through which to look at broader social indicators, Bennett-Woods considers the range of paradigmatic approaches that could be applied to assessing technology (medical, well-being, and environmental models) and employs a working conception of health grounded in social causes of poor health and health-related inequalities (therein showing sustainability in its broader sense). She gives consideration to ways in which nano-enabled technologies might operate to improve human health on a global scale, investigating in more detail the range of possible health-related applications that were traced by Howard and Wetter in Chapter 4. Central to Bennett-Woods's critique is the need to consider nanotechnology's potential environmental impacts on patients, as well as issues of accessibility (particularly costs) and the potential for greater inequity in the distribution of health care resources. She concludes by posing a set of strategically focused questions based on both the moral and practical considerations to which nanotechnology proponents must be answerable.

In "Toward Pro-Poor Nano-Innovation (Zimbabwe, Peru, and Nepal)" (Chapter 8), David Grimshaw contributes insight into the process of managing "nanotechnology dialogues." He draws upon his experience working with the United Kingdom–based nongovernment organization (NGO) Practical Action, in which he facilitated discussions between scientists and broader community stakeholders in Zimbabwe, Peru, and Nepal between 2006 and 2008. Taking a systems approach, Grimshaw documents ways in which the actions and policies surrounding nanotechnology can be used to ensure human needs are met. Using examples such as nanosensors developed for improving water quality, he mounts a case for a new approach to more appropriate scientific innovation. He believes such efforts represent "pro-poor innovation," and he pragmatically outlines its founding principles, based upon the "seven Ps": power, price, promise, poverty, pervasiveness, promiscuous utility, and paradigm.

In "Open Source Appropriate Nanotechnology" (Chapter 9), Usman Mushtaq and Joshua Pearce are primarily motivated by the question of who ultimately benefits from nanotechnology. Citing patents and flow-on costs as a significant driver of technological inequity, they consider the alternatives offered by developments in the open source movement—from open information to software and hardware. In this light, they explore emerging open source platforms for nano-innovation, as well as case studies of open nano-innovation in water, energy, and materials. In doing so, they argue that open source nanotechnology can act in support of more equitable and sustainable futures by increasing access to innovation and its outputs, ensuring local technological appropriateness, and enhancing technological oversight through the power of the crowd.

Section 4, *Governance*, considers how approaches to governance are shaping nanotechnology. A number of vital and pressing perspectives are presented, including how governance is influencing risk assessment, the nature of state regulation in India, and the intricacies and complex negotiations of global regulation. In "Nanotechnology and Risk" (Chapter 10), Fern Wickson details key concepts and critical literature central to understanding risk, especially relating to risk analysis as a decision-aiding tool. In doing so, she comprehensively introduces a range of typologies of risk. Upon evaluating the latest knowledge regarding nanotechnology's risks to environmental and human health as well as future challenges, she argues that scientific risk assessment and risk management are dominating discourses and subsequently narrowing the frame of discussion about the desirability of developing nanotechnologies. In response, Wickson details alternative decision-aiding tools that might begin to push discussions about nanotechnology toward a broader discourse of risk, thereby enabling a more deliberative negotiation of uncertainty and a more integrated consideration of social and ethical issues.

Nidhi Srivastava and Nupur Chowdhury survey the product range and depth of nanotechnology applications in the pharmaceutical sector in India in "Nanotechnology and State Regulation (India)" (Chapter 11). Their analysis provides an in-depth overview of the regulatory systems, legislation, and players in nanomedicine on the subcontinent. Following this, they provide an overview of the product safety and quality regulations that will govern the manufacture and marketing of nanomedicines in India and consider the extent to which the current regulatory framework for pharmaceutical regulation, food safety, and environmental protection is equipped to address the regulatory challenges stemming from developments within the nanotechnology and health care space. They identify critical points within the legal framework that would need to be reexamined in light of the changing characteristics of such new applications, as well as the legislative and policy work the current regime would need to undertake to develop regulatory norms to address these new challenges. They conclude by speculating what an ideal institutional framework for nanotechnology might look like.

Critical debates surrounding nanotechnology and regulation are raised by Diana Bowman and Graeme Hodge in "Nanotechnology and Global Regulation" (Chapter 12). Commencing with an outline of key regulatory concepts and methods, they survey lessons learned from regulatory reviews that have been conducted and the range of regulatory approaches available and in use. The authors build upon an evolving body of literature that deals with contemporary regulatory challenges, particularly focusing on the role that international approaches (such as framework conventions, self-regulation, and co-regulation) may play in ensuring the responsible development of nanotechnologies. In doing so, their investigation examines several current multilateral activities they believe must be in place to ensure that nanotechnologies can help drive sustainable futures.

In the concluding chapter, "Nanotechnology without Growth" (Chapter 13), we emphasize that the ideas in this book neither could nor are intended to provide a blueprint for innovation without growth. The work, research, and views of our contributors have not been drawn together with the assumption that all are in common agreement but rather that their association creates a space for new conversations, innovation at the edges, and pathways toward more equitable futures through further scientific and social exploration. Therefore in this final chapter we aim to use the collective inputs of our contributors to cultivate the grounds upon which a range of new dialogues can occur. We do so by tracing some of the new associations that have emerged between these pages, speculating as to their saliency and efficacy in regard to shaping an innovative future that is not growth dependent. By consolidating the many lines of enquiry pursued throughout the chapters, we hope to begin building pathways for further investigation and debate, so that both nanotechnology and other forms of scientific innovation can indeed become a positive impetus toward more equitable, and ultimately sustainable, futures for us all.

## References

Alexander, S. 2011. *Property beyond Growth: Towards a Politics of Voluntary Simplicity.* PhD diss., University of Melbourne. Available at http://simplicityinstitute. org/wp-content/uploads/2011/04/Property-Beyond-Growth1.pdf (accessed May 1, 2011).

Barker, T., Lesnick, M., Mealy, T., Raimond, R., Walker, S., Rejeski, D., and Timberlake, L. 2005. *Nanotechnology and the Poor: Opportunities and Risks: Closing the Gaps within and between Sectors of Society.* Washington, D.C.: Meridian Institute.

Barnett, H.J. and Morse, C. 1965. *Scarcity and Growth: The Economics of Natural Resource Availability*. Washington, D.C.: RFF Press.

Blowers, A. 1997. Environmental policy: Ecological modernisation or the risk society? *Urban Studies* 34(5–6): 845–71.

Bottero, J-Y. 2007. *Environmental Nanotechnology: Applications and Impacts of Nanomaterials*. New York: McGraw-Hill Professional.

Brundtland, G.H. 1987. *Our Common Future: Report of the World Commission on Environment and Development*. Oxford: Oxford University Press.

Brus, L.E. 1984. Electron–electron and electron–hole interactions in small semiconductor crystallites: The size dependence of the lowest excited electronic state. *Journal of Chemical Physics* 80(9): 4403–9.

Court, E., Daar, A.S., Martin, E., Acharya, T., and Singer, P.A. 2004. Will Prince Charles et al diminish the opportunities of developing countries in nanotechnology? *Nanotechweb.org*, Available at http://nanotechweb.org/cws/article/indepth/18909 (accessed July 26, 2011).

Cozzens, S. and Wetmore, J. 2010. *Nanotechnology and the Challenges of Equity, Equality and Development*. New York: Springer.

Economist, The. 2009. How to live in a bubble. *Free Exchange*, Available at http://www.economist.com/blogs/freeexchange/2009/11/how_to_live_with_bubbles (accessed January 12, 2011).

Ehrlich, P. and Holdren, J. 1972. One-dimensional ecology. *Bulletin of the Atomic Scientists* 28(5): 16, 18–27.

Eigler, D.M. and Schweizer, E.K. 1990. Positioning single atoms with a scanning tunneling microscope. *Nature* 344(6266): 524–26.

El Naschie, M.S. 2006. Nanotechnology for the developing world. *Chaos, Solitons and Fractals* 30(4): 769–73.

Ellul, J. 1967. *The Technological Society*. New York: Vintage Books.

Esteban, M., Webersik, C., Leary, D., and Thompson-Pomeroy, D. 2008. *Innovation in Responding to Climate Change: Nanotechnology, Ocean Energy and Forestry*. Yokohama: United Nations University, Institute of Advanced Studies.

ETC Group. 2008. *Downsizing Development: An Introduction to Nanoscale Technologies and the Implications for the Global South*. Geneva: UN Nongovernmental Liaison Service.

Foladori, G. and Invernizzi, N. (Eds.). 2008. *Nanotechnologies in Latin America*. Berlin: Rosa Luxembourg Foundation.

Garcia-Martinez, J. and Moniz, E.J. 2010. *Nanotechnology for the Energy Challenge*. Weinheim: Wiley-VCH.

Hall, C.A.S. and Day Jr., J.W. 2009. Revisiting the limits to growth after peak oil: In the 1970s a rising world population and the finite resources available to support it were hot topics. Interest faded—but it's time to take another look. *American Scientist* 97(3): 230–7.

Harper, T. 2003. What is nanotechnology? *Nanotechnology* 14(1): introduction.

Iijima, S. 1991. Helical microtubules of graphitic carbon. *Nature* 354(6348): 56–58.

Invernizzi, N. and Foladori, G. 2005. Nanotechnology and the developing world: Will nanotechnology overcome poverty or widen disparities? *Nanotechnology Law and Business Journal* 2(3): 101–10.

ISO Technical Committee 229. 2008. *Working Document: ISO/TC 229 N 669. Nanotechnologies—Vocabulary—Part 1: Core terms*. Available at http://www.scribd.com/doc/51619091/TC-229-N-669-JWG1-PG5–46-ISO-TC229-DTS-80004–1-Core-Terms-CD-ballot- (accessed May 16, 2011).

Jackson, T. 2009. Beyond the growth economy. *Journal of Industrial Ecology* 13(4): 487–90.

Jackson, T. 2010. *Prosperity without Growth: Economics for a Finite Planet*. London: Earthscan.

James, P. 1997. Postdependency? The Third World in an era of globalism and late-capitalism. *Alternatives* 22(2): 205–26.

Juma, C. and Yee-Cheong, L. 2005. *Innovation: Applying Knowledge in Development*. UN Millennium Project Task Force on Science, Technology, and Innovation. London: Earthscan.

Karn, B., Masciangioli, T., Zhang, W., Colvin, V., and Alivisatos, P. (Eds.). 2004. *Nanotechnology and the Environment: Applications and Implications*. American Chemical Society Symposium Series #890. New York: Oxford University Press.

Kroto, H., Heath, J., O'Brien, S., Curl, R. and Smalley, R. 1985. $C_{60}$: Buckminsterfullerene. *Nature* 318(6042): 162–63.

Latouche, S. 2009. *Farewell to Growth*. Cambridge: Polity Press.

Liao, N. 2009. Combining instrumental and contextual approaches: Nanotechnology and sustainable development. *Journal of Law, Medicine and Ethics* 37(4): 781–89

Lowe, I. 2009. *A Big Fix*. 2nd ed. Melbourne: Black Inc..

Maclurcan, D.C. 2009. Nanotechnology and the Global South: Exploratory views on characteristics, perceptions and paradigms. In *Technoscience in Progress: Managing the Uncertainty of Nanotechnology*, edited by S. Arnaldi, A. Lorenzet, F. Russo, 97–112. Amsterdam: IOS Press.

Maclurcan, D. 2011. *Nanotechnology and Global Equality*. Singapore: Pan Stanford Publishing (World Scientific).

Martínez-Alier, J., Pascual, U., Vivien, F-D., and Zaccai, E. 2010. Sustainable de-growth: Mapping the context, criticisms and future prospects of an emergent paradigm. *Ecological Economics* 69(9): 1741–47.

Mathie, A. and Cunningham, G. 2003. *Who Is Driving Development? Reflections on the Transformative Potential of Asset-Based Community Development*. Occasional Paper Series, No. 5, Coady International Institute, Antigonish, Available at http://www.coady.stfx.ca/resources/abcd/Who%20is%20Driving%20Development.pdf (accessed July 26, 2011).

Meadows, D.H., Meadows, D.L., Randers, J., and Behrens, W.W. III. 1972. *The Limits to Growth: A Report for the Club of Rome's Project on the Predicament of Mankind*. London: Earth Island.

Nieusma, D. 2010. Materializing nano equity: Lessons from design. In *Nanotechnology and the Challenges of Equity, Equality and Development*, edited by S. Cozzens and J. Wetmore, 209–30. New York: Springer.

Pachauri, R.K. and Reisinger, A. (Eds.). 2007. *Contribution of Working Groups I, II and III to the Fourth Assessment Report of the Intergovernmental Panel on Climate Change*. Geneva: IPCC.

Pieterse, J.N. 1998. My paradigm or yours? Alternative development, post-development, reflexive development. *Development and Change* 29(2): 343–73.

Polimeni, J.M., Mayumi, K., Giampietro, M., and Alcott, B. 2008. *The Jevons Paradox and the Myth of Resource Efficiency Improvements*. London: Earthscan.

Ratner, M.A. and Ratner, D. 2002. *Nanotechnology: A Gentle Introduction to the Next Big Idea*. Upper Saddle River, N.J.: Prentice Hall.

Reichel, A. and Seeberg, B. 2010. Rightsizing production: The calculus of "ecological allowance" and the need for industrial degrowth. In *Competitive and Sustainable Manufacturing, Products and Services*, edited by APMS. Berlin: Springer.

Rye Olsen, G. 1995. North-South relations in the process of change: The significance of International Civil Society. *European Journal of Development Research* 7(2): 233–56.

Salamanca-Buentello, F., Persad, D.L., Court, E.B., Martin, D.K., Daar, A.S., and Singer, P.A. 2005. Nanotechnology and the developing world. *PLoS Medicine* 2(4): 300–303.

Schumacher, E.F. 1973. *Small Is Beautiful: A Study of Economics as if People Mattered.* London: Blond and Briggs.

Scrinis, G. and Lyons, K. 2007. The emerging nano-corporate paradigm: Nanotechnology and the transformation of nature, food and agri-food systems. *International Journal of Sociology of Food and Agriculture* 15(2): 22–44.

Sellers, K. 2009. *Nanotechnology and the Environment.* Boca Raton, Fla.: CRC Press.

Simpson, R.D., Toman, M.A., and Ayres, R.U. (Eds.). 2005. *Scarcity and Growth Revisited: Natural Resources and the Environment in the New Millennium.* Washington, D.C.: RFF Press.

Slade, C.P. 2010. Exploring societal impact of nanomedicine using public value mapping. In *Nanotechnology and the Challenges of Equity, Equality and Development,* edited by S. Cozzens and J. Wetmore, 69–88. New York, Springer.

Slater, D. 2004. *Geopolitics and the Post-Colonial: Rethinking North-South Relations.* Hoboken, N.J.: Wiley-Blackwell.

Smith, G.B. and Granqvist, C-G.S. 2010. *Green Nanotechnology: Energy for Tomorrow's World.* Boca Raton, Fla.: CRC Press.

Taylor, W. 1976. Innovation without growth. *Educational Management Administration and Leadership* 4(1): 1–13.

Turner, G. 2008. A comparison of the limits to growth with 30 years of reality. *Global Environmental Change* 18(3): 397–411.

Victor, P. 2008. *Managing without Growth: Slower by Design, Not Disaster.* Cheltenham: Edward Elgar.

Wackernagel, M., Schulz, N.B., Deumling, D., Linares, A.C., Jenkins, M., Kapos, V., Monfreda, C., Loh, J., Myers, N., Norgaard, R., and Randers, J. 2002. Tracking the ecological overshoot of the human economy. *Proceedings of the National Academy of Science* 99(4): 9266–71.

Wicklein, R.C. and Kachmar, C.J. 2001. Philosophical rationale for appropriate technology. In *Appropriate Technology for Sustainable Living: 50th Yearbook of the Council of Technology Teacher Education,* edited by R.C. Wicklein, 3–17. Peoria, Ill.: Glencoe McGraw-Hill.

Wilkinson, R. and Pickett, K. 2009. *The Spirit Level: Why More Equal Societies Almost Always Do Better.* London: Penguin Books.

# Endnotes

1. The use of the words "our" and "we" throughout this book is most often for sake of ease but belies the very different circumstances that exist both between and within countries around the world, as well as the differing rights and responsibilities that can be deemed reasonable with respect to action around issues such as climate change.

2. Known also as *decroissance* (France), *decrescita* (Italy), and *decreciemento* (Spain). For a seminal work on degrowth, see Latouche (2009).
3. Throughout this book, the word *innovation* is used with particular reference to scientific and technological innovation. These are, of course, far from the only realms in which innovation can occur.
4. This assessment is made with the following assumption: "In a world of 9 billion people, all aspiring to a level of income commensurate with 2% growth on the average European Union income today" (Jackson 2009, 488).
5. Wilkinson and Pickett (2009) note: "as cars have become more fuel-efficient we have chosen to drive further. As houses have become better insulated we have raised standards of heating, and as we put in energy-saving light bulbs the chances are that we start to think it doesn't matter so much leaving them on" (219). In a similar vein, according to Polimeni et al. (2008), the doubling of food production efficiency per hectare over the past 50 years did not solve the problem of hunger because the increase in efficiency increased production and worsened hunger through a resulting increase in population.
6. As Pieterse (1998) notes, even the strongest critiques of "development" themselves arise out of modernization.
7. Speaking in the 1970s, William Taylor, director of the University of London Institute of Education, provided a speech titled "Innovation without Growth." Although he was referring to innovation within the United Kingdom's tertiary education system, he did make the comment "there is no necessary link between innovation (equated with improvement) and growth, although many of the conditions associated with growth do facilitate innovatory processes" (Taylor 1976, 6).
8. Although they remain contested, the terms "Global North" and "Global South" are seen by some (see, for example, Rye Olsen 1995; James 1997; Slater 2004) as less burdened by embedded meanings.
9. According to Wilkinson and Pickett (2009), inequality heightens competitive consumption and "governments may be unable to make big enough cuts in carbon emissions without also reducing inequality" (215).
10. An apocalyptic scenario in which self-replicating, omnivorous nanoscale robots consume the global ecosystem.
11. The scanning tunneling microscope is an instrument that uses the difference in voltage between a conducting tip and a surface to scan the surface's topography. The atomic force microscope is an instrument that uses the difference in atomic force between a cantilevered tip and a surface to map the surface's topography.
12. The ability to maneuver atoms was made famous by the 1990 manipulation of 35 xenon atoms into the letters "I.B.M." (see Eigler and Schweizer 1990).
13. Quantum mechanical computer simulation is a technique that facilitates the theoretical modeling of atoms or small molecules for the purpose of predicting the scientific characteristics of such matter. Soft x-ray lithography is a technique by which a pattern is etched onto a surface via x-rays. Chemical vapor deposition is a process by which matter, once exposed to volatile agents, will leave a material residue on a surface.
14. Quantum dots are semiconducting nanocrystals that differ in their ability to absorb and emit energy, based on the size of the crystal. Fullerenes are a class of carbon molecule that can be arranged in spherical, ellipsoidal, or cylindrical formations. A buckyball is a spherical fullerene. Nanotubes can be further disaggregated into those with single walls and those with multiple walls.

15. Here we are particularly referring to the cases of pharmaceutical and agricultural biotechnology, as well as information communications technology.
16. In essence, the instrumentalist view is a reductionist, "mechanical" vision of the relationship between science and society. From this viewpoint, poverty and social problems are largely due to a lack of technical capabilities. The contextualist view presents a holistic vision of the relationship between science and society. From this viewpoint, poverty and social problems are part of a complex web of socioeconomic trends involving systemic inequities at the global, national, and local levels.
17. For example, Sellers's (2009) exploration of the apparent paradox of using nanomaterials in environmental remediation, Karn et al.'s (2005) exploration of environmentally benign manufacturing of nanomaterials, and Smith and Granqvist's (2010) examination of energy flows in nature and how the optical properties of materials can be designed to harmonize with those flows.
18. The inspiration for this expression comes from Steb Fisher. A similar expression can be found at http://pathfindernetwork.com.au.
19. And are even more so, given our authors are largely from the Global North.
20. Some of the Global North's "most sustainable" aspirations reflected in approaches proposed for the Global South have included the "Human Development Index"; leapfrogging high-emission development paths; and reforestation (often linked to carbon offsetting schemes in the Global North).

# 2

## Nanotechnology and the Environment

David J. Hess and Anna Lamprou

### CONTENTS

## 2.1 Introduction

The advent of nanotechnology as the "next Industrial Revolution" might cause anyone with some knowledge of the environmental and health effects of previous industrial revolutions to ask some justifiably tough, skeptical questions. The promises of previous technological revolutions—a car in every garage, the peaceful atom, and better living through chemistry—have ended up generating significant environmental and health-related side effects and risks. The outcomes, in retrospect, are such that present generations would have benefited if previous generations had been more perspicacious about the regulation, design, and release of new technologies. Although precaution may be the lesson from the past, and the benefits of new technologies are often overhyped, new technologies generally involve both advantages and disadvantages, and consequently there may be little support for a political decision not to pursue at least some design variants of a proposed new technology. In this sense, nanotechnology is no different from previous generations of technologies that posed issues of both substantial societal benefit and environmental and health hazards and risks.

With respect to environmental benefits, there are various ways in which nanotechnology can contribute to products that increase energy efficiency, improve energy storage, or enable renewable energy technologies. For example, nanotechnology can contribute to the greening of the economy via applications in fuel cells, batteries, and solar photovoltaics. The combination of solar energy and hydrogen-powered fuel cells represents one way to address the challenges of intermittency associated with solar energy. Nanomaterials can also be used in electrolysis to produce hydrogen, and they generally exhibit better electronic transfer properties than bulk substances. By controlling the architecture of nanostructures, the energy conversion may become more efficient and less costly (Grätzel 1991; Wei and Zunger 1990). Nanomaterials can also play an important role in the development of methanol, which can power fuel cells. Through carbon capture and chemical conversion enabled by nanomaterials, carbon dioxide from the atmosphere or from industrial emissions can be turned into useful products like methanol, which can lower the carbon footprint of industrial processes.

More broadly in the area of energy generation and storage, nanotechnologies might prove important in improving efficiency. With respect to rechargeable batteries and capacitors, nanotechnologies are able to hold more lithium to enable batteries to have a higher charge density. Because of these energy applications, nanotechnologies could make electric vehicles more cost competitive. Nanomaterials have also been proven valuable in increasing the energy efficiency of fuel additives and insulation. They can also be used to improve fuel efficiency as catalysts, more specifically in reducing the use of platinum-group metals or even replacing them completely in surface coating and lubricants. Nanotechnology can produce very light materials, which makes transportation more efficient (Weizsäcker, Lovins, and Lovins 1998).

The use of nanotechnology to harness solar energy (nanosolar) is one example of the potential environmental benefits of nanotechnology. The advent of nanosolar could reduce the cost of solar energy significantly and rapidly, and consequently the potential environmental benefits of this type of nanotechnology are very attractive. However, the lack of information about the health-related and environmental side effects of ubiquitous nanotechnology, even nanosolar, suggests a much more unsettling picture. Using the case of solar energy as an example, in this chapter we explore how environmental social theory could be developed to shed light on complex policy issues regarding the evaluation and regulation of new technologies. We first explore the potential environmental benefits and hazards of nanosolar, followed by a consideration of the differences in strategies used to encourage more precautionary regulation.

## 2.2 Theoretical Background

In an ecological sense, most scientists now recognize that the impact of human civilizations on the global environment is unsustainable. In other words, levels of human resource consumption and waste have already exceeded the capacity of the global ecosystem to replenish and process them (Daly 1990, 1996). Unless rapid changes occur in global levels of consumption and waste, the human–ecosystem relationship will collapse (Meadows, Randers, and Meadows 2004). We cannot predict exactly when the collapse will occur, but it probably will not take the form of a single, dramatic event after which civilization descends into a dark age of rampant violence and ubiquitous political chaos (Costanza, Graumlich, and Steffen 2007). Rather, collapse will be unevenly distributed across countries, continents, ecosystems, classes, age groups, and genders. Women, children, the poor, and the elderly in the coastal areas of the poorest countries of Asia, Africa, the Pacific, and the Caribbean are most likely to suffer the worst effects of collapse. In many ways, we are already seeing the emergent signs of collapse in the rampant poverty of shantytowns, the effects of increased severe weather events on coastal populations, and other global problems and disasters.

To the general diagnosis and prognosis offered by scientists, environmental sociologists have added a political economy perspective that reframes the sustainability problem as driven by a more complex set of societal factors than the biological facts of ongoing human population growth and increasing resource consumption and pollution. The treadmill of production theory (Gould, Pellow, and Schnaiberg 2008; Schnaiberg 1980; Schnaiberg and Gould 1994) and related theories of the political economy of accumulation (Foster 2005) draw attention to the tendency for most human societies, and especially capitalist societies, to accumulate wealth and to concentrate it in the hands of elites. Profits garnered by capitalist firms tend to be reinvested in more capital-intensive production processes. This investment pattern leads to higher levels of productivity and, if those gains are passed on in the form of wages, to higher consumption for workers who remain employed in the capital-intensive industries. For workers who lose their jobs due to new efficiencies in production, the government must ensure new employment and therefore must facilitate overall job creation, a goal that generally requires policies that support economic growth. As a result, even in the absence of population growth, there is a tendency for the reinvestment of profits into innovation to lead to economic growth. In turn, economic growth is historically associated with a higher level of aggregate production and consumption, which results in the growth of environmental "deposits" of wastes and pollution into the global ecosystem and "withdrawals" of resources from the system. Eventually the ecological growth in deposits and withdrawals hits the wall of ecological limits, and the specter of various collapse scenarios emerges.

There is a way out of the dilemma. Ecological economist Herman Daly (1996) calls it a steady-state economy, in which economic growth is both limited and disentangled from environmental destruction. However, the dematerialization of the economy would require significant shifts of investment into new technologies to enable the rapid greening of a variety of industries and, to date, the shifts have not occurred. Understanding the absence of a concerted, rapid, and effective policy response by the leaders of the world's industries and governments is the second major contribution of environmental social theory to the broader, interdisciplinary discussion of the environmental crises. In an ideal world, the research of natural and social scientists about impending ecological crises and their economic foundations would be taken up immediately by elected political officials and their appointees, who would respond dramatically and swiftly with new legislation and regulations to head off future collapse.

Three basic conditions result in a huge gap between the ideal response to the environmental crisis and the actual response of policymakers. First, there is no firewall between the political field and the economic field, and consequently economic elites tend to dominate the policymaking process on issues that affect their interests. Those interests include the protection of "treadmill" industries, especially industries involved in the production and use of fossil fuels and chemicals, which benefit from political inaction on environmental policy. Second, even in the absence of treadmill industries, ongoing geopolitical rivalry among nation-states involving the ultimate sanction of warfare drives national governments toward competitive growth, because countries that occupy or aspire to positions of hegemony in the global political order require growth in order to maintain the budgets that underlie support for the military, foreign aid, and a strong economic position in general. As long as other countries are growing economically, the arms race and foreign aid race are linked to competition to attain economic growth. Third, even in the absence of the first two conditions, as long as populations are growing, national governments must maintain economic growth in order to maintain the standard of living. This third factor may be the least important of the three for a variety of reasons, including the predicted leveling off of population growth by the middle of the twenty-first century, the concentration of environmental impact in countries with lower population growth, and the capacity for economies to absorb the ecological impact of population growth through economic redistribution. Nevertheless, it remains a factor, and the concentration of population growth in the urban shantytowns of the less-wealthy countries will play a significant role in the global pattern of confronting ecological limits.

Together, the three factors result in an ongoing growth logic that is built into national economies and polities. The metaphor of treadmills in environmental sociology—of production (Gould, Pellow, and Schnaiberg 2008), accumulation (Foster 2005), consumption (Bell 2006), and, we would add, weaponry—can be generally interpreted as representing an attempt to capture two historical processes: high levels of economic growth and a lack of systemic response to changes in adaptation to the global ecology. In other words, the economic and political fields support ongoing economic and ecological growth and lack the capacity to address the ecological crises generated by the growth. However, the metaphor of a treadmill is imperfect because it does not capture the overall growth logic of the economic system with respect to the global ecosystem (see also Foster 2005). To do so, one might be better off thinking of it as an expanding treadmill in a cage: the treadmill is itself expanding, and eventually it reaches the walls of the cage of ecological limits, when the treadmill breaks down or collapses.

The metaphor of a treadmill helps capture a fundamental problem in the linkage between the economy and the global ecosystem, but it does not completely capture the dynamics of how elites respond to awareness of ecological limits. The economy is also undergoing the greening of industrial production, and the polity is undergoing a transformation of governance processes that involves the construction of a wide range of environmental regulations and reforms. The changes have been amply described in the literature in environmental sociology on ecological modernization (Mol 1995; Mol and Spaargaren 2000, 2005; Scheinberg 2003). The literature can be interpreted to claim that a new industrial revolution is taking place along ecological lines, and this interpretation ends up forcing a choice between a treadmill perspective and an ecological modernization perspective (Gould, Pellow, and Schnaiberg 2008). However, the two perspectives can be made compatible if the ecological modernization thesis is interpreted as recognition that a greening process is occurring and governance processes are changing, but the extent of such changes is highly variable across industries and countries. If interpreted in this restrictive manner, one can then recognize the coexistence of the greening of industry and governance and ongoing growth in withdrawals and deposits into the global ecosystem. Furthermore, a global perspective on sustainability, in Daly's sense, makes it possible to see that the greening of one industry and country may be associated with the export of pollution, waste, and browner industrial processes to other countries (Pellow, Schnaiberg, and Weinberg 2000; York and Rosa 2003). To date, the greening of industry and the changes in governance at a global scale have not yet addressed the fundamental issue of achieving a steady-state economy. In order for the greening of industry to be ecologically significant from the perspective of a Daly-type definition of global ecosystem sustainability, technological innovation at a global scale would have to outpace levels of absolute global growth of environmental withdrawals and sinks. To date the dematerialization of the economy generated by green technological innovation has been swamped by the overall growth of environmental sinks and withdrawals (Gould, Pellow, and Schnaiberg 2008).

Treadmill and other accumulation theories suggest a dismal diagnosis and poor prognosis regarding the capacity of technological innovation to bring about a dematerialization of the economy, and they have not offered much in the way of a therapy. Part of the appeal of ecological modernization theory is that it has analyzed policy strategies for developing cooperative relations among the state, industry, and civil society in order to move forward on pressing problems of environmental degradation. In contrast, to the extent that there is any treatment program in accumulation theories, it tends to draw attention to the role of social movements, including blue-green coalitions of labor and environmental groups, in providing the basis for a less-cooperative and more conflict-oriented strategy that contests the power of elites who have ignored warnings about environmental crisis (Gould, Pellow, and Schnaiberg 2008). Here, we suggest, is the starting point for a more interesting and fruitful debate, recast in somewhat different terms, that might move forward the field of environmental sociology as a whole. Which strategy poses a better way out of the dilemma of the treadmill: building complex partnerships among civil society organizations, the state, and industry; the confrontation, protest-oriented repertoires of the social movement sector of civil society; or some combination of both?

To be clear about the argument, from a diagnostic and prognostic perspective, the differences between a treadmill of production theory and an ecological modernization theory can be resolved. There is little doubt that greening processes and governance changes are occurring (ecological modernization), but the changes have not occurred at a sufficient pace to compensate for growth in global levels of environmental sinks and withdrawals, which are driven by the capitalist accumulation process in addition to interstate competition and, to a lesser extent, population growth (treadmill of production). If current trajectories continue, then scenarios of uneven collapse will become increasingly evident. However, the closure of the theoretical controversy on one front might also serve as a starting point for questioning the analytical focus of the field of environmental sociology on diagnosis and prognosis. In other words, recognition of closure of the debate might provide an occasion for exploring the potential of the field to contribute to the analysis of strategies of environmental policy. By linking environmental social theory to environmental policy, the field is challenged to bring its theoretical insights and empirical research findings into contact with real-world problems of pressing policy significance. In this chapter we use the case of the convergence of solar energy and nanotechnology not only as an exemplar of the complex mix of benefits and risks that nanotechnology poses but also as an opportunity to develop an analysis of how environmental social theory and environmental policy might be brought together.

## 2.3 Nanotechnology and Solar Energy

The "next Industrial Revolution" is full of promises and hype about how nanotechnology will change every aspect of human existence. The claims are at times alluring but also foreboding: Drug delivery and diagnostics will be transformed, new materials will become available at a much lower cost and higher strength, potable water will become readily available through new processes of desalinization, new systems of surveillance and chemical monitoring will become possible, and a new generation of armaments and

weaponry will emerge. The promises also extend to environmental ame-
lioration, an issue that makes an environmental sociology of nanotechnology a complicated enterprise: Nanoscale chemicals may become available to
replace the current generation of toxic, chlorinated chemicals; nanomaterials
could lead to breakthroughs in the use of fuel cells and batteries; new materials based on nanotechnology could reduce the impact of mining for metals;
and nanoscale electrical materials will be both smaller and more conductive,
leading to revolutions in electrical use and efficiency and a post-silicon era
for computing. Among the environmental benefits, we focus on solar energy,
partly because the promise here is perhaps the most appealing: solving the
problem of climate change through nanotechnology (Schmidt 2007).

Solar energy has long been the alternative energy technology preferred by
social movements, partly because its modular design can be made compatible
with decentralized and democratic ownership (Hayes 1979; Laird 2001; Reece
1979). Furthermore, from an energy perspective, solar energy is potentially
much greater than other clean or renewable energy sources, and, to date, it
has been less controversial than wind farms (Breukers and Wolsink 2007;
Firestone and Kempton 2007). Solar, wind, and related renewable-energy technologies have long been recognized as the basis for an economy that enables
the dematerialization of its energy consumption; hence, they are likely to be
essential ingredients in a transition to a steady-state economy. However, the
great problem with solar energy has been its high cost. Eventually, new materials, such as the ones that are already appearing in thin-film technology, will
bring down the costs and reduce the environmental impacts of production.
Grid parity, the point at which the price of solar energy becomes equivalent
to that of energy supplied over the grid primarily from a fossil-fuel source,
will occur sometime between 2012 and 2020. When the convergence of prices
is reached, industry analysts predict that there will be an explosion of solar
energy production and a rapid transition toward solar energy. This is not to
say that other energy sources—such as wind, geothermal, and tidal energy—
will not be important, but solar energy is different because it is a beneficiary of
the rapidly advancing innovations of photonics, materials science, and other
fields. Although perhaps not enjoying quite the rapid improvements associated with Moore's law, solar energy is likely to become much less expensive
over time, not only in relationship to fossil fuels and nuclear energy but also
in relationship to other renewable energy sources.

There are many ways that scientists are trying to achieve a revolution in
solar energy technology with nanoparticles: photosynthesis through the use
of titanium dioxide nanoparticles; nanoparticle encapsulation in polymers;
the development of calcopyrites produced as thin film photovoltaics; the use
of molecular organic solar cells; organic polymer photovoltaic systems with
nanoscale layers; the addition of single-wall nanotubes to conduct polymers
that improve efficiency; smaller size nitride semiconductors, which result in
more efficient photovoltaic systems; and photovoltaic nanoparticles coated
with thin films of polymer that can create cheap flexible solar cells. Through

a combination of the many innovations that are currently in laboratories or beginning to be tested in markets, nanotechnology becomes central to the field of solar energy by offering the potential to accelerate the decline in solar energy costs and even to help solar energy become the cheapest form of energy.

Nanotechnology and clean energy consultant Bo Varga states the potential frankly and succinctly: "Solar growth at 20 percent per year for fifty years can replace fossil fuels and nuclear and remove the causes of global warming" (2007, 1). One may argue with his assumptions and projection, but the fundamental proposition is interesting as a possible way out of some very difficult energy problems posed by climate change. There is little incentive for any country to leave oil, coal, and natural gas in the ground, and the increase in global demand will only make it more difficult to resist the temptation to drill, extract, sell, and use more fossil fuels. Carbon trading agreements could make fossil fuels less competitive with respect to renewable energy technologies, but to date the schemes in Europe have proven to be less effective than originally projected (Hansen 2009). A technological development that would bring about the widespread diffusion of a much cheaper alternative could provide an even more powerful incentive for countries and firms not to continue to use fossil fuels. Solar energy could provide that technological innovation.

To its credit, the U.S. government has recognized, albeit in a limited way, the potential of the nanosolar convergence. If one looks through the research projects funded by the Solar America Initiative, a significant number of them involve nanocrystals, nanotubes, nanowires, quantum dots, and other nanoscale materials (U.S. Department of Energy 2007). By 2011, a few companies were already bringing nanosolar technologies to market. Innovations include a printable nanocrystal technology by Solexant and a nanoparticle ink printed on thin foil by Nanosolar. The new, printable technologies enable solar photovoltaics to be produced without the glass panels that are characteristic of the older-generation, silicon-based photovoltaics.

Nanosolar convergence promises to do more than simply bring down the costs of solar and make it the preferred form of energy generation. Future scenarios include a complete redesign of energy products and technologies. Just about any surface that receives light, including clothing, could provide an opportunity for energy generation. Nano-antenna arrays can produce energy based on infrared resources. A flexible, plastic-like nanosolar cell could be sprayed onto other materials much in the way that one can spray paint onto a surface today. Stefan Lovgren, *National Geographic* correspondent and winner of the American Association of Advancement of Science journalism award, describes the following scenario: "A hydrogen-powered car painted with the film could potentially convert enough energy into electricity to continually recharge the car's battery. The researchers envision that one day 'solar farms' consisting of the plastic material could be rolled across deserts to generate enough clean energy to supply the entire planet's power needs" (Lovgren 2005). Lovgren goes on to estimate that only 0.1 percent of the Earth's surface would be needed to replace all human energy needs with this "clean and renewable" alternative.

It is impossible at this time to know how much, if any, of the promises of a nanosolar future will be realized by 2050 or 2100. In this light, while nanosolar may be new and full of possibilities, the hype surrounding it is not. Such hype is similar to that which pervades other aspects of the nanotechnology revolution and that which accompanied previous "technological revolutions" such as nuclear energy during the 1950s and 1960s, which promised to provide the world with an endless supply of cheap, clean electricity. Even today, nuclear energy advocates continue to suggest that their energy source, not solar energy, promises to solve the world's energy needs and greenhouse gas emissions problems. Nor is nuclear the only energy selected by contemporary industry to make such promises. Not surprisingly, there is also considerable hype around carbon sequestration technologies as the best choice of energy futures.

Although we have used the term *hype* to describe the futures promised by advocates of various "clean" energies, their visions of a future of ubiquitous and cheap solar energy are also expressions of a struggle among actors for dominance as important players in a highly competitive energy industry and field of funding competition. One site where this conflict plays out with special intensity is energy research funding, and in the United States there have been significant differences between Republicans and Democrats on the issue. The Republican administration of President George H. W. Bush was more supportive of research on nuclear energy and fossil fuels than renewable energy, and it attempted to cut the federal government budget for solar energy research and energy-efficiency technologies (DuBois 2008). In contrast, the Democratic administration of Barack Obama steered more resources toward solar and other renewable energy resources, and made substantial cuts in research funding for nuclear energy and fossil fuels. Even so, the budget of the Obama administration for energy research and development continued to provide higher support for nuclear energy ($495 million) and fossil fuels research ($438 million) in comparison with solar ($302 million), wind ($123 million), and geothermal energy ($55 million) (U.S. Department of Energy 2010).

## 2.4 Nanosolar Risk and Uncertainty

In addition to the parallel between the hype surrounding nuclear energy during the 1950s and the hype around nanotechnology and nanosolar today, one might draw a second parallel between the two energy sources, one that cuts deeply into the rosy futures described above. It took decades for the effects of uranium mining on local environments, the health effects of radiation exposure, the possible nightmare of terrorist attacks on nuclear energy plants, the risks of severe events such as the earthquake and tsunami that affected the Fukushima reactors, and the problem of waste disposal of spent nuclear fuel to become recognized as a bundle of negative side effects generated by nuclear technology. In a similar way, research is slowly emerging on the environmental, health, and safety (EHS) implications of nanotechnology (for reviews, see Donaldson et al. 2006; Helland et al. 2007; Lam et al. 2006; Singh and Nalwa 2007). Nongovernmental organizations such as Environmental Defense, the ETC Group, Friends of the Earth, and Greenpeace have sent warning signals about the potential for nanotechnology to repeat the mistakes of the past (Hess 2010; Lamprou 2010). Whether one makes the comparison with nuclear radiation or previous generations of materials that proved hazardous (such as asbestos and chlorinated chemicals), emerging knowledge on the EHS implications of nanotechnology suggests that society may be in the process of repeating its past mistakes by unleashing a new generation of toxic materials into the environment.

Although many of the world's industrialized powers provide some government funding for research on EHS implications of nanotechnology, several of the leading civil-society organizations in the United States, as well as prominent researchers and policymakers, have argued that the research has been systematically underfunded, whereas government support for the commercialization of nanotechnology has been much more forthcoming (Hess 2010). As a result, more is understood about the potential benefits of nanotechnology than about the risks, dangers, and unwelcome surprises.

Engineered nanoparticles are already entering the biosphere through waste streams and airborne particles. A main source of environmental exposure to engineered nanomaterials is in the waste streams from factories and research laboratories. Studies point toward the possible effects of nanoparticles on microorganisms as well as small animals such as earthworms (Brumfiel 2003; Oberdörster 2004). Studies conducted at Rice University have shown that nanoparticles could easily be absorbed by earthworms; this research suggests that it is possible for nanomaterials to move up through the food chain and reach humans (Brumfiel 2003). Another source of environmental exposure is the release of airborne nanomaterials from powders, which present especially high levels of concern because they can easily enter the human body through inhalation and become deposited in the lungs (Maynard and Kuempel 2005; Oberdörster 2000, 2004; Oberdörster et al. 2004; Oberdörster, Oberdörster, and Oberdörster 2005). Studies have shown that inhaled particles, in general—even when they have a low intrinsic toxicity to cells—may cause diseases of the lungs if the dose is of a particular strength. Diseases may arise because the immunological defenses of the lungs become overloaded if the total surface area of the affected lungs is large enough (Faux et al. 2003).

In particular, research on the toxicity of nanomaterials has characterized some risks and uncertainties associated with three commonly used nanoparticles: titanium dioxide, fullerenes ($C_{60}$), and carbon nanotubes.[1] In the case of nano-sized titanium dioxide, toxicity research shows that when they interact with cells after inhalation the nanoparticles exhibit more toxic properties than in their bulk form (Donaldson et al. 2006; Heinrich et al. 1989). In the case of fullerenes, research based on computer modeling has indicated that $C_{60}$ molecules may bind to and deform nucleotides when they come into contact with each other (Zhao, Striolo, and Cummings 2005). Recent studies have also implicated fullerenes in oxidative stress in the brains of largemouth bass and have suggested other adverse physiologic impacts on aquatic organisms (Hood 2004; Oberdörster 2004; Zhu, Oberdörster, and Haasch 2006). To date the research does not give a clear answer about how nano-$C_{60}$ behaves in aquatic environments, and until more is known about the toxicity, compounds containing nano-$C_{60}$ must be handled carefully (Lyon et al. 2005).

When it comes to carbon nanotubes, there is evidence to suggest that they may stimulate mesenchymal cell growth and cause granuloma formation and fibrogenesis (Donaldson et al. 2006). Carbon nanotubes can also be much more toxic than carbon black and quartz, and they represent a serious occupational health hazard, especially in chronic inhalation exposures (Dreher 2004; Lam et al. 2004).[2] Tests on single-wall carbon nanotubes (SWCNTs) in rats and mice showed toxicity in the form of granuloma and inflammation (Lam et al. 2004; Warheit et al. 2004). Other studies, also conducted with mice, measured the pulmonary responses to SWCNTs delivered by pharyngeal aspiration and suggested that if workers are exposed to such particles at the current permissible exposure limit, they may be at risk of developing lung lesions. The rapid fibrogenic response to aspiration of SWCNTs indicates the need for more extensive inhalation research (Shvedova et al. 2005; see also Poland et al. 2008).

The most important finding from research, with respect to the hazards of nanoparticles, remains the fact that cells and organs may have toxic responses even to normally nontoxic substances when they are exposed at a sufficient dose at the nanoscale (Borm and Kreyling 2004; Renwick, Donaldson, and Clouter 2001). Ultimately, the capacity for the kidneys to separate and discharge chemicals depends on their solubility and surface coating (Borm and Kreyling 2004). However, some particles selectively deposit in particular organs or cells, and there is a possibility that nanoparticles can penetrate cells or cross biological barriers such as the blood–brain barrier (Illum and Davis 1987). Particles smaller than one hundred nanometers (100 nm) in diameter are not only able to enter the lung interstitium and become deposited in the lungs but can also enter the bloodstream (Ferin and Oberdörster 1992; Maynard 2006; Oberdörster et al. 2002), and they can enter the liver and brain through the nerve axons (Oberdörster 2000; Oberdörster et al. 2002, 2004).

In summary, nanomaterials pose some documented risks and many unknown dangers to human health, nonhuman organisms, and the environment. Although exposure does not automatically translate into disease, the preliminary and underfunded research on the EHS dimensions of nanomaterials for humans and other animals suggests the need for a pre-cautionary approach to the regulation of nanomaterial release into the environment. But regulatory policy should also recognize that some types of nanomaterials are likely to be more dangerous than others. It may be possible that some designs of nanotechnology materials will turn out to be relatively more dangerous, in terms of human health risk, and relatively amenable to disposal that minimizes diffusion into the environment. For example, cosmetics and other personal care products that use free nanoparticles in creams that are applied directly to the skin are more likely to pose higher levels of risk due to increased contact with human bodies and a greater likelihood that the materials will degrade outside of a safe disposal process. However, even in that case we still do not know the levels of risk involved (Berube 2008). In general, free or unembedded nanomaterials are more dangerous than those that are embedded in a matrix structure or grown in a substrate. Assuming that most nanosolar products could be embedded in a matrix structure at a molecular level and placed in sealed solar panels at a product level (an assumption that may eliminate the spray-on nanosolar materials described above), this particular design of nanosolar materials may present relatively low levels of hazard and risk compared with other nanotechnological products. In other words, from a toxicological perspective, printed nanosolar on thin films may be a safe option if the materials do not degrade during use and can be recycled in a safe way. The alternate prospect of a transparent nanospray that can turn windows into photovoltaic generators may present exposure problems to both workers who use the spray and users who are exposed to the degradation of the materials due to sunlight, rain, wind, and other weather factors.

At this point, we know little about which particular kinds of nanotechnology designs can be deemed safest. The public, policymakers, and NGOs are faced with a situation of "undone science," or inadequate levels of research to provide a basis for a public-interest perspective on policy action on the risks of nanotechnology (Frickel et al. 2010). As in the case of funding for solar energy research, the budgets for EHS research on nanotechnologies have done relatively poorly during Republican administrations and only relatively better under Democratic administrations. For example, the funding level for EHS research increased from about $35 million in 2005 during the Bush administration to $117 million in 2011, during the Obama administration (Erickson 2011). The growth in the funding for nanotechnology research and product development requires a constantly increasing level of EHS research in order to keep pace, and not all funding is directly relevant to understanding human health and safety issues.

In the United States, hundreds of nanotechnology products had been released into markets by 2011, but the Environmental Protection Agency had only managed, after several years of delay, to put in place a voluntary regulatory program. The agency added to the existing and weak voluntary program an interpretation of the Toxic Substances Control Act that required a premanufacturing notice of nanoscale materials only if they were structurally different from larger chemicals that were already on the market. The classification decision is highly controversial, because nanoscale materials that are structurally similar to preexisting chemicals may have significantly different biological properties due to their size, as noted above.

In summary, the phenomenon of nanosolar presents two underlying conflicts or tensions. First, there is widespread governmental and industrial interest in nanotechnology, but solar energy has received relatively low levels of support due to competition from fossil fuels and nuclear energy. Consequently, nanosolar has not been brought to the forefront of the nanotechnology revolution. The attention it receives in the future, therefore, depends largely on a purely rationalistic assessment: how the potential for nanosolar to bring down the cost of solar energy is assessed, as well as how the importance of the rapid commercialization of renewable energy technologies is prioritized. Second, nanosolar could reduce environmental side effects resulting from human consumption by making manufacturing and electricity generation much more efficient and cost effective, although it could also generate new environmental side effects through EHS hazards and risks. Assuming that nanosolar materials are embedded in matrices or grown in a substrate, those hazards and risks would likely be concentrated in the workplace where the materials are produced and in disposal sites where sealed panels may break, matrices may degrade, and materials embedded in matrices may be released into the environment. Potentially, those problems could be addressed by extending existing models for handling toxic waste.

Given this situation, a complex response to nanosolar is in order. One would want to see much more research funding, both for the technology itself and for the EHS risks associated with different types of nanosolar design. By setting research funding goals that would allow the two strands of research to converge, it would be possible to know something about what kinds of nanosolar designs are most likely to pose minimal EHS risks to best protect workers, users, and the environment from EHS hazards posed by the new technologies.

## 2.5 Policy Strategies and Nanotechnology

Treadmill of production theorists and other accumulation theorists in environmental sociology would have no trouble explaining the rush to commercialization of nanotechnology, the government support of nanotechnology research, the relatively low levels of both solar energy research and EHS nanotechnology research, and the failure to generate a new regulatory framework for nanotechnology. Industrial interests, especially those of the fossil fuel and chemical industries, have dominated the policymaking processes that shape the destiny of nanosolar. Scientific research groups and civil society organizations are struggling to stay abreast of the problems generated by the premature release of new substances into the environment after significant investment and commercialization had already taken place. From the perspective of environmental social theory, there is little of interest in the EHS risks generated by the nanotechnology revolution. It is yet another case of capital seeking new investment opportunities and attempting to block or slow any regulatory impediments that might reduce access to those opportunities.

Although the case of nanotechnology may pose little theoretical interest for accumulation theories other than more grist for their mills, we suggest that the challenge of making sound nanotechnology policy, including nanosolar policy, does pose a greater opportunity for environmental sociology. If one begins with the assumption that a policy goal of developing a safe and responsible nanosolar industry offers potentially high societal and environmental benefits, then another analytical vista is opened up: the study of policy strategies and their effectiveness.

Those who assume that a more robust regulatory policy and higher level of EHS research would be generally beneficial would be advocating a shift in policy toward a less market-oriented and more state-interventionist approach to industrial regulation. One can then distinguish two strategies for increasing the likelihood of implementing this policy approach: an "activist" approach, modeled on historical social movements with their extra-institutional repertoires of action, such as street-based protest and civil disobedience, and an "advocate" approach, involving institutional repertoires of action that are associated with reform movements, such as working within the political field via elections, petitions, and lobbying (Hess 2007). Unfortunately, there is no good, concrete example of these two ideal types in the field of nanotechnology activism and advocacy. However, we now briefly examine the work of two organizations—the ETC Group and Environmental Defense—that have attempted to approximate the two strategies, before returning to the related theoretical implications for environmental sociology (see also Hess 2010).

The ETC Group has called for a complete moratorium on all laboratory research and commercial applications of nanotechnology until an "International Convention on the Evaluation of New Technologies" is established (ETC Group 2003a, 2003b). With respect to nanotechnology policy, the group has not engaged in extra-institutional repertoires of action characteristic of social movements, but it has built bridges with labor, consumer, and human rights groups. The primary policy tactic to date has been the circulation of petitions, publication of reports, and participation in the World Social Forum. Until now, ETC's petitions and public information campaigns have not had much policy impact. Industry has rejected the group's call for a complete moratorium until a global regulatory structure is in place, and so far the public has not taken up the NGO's warnings about the potential risks of nanotechnology. Although nanotechnology is still in a less-mature phase than biotechnology was in the late 1990s, a repeat of the public rejection of genetically modified food, especially as it occurred in Europe, has yet to happen for nanotechnology. The special place of food in cultural politics and its status as a product that is ingested on a daily basis made it easier and more tangible for social movement organizations to politicize genetically modified food. However, the story for nanotechnology is far from over, and it is possible that an anti-nanotechnology movement equivalent of that for genetically modified food could emerge (on the parallels with genetically modified food, see Sandler and Kay 2006; cf. Thompson 2008).

In contrast, Environmental Defense developed a partnership arrangement with DuPont to articulate a best-practices framework for voluntary participation by industry, and the organization helped the Environmental Protection Agency develop a voluntary program of chemical registration (Environmental Defense–DuPont Nano Partnership 2007). Over time, Environmental Defense advocates became frustrated with the voluntary approach and increasingly called for a definition of all nanomaterials as "new chemicals" under the Toxic Substance Control Act (Denison 2007c). Environmental Defense also called for a higher budget for EHS research, a separation of the nanodevelopment budget from the EHS budget, and other regulatory changes for the United States (Denison 2007a, 2007b).

From the perspective of environmental social theory, the contrasting policy approaches of the ETC Group and Environmental Defense represent relatively minor disagreements within an overall policy strategy of a shift toward higher levels of regulation of nanotechnology. However, the differences in the relative impact of the two strategies are potentially of interest for answering the kinds of questions that an environmental sociology or environmental policy might ask. In this particular case, a conflictual strategy of a call for a complete moratorium and social movement mobilization has had little impact on policy. More generally, the conflictual strategy may be an effective policy strategy for some political issues. As we have shown elsewhere in a comparative analysis of industrial opposition movements, the results often lead to partial success in the form of a partial moratorium (Hess 2007). In the case of nanotechnology, in particular, we are most likely to see partial moratoria emerging for specific particle types found in consumer products, such as nanosilver in clothing and nanoparticles in sunscreens and cosmetics. In contrast, a cooperative strategy that has involved partnerships among industry, the government, and civil society has produced some changes in the nanotechnology policy. The strategy of dialogue, partnership, and politicking through federal government institutions seems to be paying off for Environmental Defense and partner organizations, as it has showed some progress on achieving policy goals in Congress, especially relating to levels of EHS research funding.

From the perspective of treadmill theory, the work of Environmental Defense would be considered "policy tinkering." Gould and colleagues note that such approaches do have some environmental benefits and offer some potential, but they also find that such approaches do not address fundamental institutional dynamics that drive ongoing destruction of ecosystems (Gould, Pellow, and Schnaiberg 2008). Instead, they are more supportive of a social movement strategy that involves coalitions among labor, environmental, and other transnational social movement organizations. Presumably, the strategy would also maintain a focus on the fundamental goal of building a steady-state economy. Of course, the two strategies do not need to be considered in a zero-sum relationship. Having both strategies work in tandem is likely to be more effective than either strategy employed on its own, particularly because of the potential for flank effects. In other words, the threat of a public uptake of a complete moratorium on nanotechnology posed by the ETC Group may help open up political opportunities for Environmental Defense and other insider organizations that are attempting to institute industry-wide best-practice standards and achieve incremental regulatory and research funding reforms. Likewise, the process of educating legislators that Environmental Defense has undertaken may open up political opportunities for partial moratoria.

## 2.6 Conclusion

The treadmill tendency for capital to invest in new technologies that increase the efficiencies of production, as well as the tendency for governments to invest in new technologies with military and economic potential, will lead to ongoing nanotechnology innovations. Those innovations could result in the growth of nanosolar, which in turn could generate new levels of toxic exposure to workers, consumers, and the broader environment. One might argue that the rush to nanotechnology, especially outside the framework of a robust EHS research agenda and extensive regulatory oversight, is merely one more example of accumulated capital being channeled into a new industry that will generate higher profits and better weapons at the expense of a new level of environmental destruction. In a worst case scenario, ubiquitous, spray-on, plasticized solar panels will generate new energy sources along with growing levels of toxic chemical exposure.

On the other hand, from an ecological modernization theory perspective, one might counterargue that the development of nanosolar will provide a new, inexpensive source of energy that could substantially displace other, less environmentally desirable alternatives, such as nuclear energy and carbon sequestration for coal. Furthermore, by utilizing best practices in the design, manufacture, use, and disposal of the technology, spray-on nanosolar designs may end up being excluded from production, and safer nanosolar designs can be brought into existence in ways that are consistent with the emerging research of the EHS field. As a result, nanosolar might actually lead to a significant dematerialization of the economy in the sense of reduced aggregate levels of human impact of greenhouse gases on the global ecosystem.

Although playing the two frameworks against each other may be helpful in elucidating the environmental politics at play in the development of nanosolar, we are not trying to use the case of nanosolar to reopen the debate between treadmill of production theory and ecological modernization theory. In our view, the debate over diagnosis was largely resolved as follows: the greening of industry and the ecological modernization of society are occurring, but the changes are highly localized and industry specific, and to date they have not fundamentally addressed the challenge of growth in aggregate withdrawals and deposits into the global ecosystem. Even if at least half of the world's energy is produced from nanosolar by the year 2050, it is very possible that it will not be enough to enable a steady-state global economy to be achieved, because overall energy consumption may continue to grow so much that nanosolar does not displace fossil fuels enough to bring down atmospheric greenhouse gas emissions. Hence, even with a nanosolar revolution, aggregate levels of absolute withdrawals and deposits from the global ecosystem might continue to rise.

Rather than use the case of nanosolar to give new life to the controversy between treadmill of production theory and ecological modernization theory, we are instead arguing that the debate be transposed into the sociology of policy. The controversy left unanswered by the debate over diagnosis and prognosis was the crucial question of treatment, of an analysis as to which strategies for political and economic change are most likely to bring about the dematerialization of society. Even if one suspects, as we do, that the prospect of solving global greenhouse gas emissions with ubiquitous nanosolar is overblown, one might nevertheless agree that, as long as governments such as that of the United States are investing over $1 billion per year in nanotechnology research, then a priority within that investment portfolio should be nanosolar technology development coupled with EHS research that would enable the determination of how to design nanosolar in ways that minimize environmental, health, and safety risks. Although one might have some qualms about the potential toxicities of nanosolar, the other, "back to the future" energy scenarios for twenty-first-century energy production appear even less appealing, such as a return to the mid-twentieth-century world of nuclear energy or the nineteenth-century world of coal, albeit cleaned up with carbon sequestration technology. Both of these energy solutions have well-known shortcomings, including the threat of terrorist attacks, meltdowns, nuclear waste disposal, ecosystem degradation from coal mining, the unproven effectiveness and availability of underground storage, the lethal toxicity of carbon bubbles, and the cost of carbon sequestration technology. Even wind energy, which is currently cost effective at a large scale, poses problems of intermittency, storage, and transmission. As a result, an energy policy that would enable the possibility of a more rapid transition to grid parity for solar energy would seem to be a reasonable part of a balanced future energy research portfolio.

To shift environmental social theory toward the analysis of environmental policy, one could argue that no incremental or reformist policy interventions will work, and that, because more radical approaches have been taken off the table of policy debate, collapse is inevitable. In that case, retreat into marginalized, ecosocialist movement activism is probably the only viable strategy. But if one assumes that some incremental changes are possible and that they will at least mitigate the worst effects of collapse, then one has shifted the debate onto the grounds of an analysis of mainstream policy. We have developed a contrast between extra-institutional activism and institutional advocacy as ideal types. To some extent the contrast between these two different strategies for reform approximates the underlying theoretical differences in environmental social theory between the treadmill of production and ecological modernization approaches. However, the connection is probably only coincidental and contingent. For example, one might agree with the fundamental treadmill argument that, to date, the greening of industry has not had a significant impact on levels of absolute global deposits and withdrawals, and yet one might still consider an incremental, institutional advocacy policy strategy similar to that of Environmental Defense as more effective in this particular historical circumstance. Conversely, one might argue that—at least in some countries and industries—the greening of industry has significantly reduced local deposits and withdrawals into the ecosystem, yet still prefer an extra-institutional activist policy strategy as the best way to move such localized and industry-specific victories forward. More to the point, it could also be argued that a mixture of the two strategies is more likely to be effective than either solely on its own. Whichever position one takes, the broader point is that any research on this issue may help open up new vistas in environmental sociology that have been inadequately conceptualized and explored in this theoretical debate. This could help shift environmental sociology from diagnosis and prognosis to treatment, which in turn could offer more effective strategies for policy reform.

To avoid misunderstanding, it should be clear that the analysis of policy and political strategy should be distinguished from prescriptive discourse. Therefore, we are not suggesting that environmental social scientists begin proselytizing with normative statements about what should be done. Instead, we are arguing for a more empirically oriented form of social science research that seeks to understand what kinds of political strategies work best given a particular set of environmental policy goals and historical circumstances. Therefore, a key question arises as to which advocacy strategies are likely to be most effective if we wish to see a greater government role in solar research, nanosolar research, EHS research, and nanotechnology regulation. We now know that the patient is sick and possibly terminal; those debates are over. What we do not know and need to know is which therapies are most likely to help or, at a minimum, which therapies will reduce the misery as the disease of global ecosystem collapse runs its course.

## References

Bell, M. 2006. Welcome to the consumption line: Sustainability and the post-choice economy. In *Conference on Sustainability Consumption and Society*, June 2–3, 2006, Madison, Wis. Available at http://www.michaelmbell.net/suscon-program. htm (accessed February 15, 2011).

Berube, D. 2008. Rhetorical gamesmanship in the nano debates over sunscreens and nanoparticles. *Journal of Nanoparticle Research* 10(1): 23–37.

Borm, P. and Kreyling, W. 2004. Toxicological hazards of inhaled nanoparticles: Potential implications for drug delivery. *Journal of Nanoscience and Nanotechnology* 4(5):521–31.

Breukers, S. and Wolsink, M. 2007. Wind power implementation in changing institutional landscapes: An international comparison. *Energy Policy* 35(5): 2737–50.

Brumfiel, G. 2003. Nanotechnology: A little knowledge. *Nature* 424(6946): 246–48.

Costanza, R., Graumlich, L., and Steffen, W., eds. 2007. *Sustainability or Collapse? An Integrated History and Future of People on Earth*. Cambridge, Mass.: MIT Press.

Daly, H. 1990. Toward some operational principles of sustainable development. *Ecological Economics* 2(1): 1–6.

Daly, H. 1996. *Beyond Growth: The Economics of Sustainable Development*. Boston, Mass.: Beacon Press.

Denison, R. 2007a. Questions for the record to Dr. Richard A. Denison, Senior Scientist, Environmental Defense. U.S. House of Representatives Committee on Science and Technology. Available at http://www.edf.org/documents/7347_DenisonQFRresponsesFINAL.pdf (accessed February 15, 2011).

Denison, R. 2007b. Statement of Richard A. Denison, Ph.D., Senior Scientist, Environmental Defense, before the U.S. House of Representatives Committee on Science and Technology. Available at http://www.edf.org/documents/7287_DenisonTestimony_10312007.pdf (accessed February 15, 2011).

Denison, R. 2007c. Statement of Richard A. Denison, Ph.D., Senior Scientist: US EPA's public meeting on the development of a voluntary nanoscale materials stewardship program. Available at http://www.edf.org/documents/6749_EnvironmentalDefenseStmt2007NanoHearing.pdf (accessed February 15, 2011).

Donaldson, K., Aitken, R., Tran, L., et al. 2006. Carbon nanotubes: A review of their properties in relation to pulmonary toxicology and workplace safety. *Toxicological Sciences* 92(1): 5–22.

Dreher, K. 2004. Health and environmental impact of nanotechnology: Toxicological assessment of manufactured nanoparticles. *Toxicological Sciences* 77(1): 3–5.

DuBois, D. 2008. FY 2009 budget means big cuts for efficiency, renewables. *Energy Priorities Magazine*, February 11. Available at http://energypriorities.com/entries/2008/02/fy_2009_budget_energy.php (accessed February 15, 2011).

Environmental Defense–DuPont Nano Partnership. 2007. *Nano Risk Framework.* Available at http://www.edf.org/documents/6496_Nano%20Risk%20 Framework.pdf (accessed February 15, 2011).

Erickson, B. 2011. Nanotechnology investment. *Chemical and Engineering News* 88(15): 22–24.

ETC Group. 2003a. *The Big Down: Atomtech—Technologies Converging at the Nanoscale.* Available at http://www.etcgroup.org/en/issues/nanotechnology (accessed February 15, 2011).

ETC Group. 2003b. *No Small Matter Ii: the Case for a Global Moratorium.* Occasional Paper Series 7, pp.1–20. Available at http://www.etcgroup.org/en/issues/ nanotechnology (accessed February 15, 2011).

Faux, S., Tran, C., Miller, B., and Jones, A. 2003. *In Vitro Determinants of Particulate Toxicity: The Dose-Metric for Poorly Soluble Dusts.* Norwich, U.K.: Institute of Occupational Medicine for the Health and Safety Executive.

Ferin, J. and Oberdörster, G. 1992. Translocation of particles from pulmonary alveoli into the interstitium. *Journal of Aerosol Medicine* 5(3): 179–87.

Firestone, J. and Kempton, W. 2007. Public opinion about large offshore wind power: Underlying factors. *Energy Policy* 35(3): 1584–98.

Foster, J. 2005. The treadmill of accumulation. *Organization and Environment* 18(1): 7–18.

Frickel, S., Gibbon, S., Howard, J., et al. 2010. Undone science: Social movement challenges to dominant scientific practices. *Science, Technology, and Human Values* 35(4): 444–73.

Gould, K., Pellow, D., and Schnaiberg. A. 2008. *The Treadmill of Production: Injustice and Unsustainability in the Global Economy.* Boulder, Colo.: Paradigm.

Grätzel, M. 1991. The artificial leaf, molecular photovoltaics achieve efficient generation of electricity from sunlight. *Comments on Inorganic Chemistry: A Journal of Critical Discussion of the Current Literature* 12(2): 93.

Hansen, G. 2009. *Storms of My Grandchildren: The Truth about the Coming Climate Catastrophe and the Last Chance to Save Humanity.* New York: Bloomsbury.

Hayes, R. 1979. *Blueprint for a New America.* Washington, D.C.: Solar Lobby.

Heinrich, U., Muhle, H., Hoymann, H.G., and Mermelstein, R. 1989. Pulmonary function changes in rats after chronic and subchronic inhalation exposure to various particulate matter. *Experimental Pathology* 37(1–4): 248–52.

Helland, A., Wick, P., Koehler, A., et al. 2007. Reviewing the environmental and human health knowledge base of carbon nanotubes. *Environmental Health Perspectives* 115(8): 1125–31.

Hess, D. 2007. *Alternative Pathways in Science and Industry.* Cambridge, Mass.: MIT Press.

Hess, D. 2010. The environmental, health, and safety implications of nanotechnology: Environmental organizations and undone science in the United States. *Science as Culture* 19(2): 181–214.

Hood, E. 2004. Fullerenes and fish brains: Nanomaterials cause oxidative stress. *Environmental Health Perspectives* 112(10): A568–A568.

Illum, L. and Davis, S. 1987. Targeting of colloidal particles to the bone-marrow. *Life Sciences* 40(16): 1553–60.

Laird, F. 2001. *Solar Energy, Technology Policy, and Institutional Values.* Cambridge: Cambridge University Press.

Lam, C., James, J. McCluskey, R., and Hunter, R. 2004. Pulmonary toxicity of single-wall carbon nanotubes in mice 7 and 90 days after intratracheal instillation. *Toxicological Sciences* 77(1): 126–34.

Lam, C., James, J., McCluskey, R., et al. 2006. A review of carbon nanotube toxicity and assessment of potential occupational and environmental health risks. *Critical Reviews in Toxicology* 36(3): 189–217.

Lamprou, A. 2010. *Nanotechnology Regulation: Policies Proposed by Three Organizations for the Reform of the Toxic Substances Control Act.* Available at http://www.chemheritage.org/research/policy-center/publications/studies-in-sustainability.aspx (accessed February 15, 2011).

Lovgren, S. 2005. Spray-on solar-power cells are true breakthrough. *National Geographic News*, January 14. Available at http://news.nationalgeographic.com/news/2005/01/0114_050114_solarplastic.html (accessed February 15, 2011).

Lyon, D.Y., Fortner, J.D., Sayes, C.M., Colvin, V.L., and Hughes, J.B. 2005. Bacterial cell association and antimicrobial activity of a $C_{60}$ water suspension. *Environmental Toxicology and Chemistry* 24(11): 2757–62.

Maynard, A. 2006. Nanotechnology: Assessing the risks. *Nano Today* 1(2): 22–33.

Maynard, A. and Kuempel, E. 2005. Airborne nanostructured particles and occupational health. *Journal of Nanoparticle Research* 7(6): 587–614.

Meadows, D., Randers, J., and Meadows, D. 2004. *Limits to Growth: The Thirty-Year Update.* White River Junction, Vt.: Chelsea Green, 2004.

Mol, A. 1995. *The Refinement of Production: Ecological Modernization Theory and the Chemical Industry.* Utrecht, The Netherlands: International Books.

Mol, A. and Spaargaren, G. 2000. Ecological modernisation theory in debate: A review. *Environmental Politics* 9(1): 17–49.

Mol, A. and Spaargaren, G. 2005. From additions and withdrawals to environmental flows: Reframing debates in the environmental social sciences. *Organization and the Environment* 18(1): 91–108.

Oberdörster, G. 2000. Pulmonary effects of inhaled ultrafine particles. *International Archives of Occupational and Environmental Health* 74(1): 1–8.

Oberdörster, G, 2004. Manufactured nanomaterials (fullerenes, $C_{60}$) induce oxidative stress in brain of juvenile largemouth bass. *Environmental Health Perspectives* 112(10): 1058–62.

Oberdörster, G., Oberdörster, E., and Oberdörster, J. 2005. Nanotoxicology: An emerging discipline evolving from studies of ultrafine particles. *Environmental Health Perspectives* 113(7): 823–39.

Oberdörster, G., Sharp, Z., Atudorei, V., et al. 2002. Extrapulmonary translocation of ultrafine carbon particles following whole body inhalation exposure of rats. *Journal of Toxicology and Environmental Health: Part A* 65(20): 1531–43.

Oberdörster, G., Sharp, Z., Atudorei, V., et al. 2004. Translocation of inhaled ultrafine particles to the brain. *Inhalation Toxicology* 16(6/7): 437–45.

Pellow, D., Schnaiberg, A., and Weinberg, A. 2000. Putting the ecological modernization theory to the test: The promises and performances of urban recycling. *Environmental Politics* 9(1): 109–37.

Poland, C., Duffin, R., Kinloch, I., et al. 2008. Carbon nanotubes introduced into the abdominal cavity of mice show asbestos-like pathogenicity in a pilot study. *Nature Nanotechnology* 3(7): 423–28.

Reece, R. 1979. *The Sun Betrayed: A Report on the Corporate Seizure of U.S. Solar Energy Development.* Boston: South End Press.

Renwick, L., Donaldson, K., and Clouter, A. 2001. Impairment of alveolar macrophage phagocytosis by ultrafine particles. *Toxicology and Applied Pharmacology* 172(2): 119–27.

Sandler, R. and Kay, W.D. 2006. The GMO-nanotech (dis)analogy? *Bulletin of Science and Technology* 26(1): 57–62.

Scheinberg, A. 2003. The proof of the pudding: Urban recycling in North America as a process of ecological modernization. *Environmental Politics* 12(4): 49–75.

Schmidt, K. 2007. Nanofrontiers: Visions for the future of nanotechnology. Washington, D.C.: Wilson Center. Available at http://www.nanotechproject.org/publications/archive/nanofrontiers_visions_for_future/ (accessed July 27, 2011).

Schnaiberg, A. 1980. *The Environment*. Oxford: Oxford University Press.

Schnaiberg, A. and Gould, K. 1994. *Environment and Society*. New York: St. Martin's Press.

Shvedova, A., Kisin, E., Mercer, R., et al. 2005. Unusual inflammatory and fibrogenic pulmonary responses to single-walled carbon nanotubes in mice. *American Journal of Physiology: Lung Cellular and Molecular Physiology* 289(5): L698–708.

Singh, S. and Nalwa, H. 2007. Nanotechnology and health safety: Toxicity and risk assessments of nanostructured materials on human health. *Journal of Nanoscience and Nanotechnology* 7(9): 3048–70.

Thompson, P. 2008. Nano and bio: How are they alike? How are they different? In *What Can Nanotechnology Learn from Biotechnology? Social and Ethical Lessons for Nanoscience from the Debate over Agrifood Biotechnology and GMOs*, edited by K. David and P. Thompson, 125–55. New York: Elsevier/Academic Press.

U.S. Department of Energy. 2007. *Future Generation Photovoltaic Devices and Processes Selections*. Solar Energy Technologies Program. Available at http://www1.eere.energy.gov/solar/pdfs/next_generation_pv_prospectus.pdf (accessed July 20, 2010).

U.S. Department of Energy. 2010. *FY 2011 Congressional Budget Request: Budget Highlights*. Available at http://www.mbe.doe.gov/budget/11budget/Content/FY2011Highlights.pdf (accessed February 15, 2011).

Varga, B. 2007. Nano solar news: global warming and 2015 cost trends. *Nanotechnology Now*, March 14. Available at http://www.nanotech-now.com/columns/?article=038 (accessed February 15, 2011).

Warheit, D., Laurence, B. R., Reed, K. L., et al. 2004. Comparative pulmonary toxicity assessment of single-wall carbon nanotubes in rats. *Toxiological Sciences* 77(1): 117–25.

Wei, S. and Zunger, A. 1990. Band-gap narrowing in ordered and disordered semiconductor alloys. *Applied Physics Letters* 56(7): 662.

Weizsäcker, E.U.V., Lovins, A.B., and Lovins, L.H. 1998. *Factor Four: Doubling Wealth—Halving Resource Use: The New Report to the Club of Rome*. London: Earthscan.

York, R. and Rosa, E. 2003. Key challenges to ecological modernization theory. *Organization and Environment* 16(3): 273–88.

Zhao, X., Striolo, A., and Cummings, P. 2005. $C_{60}$ binds to and deforms nucleotides. *BioPhysical Journal* 89(6): 3856–62.

Zhu, S., Oberdörster, E., and Haasch, M.L. 2006. Toxicity of an engineered nanoparticle (fullerene, $C_{60}$) in two aquatic species, Daphnia and Fathead Minnow. *Marine Environmental Research* 62 (Supplement 1): S5–S9.

## Endnotes

1. $C_{60}$ is a spherical carbon molecule in the form of twenty hexagons and twelve pentagons, named "fullerene" after Buckminster Fuller.
2. Carbon black is a material produced from the incomplete combustion of petroleum products that is used as a pigment and reinforcement in rubber and other products.

# 3

# Nanotechnology and Traditional Knowledge Systems

Ron Eglash

## CONTENTS

## 3.1 Introduction

Traditional knowledge systems, in particular those of indigenous societies with hunter-gatherer or horticultural economies, have made a surprising impact on many disciplines surrounding science and technology. The pharmaceutical industry, for example, has long used "ethnobotany"—the study of indigenous utilization of plants—to help discover biologically active molecules. This is not entirely counterintuitive; we can imagine many centuries in which an exhaustive trial-and-error search by indigenous communities resulted in the discovery of medically useful preparations. Such common-sense assumptions, however, leave us unprepared to appreciate the aspects of indigenous knowledge that go beyond mere accident, nor do they prepare us for indigenous practices in nonorganic arenas. In this essay I present case studies in which indigenous knowledge has produced parallels to particular artifacts or processes in the high-tech world of nanoscience. Not only do these cases provide evidence that sophisticated traditional knowledge can extend beyond the organic world, but they also help us understand the contributions of indigenous practices in contemporary science and technology.

The uses of curare in surgery, quinine in malaria prophylaxis, and other traditional medical applications of indigenous discoveries now pale in comparison to the massive efforts in "bioprospecting," in which large-scale scientific programs launch a comprehensive search for biologically useful compounds (often in ecosystems of indigenous communities). The legal and ethical issues have become so significant that the American Association for the Advancement of Science has instituted a project on indigenous knowledge and intellectual property (IP) rights, and the concerns have spread to other ethnoscience disciplines (Hansen and VanFleet 2003; Nicholas and Bannister 2004).

As Watson-Verran and Turnbull (1995, 116) note, "indigenous peoples have been frequently portrayed as closed, pragmatic, utilitarian, value laden, indexical, context dependent, and so on, implying that they cannot have the same authority and credibility as science." A key challenge to this dismissal of traditional knowledge has been the acknowledgment that Western science is also local and value laden. Scott (1996), for example, notes that all knowledge systems make use of "root metaphors" to provide a cohesive framework. A root metaphor is a fundamental analogy that guides our thinking across phenomena that are otherwise hard to conceptualize. For example, our understanding of electrical current uses the root metaphor of fluid flow—we know that the movement of electrical charge is not really a fluid, but we can conceptualize it by talking about electrical "resistance" as if it were a blockage in a fluid pipe, electrical "capacitance" as a tank with a certain fluid capacity, and so forth (hence electrical "current"). Scott shows that the root metaphor of personhood in native conceptions of particular species allows the construction of (what we could translate as) ecologically sustainable natural resource management. In a similar example, Langdon (2007) provides archaeological evidence suggesting that as the indigenous population of what is now coastal southern Alaska increased, salmon populations also increased. Langdon attributes this to an intertidal fishing practice that increased salmon habitats, guided in part by the personhood root metaphor. One need not partake of a mystical or religious point of view here; rather we can think of the personhood root metaphor as one way of understanding the "agency" of nonhuman systems, which has been an important development in recent social studies of science and technology (cf. Pickering 1995), and an ancient practice for indigenous societies.

Indigenous knowledge systems cannot be automatically dismissed as merely nontheoretical, unintentional, and unconscious, just as they cannot be automatically valorized as transparent equivalents to Western science. Skeptical questions and falsifiable hypothesis testing are crucial to the documentation of traditional knowledge; both deliberate charlatanism as well as well-meaning efforts leaning on pseudoscience or misinformation can destroy its scholarly value (Oritz de Montellano1993; Martel 1994; Restivo 1985). The best approach is not to claim wholesale analogies but rather to show the various connections between the body of knowledge in its original and historical contexts and their parallels (and differences) in contemporary science and technology—illuminating both in the process.[1]

The epistemological status of traditional knowledge is critical to preventing appropriations of indigenous heritage, which also makes its use possible in a contemporary context—as in the case of nanotechnology. Another important area of application for traditional knowledge has been in underrepresented minority student education. Many teachers in science, technology, and math have turned to cultural connections that often include traditional knowledge from ancient state societies, such as Egyptian, Mayan, and Hindu, as well as those of smaller-scale tribal or band societies. Again, this application hinges on the epistemology of these practices; unless they have the status of knowledge, they cannot be used to contest primitivist or ethnocentric portraits of non-Western culture. In this chapter I first map a number of connections between traditional knowledge and nanotechnology, thereby arguing that a more careful examination of the social histories of these artifacts and practices can shift our understanding of their epistemological status. I then detail some of the new and significant possibilities emerging for their application, especially in the areas of education and further research development, with the caveat that such practices should be done with the consent and legal safeguards that protect indigenous intellectual property rights and the well-being of their communities.

## 3.2 Connections to Nanotechnology in the Traditional Knowledge of Ancient State Societies

### 3.2.1 Case study: Wootz and Damascus Steel

One of the most spectacular connections between nanotechnology and culture has been the discovery of the famed Damascus steel, used in Middle Eastern sword making from about 1100 to 1700 CE. In ordinary steel production of that period, sharpness and strength would be opposing tradeoffs: Increasing carbon content for sharpness would make a sword more brittle, and decreasing carbon content for strength would prevent it from holding an edge. We now know that it was a special type of steel ("wootz") from India—developed perhaps as early as 300 BCE—that was used to forge the blades. Indian metallurgists used ores from particular mines that included alloying trace elements such as vanadium and molybdenum (Verhoeven, Pendray, and Dauksch 1998). The name "Damascus steel" may have originated in association with the forging of blades in Damascus, Syria, but another possibility is that it was named after the characteristic pattern of wavy lines seen on the blade (in Arabic, "damas"). The disappearance of wootz steel in the eighteenth century is attributed to the diminishing supply of Indian ores with the proper trace elements. Bladesmiths continued to mimic the wavy line pattern by forge welding alternating sheets of high- and low-carbon steels, but the extraordinary material properties of wootz were no longer present. It has been known for some time that the wavy pattern in blades of wootz steel origin was due to bands of iron carbide particles ("cementite"). But cementite is typically brittle; somehow the trace elements, together with the particular heat treatments, were preventing the cementite from weakening the blade. Recently, high-resolution transmission electron microscopy was used to examine a sample of Damascus saber steel from the seventeenth century; it showed the presence of carbon nanotubes as well as cementite nanowires (see Reibold et al. 2006).

This example illustrates that traditional knowledge can include manipulation or use of material properties relevant to nanotechnology. But if we stop there, we leave the impression that wootz was simply an interesting artifact from the ancient past, and science merely tells us what the ancients failed to understand. Such a view leaves out the active role that wootz has played throughout the history of metallurgic science. Scientific analysis of wootz is nothing new; Europeans have long been aware that there was something special about it, and this mystery has inspired a great deal of important metallurgical research. Michael Faraday, for example, is best known for his foundational work in electrical and chemical physics, but prior to those experiments he sought to discover the secret of wootz (not a surprising move, given his father's employment as a blacksmith). His study proved that the wavy pattern on wootz blades was due to an inherent crystalline structure and not a mechanical mixture of substances. Faraday's later experiments with metallic colloids, in which he suggested that size differences in extremely small metal particles could produce color changes, has been cited as the birth of nanoscience (Edwards and Thomas 2007).

Wootz experimentation has since continued over centuries and continents, such as by Giambattista in Italy (1589), Reaumur in France (1722), Bergman in Sweden (1781), Anossoff in Russia (1841), and Smith in the United States (1960), among many others. In their review of this history, two professors of materials science in India, Sharada Srinivasan and Srinivasa Ranganathan, concluded that several important innovations in metallurgical science—most strikingly the role of carbon in steel—have been associated with wootz research. They also pointed out that discoveries in this historical trajectory are still ongoing. For example, Reaumur proposed that the properties of steel are determined at several scales, from microscopic "grains" to a hypothesized nanoscale of "periodic spheres." At Massachusetts Institute of Technology in the 1960s, Cyril Smith, sometimes referred to as a "philosopher-metallurgist," recovered the work of Reaumur, translating it into the modern idiom of a multiscale architecture where crystalline, molecular, and atomic processes have mutual influence on each other. Professor of material science Greg Olsen, inspired by Smith as a student, later developed software to "design" steel using this model of multiscale processes. Recently, Olsen's work was celebrated for its mixture of humanities and technology, as he hired bladesmith Richard Furrer, an expert in the reproduction and use of wootz steel, to make a "mythic" blade using his computationally designed steel, Ferrium C69 (Davis 2001). Therefore, in short, wootz steel as an example of nanotechnology in traditional knowledge is not merely an matter of historical curiosity but rather a "boundary object" (Star and Griesemer 1989) through which Western and non-Western metallurgists have maintained a dialogue over the past 400 years—one still relevant today.[2]

### 3.2.2 Case Study: Maya Blue Pigment

Another remarkable example of "retrospective" nanotechnology is Maya blue pigment. First formally identified by Harvard archaeologist R. E. Merwin at Chichén Itzá in the 1930s, it is notable for its stability, maintaining a brilliant blue color despite centuries of exposure to heat and moisture in a tropical climate. Why this mixture of indigo and white clay (palygorskite) did not fade was a mystery. Miguel José-Yacamán, a materials scientist then at the University of Mexico, proposed that nanosized channels in the palygorskite protected the indigo and metal combination (José-Yacamán et al. 1996). Recent evidence suggests that the indigo is actually embedded in surface grooves, rather than interior channels (Chiari et al. 2008) and that the carbonyl oxygen of the indigo is bound to a surface aluminum ion (Polette-Niewold et al. 2007). José-Yacamán found an almost identical composition in eight paint samples, even though they came from sites dozens of kilometers apart. He concluded that there was a remarkable level of "quality control" in the paint production.

Again this retrospective view—remarkable as it is—is not the end of the story. Maya blue is not only resistant to heat and moisture, it is also resistant to biocorrosion, mineral acids, and alkalis (Sanchez del Rio 2006). Researchers at the University of Texas in El Paso noted that, since indigo could be substituted by other organic dyes, its chemistry not only offered a new class of organic/inorganic complexes for research but also exciting new possibilities for application, since current pigments are mostly based on either environmentally unfriendly heavy metals or strategic metals that are in short supply (Polette-Niewold et al. 2007). The researchers recently formed a private company, Mayan Pigments Inc., and have already received National Science

Foundation funding for their research as well as contracts with industry for their services.

Finally, it is important to note that although popular representations limit the use of Maya blue as an artifact from the precolonial past, Maya blue pigment continued to be used even after colonization. It was applied in the sixteenth century in Catholic convents in Mexico; the best examples are in the paintings of Native American Juan Gerson in Tecamachalco. Although its use in Mexico apparently ended after that point, in Cuba its use continued up to 1830 (Chiari et al. 2000).

## 3.3 Connections to Nanotechnology in the Traditional Knowledge of Nonstate Indigenous Societies

### 3.3.1 Case Study: The Obsidian Blade

Although indigenous societies that did not constitute a state typically lacked the labor specialization we associate with knowledge production, they still managed to refine their use and manipulation of materials over many centuries. Perhaps one of the best known is obsidian tools, which reached a highly sophisticated state of craft among Native Americans. Obsidian is a glassy mineral used in many indigenous cultures to produce blades for arrows, spears, and knives. Anthropologist Payson Sheets of the University of Colorado, Boulder, was excavating obsidian glass blades in El Salvador during the early 1970s. Sheets investigated the blades' cutting properties, replicating the fracturing process used in ancient indigenous cultures of Central America. Using an electron microscope, he compared the cutting edges of the obsidian blades to those of modern disposable steel scalpels and to diamond scalpels, the sharpest surgical tools available. The obsidian blades turned out to be two to three times sharper than diamond scalpel blades—at their smallest only 3 nanometers across—but at 1/100th the cost, and have since gone into commercial production (Sheets 1989). One study comparing wound healing using obsidian and steel scalpels found that the extremely thin edges of obsidian create statistically significant wound healing advantages (Disa, Vossoughi, and Goldberg 1993).

A significant contradiction exists between the popular press reports of this technology and the actual history of native flint knapping (that is, sculpting a stone into a blade). For example, the Michigan *University Record* (September 10, 1997) titled its article "Surgeons use Stone Age technology for delicate surgery" and ended with a comment about "our Paleolithic ancestors." While the contrast between the prehistoric and modern works well to grab readers' attention, it misleads the popular audience into thinking that this technology stopped advancing when glaciers receded. Flint knapping was widely used by Native American groups well into the nineteenth century (and in some cases beyond). Nonnative admiration for obsidian tools was dramatically increased when flint knapping took hold as both a hobby and aid to professional archaeologists (such as Payson Sheets) wishing to reconstruct original methods to better understand these tools. Some of the most highly crafted projectile points are still beyond the skill level of all but a handful of dedicated artisans. Computational models for flint knapping have been applied to fracture mechanics in dentistry, metal fatigue, tribology, and other areas of contemporary concern (cf. Fonseca, Eshelby, and Atkinson 1971). Again, far from an irrelevant artifact of interest only to antiquarians, this body of traditional knowledge and practice is still a resource for contemporary technological development.

### 3.3.2 Case Study: Piezoelectricity

Another mineral-based case of indigenous connections to nanoscience can be found in the Native American use of quartz crystals to generate flashes of light for ceremonial purposes. Piezoelectric actuators (in the form of tiny polycrystalline ferroelectric ceramic materials) are often used in nanopositioning systems: Energy inputs create mechanical movement. The Native American use is the reverse: The light is energy output generated by the piezoelectric effect of mechanical stress on the crystals. This is accomplished by placing the crystals in translucent rattles. Again, the popular representations are at odds with the actual practice. The phenomenon first received wide coverage when it was described in a Wikipedia page on the Uncompaghre Ute Indians (Colorado–Utah area) and quickly became a standard "human interest" component for popular accounts of the piezoelectric effect. *Plenty* magazine, for example, reported that "Thousands of years ago, the Ute Indians of Colorado cleverly filled rattles with pieces of quartz that glowed when shaken together to create the world's first flashlight, no batteries required" (Clark 2007).

However, a review of the literature on Ute traditions has not revealed any mention of it. Two ethnographers contacted by this author have confirmed reports among Southwestern groups (Eastern Pueblo and Northern Ute), but it appears that this was adopted in the twentieth century through "pan-Indian" syncretism—that is, the blending of various native cultures that occurred after (and partly in oppositional response to) colonization (Cornell 1988). The only well-documented reports of traditional use of the piezoelectric effect appear to be in the Lakota *yuwipi* ("stone power") ceremony. Descriptions of this ceremony typically report sparks at the crucial moment when all light is extinguished; the sound of rattles accompanied by these blue flashes of light is said to indicate the presence of spirits (cf. Powers 1986).

An Assiniboine spiritual leader in northern Montana (culturally of close relation to the Lakota) described how a medicine man would send a request to ants—"Please mine white quartz"—and then return a week later to

anthills to gather the stones, which he said were valued because they produce flashes of light in the dark during the *yuwipi* ceremony (Mayor 2005).
Two aspects of this story are particularly significant in making the case for
indigenous knowledge of the piezoelectric effect. First, this shows how the
root metaphor of personhood in nonhumans (in this case, ants) can work for
traditional knowledge concerning inorganic physical phenomena. Ants are
spiritually significant because they connect the subterranean world, identified with sacred origins, and the world that humans inhabit. Mayor notes
that in addition to sorting out quartz for the *yuwipi* ceremony, ant mounds
also sort out tiny fossils that have ceremonial use among the Sioux as well
as the Cheyenne. He also describes similar use of ant mounds by paleontolo-
gists, who found them so useful in sorting out tiny fossils that they would
bring soil samples from other areas to ant mounds to be sorted for them, and
even packed ant mounds in crates and had them shipped to sites. Pickering
(1995) would describe such scientific phenomena as a "mangle"—capturing
nonhuman agency in ways that change both scientific practice and the nonhumans. Mangle here seems an apt way to identify the Native American root
metaphor as well, although they would likely reject Pickering's use of the
term "capturing" as identifying a more European-American attitude than
the collaborative approach they emphasize with personhood.

A comparison between Native American and European histories of piezoelectricity is also illuminating. Katzir (2006) notes, "Various references in
ancient and mediaeval literature suggest the possibility that the phenomenon was observed in the West long before. However even if the attraction
of tourmaline was known before … it was forgotten and had no practical
tradition. No one knew how to identify the stone or stones mentioned in the
books" (24).

This stands in contrast to the Native American practice, which had both
practical application (*yuwipi*) and a systematic method for identifying the
particular stones that were most effective in producing the piezoelectric
effect (searching ant mounds). The books to which Katzir refers are ancient
Greek descriptions of pyroelectricity, a related phenomenon in which crystals create an electrical charge when heated. This was rediscovered by Dutch
gem cutters in the early eighteenth century, but it was not until 1880 that
Jacques and Pierre Curie found that applying mechanical stress to crystals
could also create an electrical charge.

Thus, while the popular accounts (as in the case of obsidian) portray the
Native American practice in terms of a frozen prehistoric past, this case of
indigenous knowledge shows a dynamic history. Not only did the Native
American observations of the piezoelectric effect predate by many centuries
that of the European discovery, but more importantly their root metaphor
proved more reliable in allowing others to replicate the phenomenon than
did that of the ancient Greeks, who failed to transmit the ability to replicate a
similar phenomenon (the pyroelectric effect) to later generations. One might
defend the Greeks as having a difficult cultural barrier in communicating

with Dutch and French scientists, centuries later, but the Ute adopters and the Lakota originators were also from widely disparate cultures.

## 3.4 A Whole World of Scientific Knowledge

Both examples of exploiting the nanoscale properties of obsidian and quartz are merely the tip of the iceberg for the intimate knowledge of physical and chemical phenomena that indigenous societies have accumulated through centuries of experimentation. The Gwich'in Athabascan tribe in Alaska, for example, has over sixty artifacts/products they produce from the birch tree (Engbloom-Bradley 2006). They not only differentiate use between botanical parts of the tree (bark, roots, and so on) but also more subtle variations; for example, they use the north side of the tree to make arrows because of its greater hardness, and the south side to make bows because of its greater flexibility. Plant extracts such as resins and saps are the most common indigenous encounters with chemistry; traditional uses include adhesives for crafting artifacts, waterproofing for containers, incense for religious ceremonies, medicinal compounds, and many other applications (Langenheim 2003).

As in the previous examples of wootz and obsidian, use of plant extracts by indigenous populations has also been part of a long-term conversation with Western science. Imagine what the history of technology would have been without rubber, a plant extract introduced to Europeans by South American Indians. Synthetics did not simply replace these natural extracts, since some required plant extracts as part of their production (for example, the addition of camphor to make celluloid), and many of these plant extracts are still in use today (for example, shellac). As Peters (1994) notes, resin tapping "probably comes the closest to conforming to the ideal of sustainable non-timber forest product extraction" and shows great promise for linking indigenous livelihoods with forest conservation. One exemplar in this case is investigation of spiniflex resin through collaboration between the indigenous people of the Myuma group in northwest Queensland and the Aboriginal Environments Research Centre at the University of Queensland's School of Architecture. The Research Centre's director, Paul Memmott, notes that the project includes experts in botany, bio-nano-engineering, chemistry, and architecture, as well as Aboriginal community members ranging from elders with traditional knowledge to postgraduate student Malcolm Connolly, who conducts experiments in harvesting and regrowth of the plant.

One special category in the relation between nanoscale effects and indigenous knowledge is the contrast between our normal expectations of physical phenomena and the counterintuitive physics enabled by certain specific nanoscale structures. For example, the indigenous descriptions of the

Lakota *yuwipi* ceremony specifically mention the blue color of the sparks, differentiating them from sparks generated by combustion. This special category—indigenous knowledge that derives from counterintuitive physics—is particularly important because of its potential for establishing the kinds of cultural connections that could be applied to educational contexts. In the West we think of magic as something requiring illusion or fakery, something hidden up the sleeve or done with mirrors. But many traditional examples of African "magic" are performed openly and depend on counterintuitive physics. Some of these are related to nanoscale phenomena. For example, anthropologist Paul Stoller reported that, during his apprenticeship to a Songhay sorcerer, he witnessed his teacher spread a fine powder on the surface of a bowl of water and then retrieve an item from the bottom of the bowl without getting his fingers wet (Stoller 1987). Such demonstrations of surface tension are common in contemporary science classes (where lycopodium powder is typically used). Another example occurs in the Bayaka society of central Africa, where fluids from a luminescent fungus are used as body paint during a nighttime ritual (Sarno 1993).

## 3.5 Connections to Nanotechnology in Traditional Knowledge at the Macroscale

Certain macroscale structures found in traditional knowledge systems also offer important cultural connections to nanoscale phenomena. The relation between fullerene molecules and the geodesic domes of Buckminster Fuller is well known. Less well known is the use of similar structures in baskets and other indigenous artifacts. Paulus Gerdes, professor of mathematics in Mozambique, has studied the use of hexagonal weaving patterns in Africa and indigenous cultures elsewhere (such as in India, Brazil, and Malaysia) and has investigated their use in modeling fullerenes (Gerdes 1998). As a flat sheet, the hexagonal weave resembles the structure of graphite; rolled into a cylinder, it resembles a nanotube. Gerdes notes that weavers introduce a pentagonal weave when they need a corner. Figure 3.1(a) and Figure 3.1(b) show a Malaysian sepak ball in which the pentagons and hexagons tile the surface, creating the same fundamental structure as that of a $C_{60}$ fullerene molecule (that is, both are truncated icosahedrons). Gerdes's work demonstrates how all the fullerene structures can be generated using this indigenous weaving technique. This has important applications in education, as we will see in Section 3.6.

Other macroscale connections can be found under the rubric of self-organization, which is used in both nanotechnology (molecular self-assembly) and indigenous social organization. A particularly vivid example of indigenous

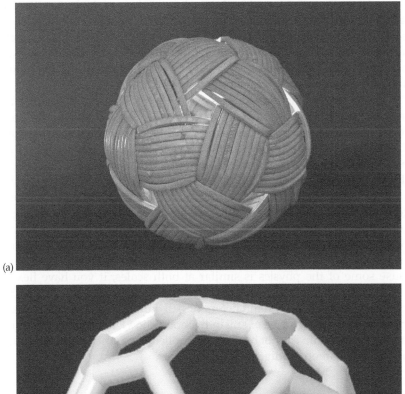

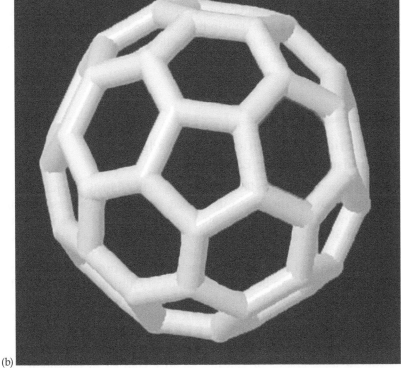

**FIGURE 3.1**
(a) Malaysian sepak ball (b) $C_{60}$ fullerene molecule.

self-organization can be found in the fractal structure of African settlement patterns, where consistent geometric patterns occur over several magnitudes of scale (Eglash 1999). Coppens (2009) provides an extensive list of fractal nanostructures, some with implications for improving environmental sustainability, such as distribution of oxygen over a hydrogen fuel cell (which would only produce water vapor as a by-product).

It is important to distinguish between the previous examples that are based on knowledge of physical properties—wootz steel, Lakota rattles, and so forth—and these examples of macroscale structures. There is no evidence that any indigenous group had knowledge of nanoscale geometry, and the pop-culture texts that claim such knowledge for ancient Hindus or Zen Buddhists are detrimental to scholarly work (see critique in Restivo 1983). Macroscale structures are significant because they show indigenous knowledge of relations between geometry and physical properties (for example, the structural integrity afforded by a hexagonal mesh). These relations can also apply at the nanoscale—not because of indigenous scanning electron microscopes, mystical cosmic knowledge, or ancient astronauts but simply because some of the physics is similar at both scales; if you have flexible joints, the only way to make a rigid structure is to use triangles. That is true whether the triangle is 10 nanometers or 0.01 kilometers. The macro/nano correlation is not itself a part of any indigenous knowledge, but that does not mean the relationship cannot be usefully applied to nanoscale science education, especially among indigenous descendants.

## 3.6 Applications of Traditional Knowledge to Nanoscale Science Education

Research by many scholars indicates that some of the statistically poor record of achievement and participation in science, technology, engineering, and math disciplines by African American, Latino, and Native American youth in the United States can be attributed to cultural barriers. One barrier can be found in myths of genetic determinism, which lowers expectations for minority students and thus becomes a self-fulfilling prophecy (Hoberman 1997). Another barrier can be found in myths of cultural determinism. For example, African American students sometimes perceive a forced choice between black identity and high scholastic achievement (Ogbu and Simons 1998). Many high-achieving African American students report that they have been accused of "acting white" by their peers (Austen-Smith and Fryer 2005). Similar assessments of cultural identity conflict in education have been reported for Native American, Latino, and Pacific Islander students (Kawakami 1995; Lipka and Adams 2004; Lockwood and Secada 1999).

Cultural connections to science and technology can be important resources for defeating these barriers. Myths of genetic determinism and myths of cultural determinism can be contradicted by evidence for sophisticated bodies of knowledge from the heritage cultures of these students. Again, it is important to note that the epistemological status of the traditional practice is critical to this use; merely showing that one can carry out a nanoscience analysis of an indigenous material is far weaker than showing an indigenous knowledge system that makes intentional use of a nanoproperty. Moreover, the cultural connection offers the opportunity to discuss not only the cultural context for indigenous innovation but also the social implications of contemporary practice. For example, students who insist that only the Western version offers a complete understanding can be asked to consider the salmon case cited in the opening of this chapter. If it's true that only the Western version is a complete understanding, then why are Western societies destroying their salmon populations (Krkosek 2010), in comparison to the indigenous sustainable harvest?

Case studies have shown effective culturally situated learning for minority students when using the epistemologically stronger examples of intentional knowledge (cf. Lipka and Adams 2004). In one recent U.S. study, black middle-school student responses to nanotechnology education were less engaged than those of white peers; researchers concluded that experiences that were more "connected to students' lives" would stand the best chance of addressing this disparity (Jones et al. 2007).

With others, my own research results in math education also support this culturally situated framework. Culturally Situated Design Tools (CSDTs) are web applets (http://www.csdt.rpi.edu/) based on ethnomathematics—in particular, the mathematical knowledge embedded in cultural designs such as cornrow hairstyles, Native American beadwork, Latino percussion rhythms, urban graffiti, and others. These tools allow students to use underlying mathematical principles to simulate the original cultural designs, create new designs of their own invention, and engage in specific math inquiries. Preliminary evaluations with minority students indicate statistically significant increases in math and computing education achievement, as well as attitudes toward technology-based careers (Eglash et al. 2006; Eglash and Bennett 2009; Eglash et al. forthcoming).

Surprisingly, we did not see a strong correlation between design tool selection and heritage identity. Minority students who had been trained in the use of all the tools and were allowed to select any tool for their final project did not show an overwhelming preference for designs from their own ethnic group. On the other hand, we often saw cases of "appropriation"—African American students using the Native American beadwork tool to create something similar to graffiti tags, or Latino students using the graffiti tool (based mainly on samples from New York City) but creating artwork that specifically reflected Latino cultural origins. Such appropriation fits well with the recent studies on the formation of cultural identity by contemporary youth, which replaces older models of ethnicity as a static given with portraits of youth in the process of actively constructing their identity, often in terms of hybridity and syncretism (Pollack 2004).

This observation on the importance of creativity and flexibility in culture-based education frameworks creates a challenge for culturally situated nanoscience education, which may lend itself less to design or other potentially expressive activities. In February 2008, with others, I conducted a brief workshop with African American students at Rensselaer Polytechnic Institute using Gerdes's African weaving approach. The website includes video clips of Baka women in Cameroon weaving a basket using a hexagonal lattice, visual content showing the connection between the woven artifacts and molecular structures, and finally instructions for creating a $C_{60}$ fullerene molecule model using paper hexagonal weaves. The workshop successfully competed for students with other programs offered at the time, drawing approximately fifty African American youth, with a majority of girls (possibly due to the traditionally gendered nature of weaving). A longer Nanoscale Technology and Youth summer program was offered in 2009 at State University of New York, Albany. It offered middle school students the opportunity to take several different nanotechnology workshops. At the end, the students selected images from our African weaving workshop over the other workshops as the image for their T-shirts. Both the 2008 and 2009 sessions indicated that the African weaving approach to creating a $C_{60}$ fullerene molecule model was sufficiently engaging to compete with other activities. The current technique requires multicolored strands with numbers at specific intersections; this method is probably not optimal in terms of simplicity, and we suspect that it discourages the students from using the weaving technique creatively, but clearly there is a strong potential for sparking interest through a cultural connection to nanotechnology.

Another tool in the CSDT suite offering a nanoscale connection is that of Anishinaabe Arcs, which offers simulations of wigwams, long houses, canoes, and other structures based on wooden arcs. The arcs are placed in very specific geometric relationships, such that a strong case can be built for indigenous geometric knowledge. The building materials are carefully selected according to factors such as terms of species and time of year to obtain the maximum structural characteristics (for example, elasticity and strength) as they must be placed into tension to form arcs. We tested our prototype with Anishinaabe students in a summer camp run by the Native American Studies center at Northern Michigan University. In contrast to the African weave, students were highly creative with this tool and produced a wealth of different forms (see Figure 3.2).

Finally, we tested Maya blue with a group of Latino and African American students during an after-school program (Figure 3.3). Students created their own Maya blue mix, heated it in a kiln, and painted it on white tiles. Although the small sample size (eight students) prevented statistical significance,

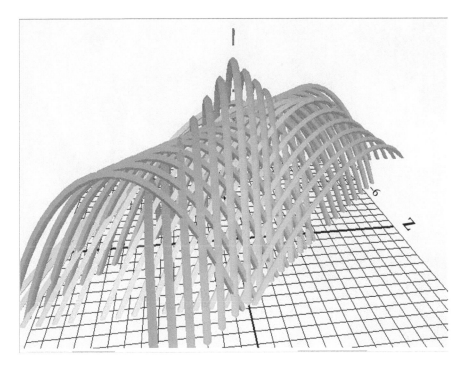

**FIGURE 3.2**
Design created by a Potawatomi student with the Anishinaabe Arcs tool.

pre- and post-test contrasts showed that "preventing pollution" became more strongly associated with nanotechnology after the workshop.

## 3.7 Applications of Traditional Knowledge to Intellectual Property Rights of Nanotechnology

In several cases, groups opposed to foreign ownership of traditional knowledge have used the legal system to prevent the misappropriation of patents (Ruiz 2002; Hansen and VanFleet 2003; Nicholas and Bannister 2004). The best publicized cases are those of the turmeric patent (U.S. Patent No. 5,401,504), the neem (*Azadirachta indica*) patents (over forty in the United States), and the ayahuasca (*Banisteriopsis caapi*) patent (U.S. Plant Patent No. 5,751). Five fundamental concerns are put forward by protestors.

1. That a patent unjustly appropriates the intellectual resources of the culture that created the knowledge.
2. That nontraditional use may offend indigenous cultural or spiritual sensibility or disrupt the social order.

**FIGURE 3.3**
Students prepare Maya blue by mixing indigo with palygorskite clay.

3. That a patent could block that culture from further development of its traditional knowledge or even its use.

4. That traditional knowledge properly belongs in the public domain, whereas a patent would privatize this knowledge.

5. That traditional knowledge practices serve an important role in protecting species, ecosystems, and landscapes.

In 1992 the United Nations Convention on Biological Diversity (CBD) used the fifth reason, protection of the ecosystem, to introduce the first regulations involving traditional knowledge. The CBD, coupled with organizing efforts by various indigenous groups and their advocates (such as the Declaration of Belem and the Indigenous Peoples Earth Charter), eventually led to an investigation by the World Intellectual Property Organization (WIPO), which broadened the rationale beyond the fifth to include Nos. 1 to 3. The WIPO established the Intergovernmental Committee on Intellectual Property and Genetic Resources, Traditional Knowledge, and Folklore. This committee is primarily focused on "negative protection," that is, a view of traditional knowledge as "prior art." For example, Subsection 102(f) of the U.S. Patent Act (35 U.S.C.) specifies that a patent will not be awarded when the applicant was not the original inventor. Thus any traditional knowledge—whether published or unpublished, whether in the United States or abroad, and which proves that the applicant is not the inventor—could be a basis for rejecting the application. A second form of "positive protection" concerns protective legal rights over traditional knowledge as part of cultural self-determination. In U.S. law, for example, many Native American tribes have retained their sovereign rights (albeit only in the wake of genocidal policy and extensive legal battles) and have maintained that their knowledge systems fall under these sovereign rights as much as land claims do (Brown 2003).

These strategies have had some success in defeating patent misappropriation (sometimes referred to as "biopiracy"). For example, in 1986 American scientist Loren Miller obtained a U.S. patent on a strain of the ayahuasca vine, which had been used by traditional healers in the Amazon for many generations. In 1999, Antonio Jacanamijoy, leader of a council representing more than 400 indigenous tribes in South America, achieved a rejection of Miller's ayahuasca patent by the U.S. Patent Office. Similarly, a U.S. patent for the use of turmeric in wound healing was awarded to the University of Mississippi Medical Center in 1995, despite its much publicized use as a traditional medicine in India. A complaint was filed by India's Council of Scientific and Industrial Research, and the U.S. patent office revoked the University of Mississippi's patent in 1997. In 1995 the U.S. Department of Agriculture and a pharmaceutical research firm were awarded a patent on an antifungal agent from the neem tree, which was used in traditional medicine in India. Following widespread public outcry, legal action was pursued by the Indian government, and the patent was eventually overturned in 2005.

Could the traditional knowledge involving material nanoproperties—wootz steel, Maya blue, obsidian blades—play a similar role in protecting the intellectual property (IP) rights of indigenous groups? At least two barriers are at play in these cases that were not prominent for the biopiracy cases. First, recall that the origins of IP protection of traditional knowledge in the United Nations CBD were purely based on ecological impact; it was only later that other factors were added. Since these "indigenous nanotechnology" cases are more focused on material properties than ecological properties, they may not fit the CBD protections and hence the subsequent protections built on that foundation. Second, these nanoproperty cases are often more difficult to connect to current populations, at least in cases in which they are no longer in use. However, neither of these barriers is absolute. First, as nanotechnology becomes increasingly blurred with biotechnology, examples such as the Bayaka use of fungi will increasingly link biopiracy with what could perhaps be termed "nanopiracy." Second, cases such as Maya blue are attractive precisely because they may provide environmentally preferable alternatives. Finally, as we have seen above, the popular descriptions of these nanoscience aspects of traditional knowledge often misleadingly present a portrait of knowledge frozen in an ancient or even prehistoric past, when in fact many of these technologies have been in dynamic play at least until the nineteenth century (if not beyond) and thus are more connected to current populations than it might appear at first glance.

However, it may be that traditional knowledge in the nanosciences will have an impact not as "positive protection" in terms of the sovereign rights of cultural self-determination but rather as "negative protection," constituting "prior art" that prevents patents from issuing at all. In such cases, the motivation might be more aligned with No. 4, traditional knowledge as public domain knowledge. The reasons for this are well explicated by the ETC Group's 2005 report, *Nanotech's "Second Nature" Patents: Implications for the Global South*. They note, for example, that U.S. patent 5,897,945 on nanoscale metal oxide nanorods covers not just one metal oxide but also oxides selected from any of thirty-three chemical elements (such as nanorods comprised of titanium, nickel, copper, zinc, or cadmium). This is nearly one third of all chemical elements in the Periodic Table in a single patent claim. Furthermore, many of these nanotechnology patents are assigned to all of the major patent classes, including electricity, human necessities, chemistry/metallurgy, performing operations and transportation, mechanical engineering (lighting, heating, weapons, blasting), physics, fixed construction, and textiles and paper. Despite the legal restrictions preventing patents on "natural phenomena," the breadth of nanotechnology patents at this (literally) elemental level, as well as the breadth of their application, suggests that many fundamental aspects of nature itself could become privatized as intellectual property.

The ETC Group concludes by noting that "patent claims on nano-scale formulations of traditional herbal plants are providing insidious pathways to monopolize traditional resources and knowledge" and recommends that protection take place by adding a nanotechnology component to the UN's CBD. However, keeping in mind that the CBD was for the purpose of environmental sustainability, such a strategy may only support those cases of indigenous nanoscience that fit under the biopiracy rubric. A broader commitment to the protection of indigenous knowledge of nanoproperties, including those of purely inorganic origin (such as wootz) or organic–inorganic hybrids (such as Maya blue) would provide better protection.

## 3.8 Conclusion

Traditional knowledge of nanoproperties, just like traditional knowledge of medicine, includes unique innovations that have already contributed to contemporary science and technology. But popular representations tend to portray them as curious artifacts frozen in an ancient past and only relevant as proof that science can reveal what the ancients failed to understand. The actual histories and contexts of this new class of traditional knowledge show a much more dynamic, vibrant set of practices that have, in some cases, provided important dialogues with the development of Western science and in other cases have applications that may lie in the future. Making these indigenous innovations available in a just and responsible manner—either as a component of culturally responsive education or as protection against misappropriated intellectual property rights—hinges on properly understanding and representing their epistemological status.

# References

Austen-Smith, D. and Fryer, R. 2005. An economic analysis of "acting white." *Quarterly Journal of Economics* 120:551–83.

Brown, M. 2003. *Who Owns Native Culture?* Cambridge, Mass.: Harvard University Press.

Chiari, G., Giustetto, R., Druzik, J., Doehne, E., and Richiardi, G. 2008. Pre-Columbian nanotechnology: Reconciling the mysteries of the Maya blue pigment. *Applied Physics A: Materials Science and Processing* 90:3–7.

Chiari, G., Giustetto, R., Reyes-Valerio, C., and Richiardi, G. 2000. Maya blue pigment: A palygorskite-indigo complex. *XXX Congresso Associazione Italiana di Cristallografia* 48(1): 115.

Clark, J. 2007. Back to the Future. *Plenty* (17 September). Retrieved from http://www.plentymag.com/magazine/back_to_the_future.php?page=2 on 4-1-09 (accessed July 27, 2011).

Coppens, M. 2009. Nature inspired chemical engineering: A new paradigm for sustainability. In *Confluence: Interdisciplinary Communications 2007/2008*, edited by Willy Østreng, 102–6. Oslo: Centre for Advanced Study.

Cornell, S. 1988. *The Return of the Native: American Indian Political Resurgence*. New York: Oxford University Press.

Davis, E. 2001. Forging the dragonslayer. *Wired* 9(2): 136–43.

Disa, J., Vossoughi, J., and Goldberg, N. 1993. A comparison of obsidian and surgical steel scalpel wound healing in rats. *Plastic and Reconstructive Surgery* 92:884–87.

Edwards, P.P. and Thomas, J.M. 2007. Gold in a metallic divided state—From Faraday to present-day nanoscience. *Angewandte Chemie International Edition* 46:5480–86.

Eglash, R. 1999. *African Fractals: Modern Computing and Indigenous Design*. New Brunswick, N.J.: Rutgers University Press.

Eglash, R. and Bennett, A. 2009. Teaching with hidden capital: Agency in children's mathematical explorations of cornrow hairstyle simulations. *Children, Youth, and Environments* 19(1).

Eglash, R., Bennett, A., O'Donnell, C., Jennings, S., and Cintorino, M. 2006. Culturally Situated Design Tools: Ethnocomputing from field site to classroom. *American Anthropologist* 108:347–62.

Eglash, R., Krishnamoorthy M., Sanchez J., and Woodbridge, A. Forthcoming. Fractal simulations of African design in pre-college computing education. *ACM Transactions on Computing Education*.

Engbloom-Bradley, C. 2006. Personal communication, September 2.

ETC Group. 2005. *Special Report: Nanotech's "Second Nature" Patents: Implications for the Global South*. Available at http://www.etcgroup.org/upload/publication/pdf_file/54 (accessed May 22, 2007).

Fonseca, J.G., Eshelby J.D., and Atkinson, C. 1971. The fracture mechanics of flint-knapping and allied processes. *International Journal of Fractal Mechanics* 7(4): 421–33.

Fujimura, J.H. 1992. Crafting science: Standardised packages, boundary objects, and translations. In *Science as Practice and Culture*, edited by A. Pickering, 168–211. Chicago: University of Chicago Press.

Gerdes, P. 1998. Molecular modelling of fullerenes with hexastrips. *Chemical Intelligencer* 40:41–45.

Handy, R.L. 1973. The igloo and the natural bridge as ultimate structures. *Arctic* 26(4): 276–81.

Hansen, S. and VanFleet, J. 2003. *Traditional Knowledge and Intellectual Property: A Handbook on Issues and Options for Traditional Knowledge Holders in Protecting their Intellectual Property and Maintaining Biological Diversity*. Washington, D.C.: American Association for the Advancement of Science.

Harding, S. 2006. *Science and Social Inequality: Feminist and Postcolonial Issues*. Champaign: University of Illinois Press.

Hoberman, J.M. 1997. *Darwin's Athletes: How Sport Has Damaged Black America and Preserved the Myth of Race*. New York: Houghton Mifflin.

Huerta, S. 2006. Structural design in the work of Gaudi. *Architectural Science Review* (December).

Jones, M., Tretter, T., Paechter, M., Kubasko, D., Bokinsky, A., Andre, T., and Negishi, A. 2007. *Journal of Research in Science Teaching* 44:787–99.

José-Yacamán, M., Rendon, L., Arenas, J., and Carmen Serra Puche, M. 1996. *Science* 273:223.

Katzir, S. 2006. *The Beginnings of Piezoelectricity: A Study in Mundane Physics*. Dordrecht: Springer.

Kawakami, A.J. 1995. *A Study of Risk Factors among High School Students in the Pacific Region*. Honolulu: Pacific Resources for Education and Learning.

Krkosek, M. 2010. Sea lice and salmon in Pacific Canada: Ecology and policy. *Frontiers in Ecology and the Environment* 8:201–9.

Langdon, S.J. 2007. Sustaining a relationship: Inquiry into the emergence of a logic of engagement with salmon among the southern Tlingits. In *Native Americans and the Environment: Perspectives on the Ecological Indian*, edited by M.E. Harkin and D.R. Lewis, 233–73. Lincoln: University of Nebraska Press.

Langenheim, J.H. 2003. *Plant Resins: Chemistry Evolution Ecology and Ethnobotany*. Portland, Ore.: Timber Press.

Lipka, J. and Adams, B. 2004. *Culturally Based Math Education As a Way to Improve Alaskan Native Students' Math Performance*. ACCLAIM Working Papers #20. Available at http://acclaim.coe.ohiou.edu/rc/rc_sub/pub/3_wp/LipkaAdams20.pdf.

Lockwood, A.T. and Secada, W.G. 1999. Transforming education for Hispanic youth: Exemplary practices, programs, and schools. *NCBE Resource Collection Series*, no. 12.

Martel, E. 1994. How not to teach ancient history. *American Educator* (Spring):33–37.

Mayor, A. 2005. *Fossil Legends of the First Americans*. Princeton, N.J.: Princeton University Press.

Nicholas, G. and Bannister, K. 2004. Copyrighting the past? Emerging intellectual property rights in archaeology. *Current Anthropology* 45:327–50.

Ogbu, J. and Simons, H. 1998. Voluntary and involuntary minorities: A cultural-ecological theory of school performance with some implications for education. *Anthropology and Education Quarterly* 29:155–88.

Oritz de Montellano, B. 1993. Melanin, Afrocentricity, and pseudoscience. *Yearbook of Physical Anthropology* 36:33–58.

Peters, C.M. 1994. *Sustainable Harvest of Non-Timber Plant Resources in Tropical Moist Forest: An Ecological Primer*. Washington, D.C.: Biodiversity Support Program-WWF.

Pickering, A. 1995. *The Mangle of Practice: Time, Agency, and Science*, 1–27, 37–67. Chicago: University of Chicago Press.

Polette-Niewold, L., Manciu F., Torres B., Alvarado M., and Chianelli R. 2007. Organic/inorganic complex pigments: Ancient colors Maya blue. *Journal of Inorganic Biochemistry* 101:1958–73.

Pollock, M. 2004. Race bending: "Mixed" youth practicing strategic racialization in California. *Anthropology and Education Quarterly* 35:30–52.

Powers, M. 1986. *Oglala Women: Myth, Ritual, and Reality*. London: University of Chicago Press.

Reibold, M., Paufler, P., Levin, A,. Kochmann, W., Pätzke N., and Meyer, D. 2006. Materials: Carbon nanotubes in an ancient Damascus sabre. *Nature* 444:286.

Restivo, S. 1985. *The Social Relations of Physics, Mysticism and Mathematics*. Dordrecht: Pallas Paperbacks, Reidel.

Ruiz, M. 2002. *The International Debate on Traditional Knowledge As Prior Art in the Patent System: Issues and Options for Developing Countries*. Washington, D.C.: CIEL.

Sanchez del Rio, M., Martinetto, P., Reyes-Valerio, C., Dooryhée, E., Peltier, N., and Suárez M. 2006. *Archaeometry* 48:115–30.

Sarno, L. 1993. *Song of the Forest*. New York: Houghton Mifflin.

Science for the West, myth for the rest? The case of James Bay Cree knowledge construction. In *Naked Science: Anthropological Inquiries into Boundaries, Power and Knowledge*, edited by Laura Nader, 69–86. London: Routledge.

Sheets, P. 1989. In *Applying Anthropology: An Introductory Reader*, edited by A. Poloefsky and P. Brown, 113–15. Mayfield: Mountain View.

Star, S.L. and Griesemer, J.R. 1989. Institutional ecology, "translations" and boundary objects: Amateurs and professionals in Berkeley's Museum of Vertebrate Zoology, 1907–39. *Social Studies of Science* 19(4): 387–420.

Stoller, P. 1987. *In Sorcery's Shadow: A Memoir of Apprenticeship among the Songhay of Niger*. Chicago: University of Chicago Press.

Thalos, M. 2008. Two conceptions of collectivity. *Journal of the Philosophy of History* 2:83–104.

Verhoeven, J., Pendray A., and Dauksch, W. 1998. The key role of impurities in ancient Damascus steel blades. *Journal of Metals* 50:58–64.

Watson-Verran, H. and Turnbull, D. 1995. Science and other indigenous knowledge systems. In *Handbook of Science and Technology Studies*, edited by S. Jasanoff, G. Markle, T. Pinch, and J. Petersen, 115–39. Thousand Oaks: Sage.

### Endnotes

1. As Harding (2006) notes, we cannot afford a "tolerant pluralism" that leaves us without a critical apparatus for either scientific or social issues, nor can we simply fall back on a universalism that makes Western science the monolithic repository of rationality and truth.

2. Although space does not permit a full discussion here, it is worth noting that wootz has not only empowered Western science but also Western pseudoscience in the form of rhondite steel. Despite its entry in Wikipedia and various media reports about patent claims and commercial applications, this author's investigation suggests that rhondite steel does not exist. Just as Bloor points out that we need to avoid the asymmetry that occurs if we only investigate failed science, we also need to avoid the reverse asymmetry of only investigating connections to successful science when discussion traditional knowledge.

# Section II

# Capacity

# 4

## Nanotechnology and Geopolitics: There's Plenty of Room at the Top

**Stephanie Howard and Kathy Jo Wetter**

### CONTENTS

## 4.1 Introduction

The capacity to manipulate matter at the atomic level is expected to usher in an era of unprecedented technological innovation that will provide answers to the current global crises in food, fuel, finance, Fahrenheit, and flu.

Visions of the new economic opportunities arising from the nanoscale level have inspired state technology-resource mobilization not seen since the Cold War space race (Wilsdon and Keeley 2007; Lok 2010; Organisation for Economic Cooperation and Development [OECD] 2010a). A decade into the nano race, however, the technology remains largely a research and development (R&D) enterprise living on promises, and the products that have made it to market thus far—among them, stain-resistant trousers, antibacterial door handles, and miniaturized components for electronics—are modest in light of the great transformations that had been predicted.

In this chapter, we survey the emerging geopolitical landscape of nanotechnology and how the technologies and their ownership, control, and governance are evolving. Features of the "nanotechnology revolution" include its global scope (with many Global South governments shadowing the enthusiasm of Global North governments) and the central role of governments in financing and facilitating the commercialization of nanoproducts and systems. Analysis underpinning state investment in nanotechnologies has been simplistic, apparently aimed to provide justification for ongoing use of the public purse to support national nanoaspirations. Preoccupation with the technologies as engines of economic growth also appears to be leading to nanotechnology favoritism, obscuring the potential for other, potentially less risky, and more environmentally sustainable systems and approaches to achieve equal or greater gains.

Like no other, the twentieth century, in which many of our current crises originated or accelerated, has taught the lesson that it is easier to unleash a technology than to control its path. Nanotechnology's first decade as a state technology program (2000–2009) was one of global turmoil, marked by the increased probability of dangerous climate change arising from industrial activity, unprecedented levels of poverty and hunger, biodiversity loss that proceeded unchecked, and a global economic crisis. Whether the stellar achievements predicted for nanotechnologies and the move to a nanoeconomy will usher in the changes needed to restore a world of vastly compromised ecological health—in which more than half the human population and a great proportion of nonhuman species have been forced to make do on the margins—depends on governance. We discuss the course of the responsible governance culture that many governments have adopted. Although early statements contain commendably inclusive commitments, thus far, governments have been responsive to narrow research and commercial interests, rendering the wider community passive recipients of technological developments with far-reaching consequences. Meanwhile, regulatory discounts, a preference for voluntary approaches, and the drive to commercialize nanotechnologies well before their potential for harm can be identified, assessed, or tracked place them within the fold of earlier, poorly governed technology introductions, where risks were unanticipated, downplayed, or ignored by operators and governments, generating enduring legacies of harm.

Although the nanotechnology race is, at present, accompanied by a bureaucratic race to the bottom, there is plenty of room at the top. It is still possible for governments to act to bring about the enlightened, participatory innovation cultures many have notionally committed to, and to place new technologies in the service of the wider communities—human and nonhuman.

## 4.2 State-Sponsored Technology Revolution and Corporate Hitchhikers

In what is now logged as a seminal moment in nanotechnology's short history, former U.S. president Bill Clinton's millennial speech conjuring new capabilities at the nanoscale sketched out a blueprint—the National Nanotechnology Initiative (NNI)—for reshaping the world "atom by atom" (U.S. National Science and Technology Council 1999). The launch of the NNI in 2000 unleashed U.S. federal funding of $12 billion and prompted other countries to similarly commit to underwriting the development of nanotechnologies.

The vision of an economy (re)built upon nanotech's cross-sectoral platform has overwhelmed state conceptions of the future; technologically speaking, there is no competing vision. For Berlin, Moscow, Kuala Lumpur, Washington, D.C., and Johannesburg, future economic competitiveness and industrial growth are believed to depend upon dominance in nanotechnologies.[1] While Global North countries are focused on shoring up their position or gaining new advantage—including military superiority—countries of the Global South must become nano capable, they are told, in order to make the transformation into modern, industrialized economies. Reflecting the pivotal position nanotechnologies have acquired, they have been dubbed a "bearer of the future" in Brazil's state planning (see Guillermo and Invernizzi 2008), embody Malaysia's aspirations "to build a high-income economy dependent on high-value sectors of growth" (Government of Malaysia 2010), and are considered sufficiently vital to Mexico's future that a National Emergency Program was deemed necessary by parliamentarians (Foladori 2006).

According to a survey by Maclurcan (2005), by 2005 the number of countries engaging with nanotechnology R&D on a national level had grown to sixty-three. A further thirteen expressed "country-level interest," and sixteen countries reported dedicated research activities (in the absence of a national-level program).

By the end of 2009, state investment in nanotech R&D around the world had reached over $50 billion (Cientifica 2009). That year, the European Union (EU) led global state investment, accounting for over a quarter of public funds (European Commission 2009c; Hellsten 2008). Driving EU activity at the member-state level is Germany, one of the world's largest chemical

## GAME-CHANGING INVESTMENTS: MILITARY NANO R&D INVESTMENT

Projected military applications of nanotechnologies range from lighter, more efficient military battle suits, ubiquitous surveillance, more lethal weapons, and enhanced super "war fighters" engineered through the convergence of digital, bio-, nano-, and computing sciences, with the potential "to radically change the balance of power" (Ibrügger 2005, 6).

Military aspirations have commanded a considerable share of government funding, particularly in the United States, which is understood to be making the largest investment in military nano R&D, by one estimate accounting for 90 percent of global investment by states (Altmann 2009). Concern about the extent to which military interests are driving the nanotech research agenda have been raised by the United Nations Educational, Scientific and Cultural Organization (UNESCO) (Khan 2011).

The U.S. Department of Defense (DOD) has invested in "submicron technologies" since the 1980s, and, until 2010, has commanded $3.4 billion—around 30 percent of the total NNI funds—to develop military applications (U.S. Department of Defense 2007; U.S. Government 2010). This compares with just over a quarter to the National Science Foundation, 18 percent to the Department of Energy, and around 15 percent to the Department of Health and Human Services/National Institutes of Health (U.S. Government 2010). Over the past two years, reprioritization of the NNI budget has seen the military's share of the federal budget for nano R&D reduced in favor of other sectors (such as energy, health, and nanomanufacturing) (Service 2010; U.S. Nanoscale

economies, which leads on a number of measures, including funding (€441.2 million from all public sources in 2009), intellectual property (IP) activity, publications, and general technology capability (Federal Ministry of Education and Research 2009; President's Council of Advisors on Science and Technology [PCAST] 2010). That same year, Russia's investment (23 percent of global state funding) trumped that of the United States (19 percent). Asian countries make up the remainder of governments investing more than 1 percent of the global total (Japan 12 percent; China 10 percent; Korea 4 percent, and Taiwan 1 percent) (Cientifica 2009).

However, the rise of aspiring technology powers in the investment league table—reflecting a more general eastward shift in the gravitational center of innovation (Wilsdon and Keeley 2007; Adams and Pendlebury 2010)—has significantly quickened the pace of the nanotechnology race. Russia's late but massive 2007 investment (estimated to be around US$10–11 billion) has put the country second only to the EU block (Elder 2007; Vahtra 2010). Meanwhile, nanotechnology commanded a greater portion of the R&D budget in China

Science, Engineering, and Technology Subcommittee, National Science and Technology Council [NSET] 2010, 2011). At the time of writing, the White House has proposed that military nano R&D receive 17 percent of NNI funds in the 2012 fiscal year—a significant cut, but one that would still leave the military establishment some $368 million to spend (NSET 2011).

Between January 2000 and December 2009, U.S. military institutions (the Army Research Office, the Office of Naval Research, the Defense Advanced Research Projects Agency [DARPA], the U.S. Navy, and the Air Force Office of Scientific Research) have funded and secured a share in the rights to 195 nanotechnology inventions at the U.S. Patent and Trademark Office (USPTO), with a further 151 applications pending. (Data is from a review of patents awarded and applications filed at the USPTO over the period January 1, 2000, to December 31, 2009. The database search was conducted in February 2010 and may not represent the full number of applications, in particular, due to the lag between filing and online publishing by the patent office.)

The United Kingdom, Netherlands, Sweden, France, Israel, India, China, and Iran are also said to be investing public funds in military research, but few are believed to be spending more than $10 million a year (Altmann 2009). Russia has also declared military applications to be a priority of its nanotech R&D agenda (RIA Novosti 2007), an interest it punctuated with the 2007 televised detonation of what the Kremlin declared to be the world's first "nanobomb"—a fuel–air explosive with reportedly nano features—dubbed "the Father of All Bombs" (Elder 2007; BBC 2007; RIA Novosti 2007).

than in the United States in 2009, and the country's investment in nanotechnology R&D (reports range from $180 million to $510 million in 2008) places it third after the EU and Russia, when adjusted for purchasing power parity (PCAST 2010). By at least one account, public funding is yet to peak as more states enter the arena (Spinverse 2010). Nevertheless, the rate of investment that marked the first half of the nanotech decade has dropped sharply (Cientifica 2009).

The entrance of Russia, China, and other "emerging economies" (such as India and Brazil) to the nanotechnology race has seen the share of global spending by Japan, the EU, and the United States drop from 85 percent in 2004 to just 58 percent in 2009 (Cientifica 2009). However, some relief is offered OECD countries in accounts of national investment that include both private and public sources (PCAST 2010; see below). Nevertheless, Brussels, Tokyo, and Washington are reportedly nervous. The United States' assumed dominance across the board of technology-capture metrics (including scientific publications, IP, investment, education, technology markets, and commercialization) is considered tenuous (Sargent 2011), prompting frequent calls for the release of further funds to help shore up its technology rankings (Nordan 2008; Hobson 2009; PCAST 2010; Tour 2011). Similar concerns have been expressed in the European Union (European Commission 2009c; Council of the European Union 2007).

While investment by the private sector is often less transparent and therefore more difficult to calculate, market analysts agree that corporate investment in nanotech R&D now outstrips government spending (Lux Research 2008a; Cientifica 2008).

By 2010, U.S. corporations had invested around $2.75 billion into R&D (under a quarter of the total public investment), with around 50 percent of this in electronics and information technology (IT), 37 percent in materials and manufacturing, 8 percent in health care and life sciences, and 4 percent in the energy and environment sector (Lux Research figures cited in PCAST 2010).

According to industry consultancy Cientifica, the global semiconductor industry has seen the largest share of corporate R&D investment in nanotechnologies, with a total of $55 billion during the 2005–2010 period. Pharmaceutical and health care is the fastest growing sector of corporate R&D investment, while the food sector's investment is the lowest by a considerable margin, with just $22 million in 2010 (Cientifica 2008).

## 4.3  Ill-Conceived Revolution

### 4.3.1  Intoxicating Nanotechnologies: The Need for Sober Thinking

A decade into the nanotechnology race, considerable hype surrounds the technologies' media accounts, market forecasts, consultancy reports, and corporate and research institution press releases. The hype is not exclusively the purview of industry or research institution public relations (PR) departments. Governments are also generating breathless headlines: Nanotechnologies are "crucial" for European citizens' quality of life, says the Council of the European Union (Council of the European Union 2007, 1); "humankind's great hope," according to a South African government minister (Government of South Africa 2009); and, as a Korean minister put it, a "jumbo-size hope for mankind" (Government of Korea 2006).

In the nanotechnology lexicon, "ubiquitous" is, itself, ubiquitous, and repeated to assert the prediction that there is no area of the economy or social activity that nanotechnologies will not reach or transform for the better. Yet, as of this writing, the technologies exist largely in the future/conditional tense, with applications more likely to be theoretical and serve as fund-raising slogans than actual artifacts (Davies, Macnaghten, and Kearnes 2009). Despite this, nanotechnologies are still heralded as a boon for economic growth (PCAST 2010), a key to combating climate change (Cientifica 2007; Saxl 2009), the basis of a new "green revolution" in agriculture (Sastry et al. 2007), and a means for economies to climb out of recession (Ray 2009; Rickett 2009).

Enthusiasts are to be expected. Yet the hyperbole that crowds public and political forums makes for an ill-considered technology drive. As discussed in the remainder of Section 4.3, a less-than-rigorous economic analysis underpins the case for a nanoeconomy and vastly reduces the likelihood that it could bring about the sea changes required to bring industrialized economies to live within their ecological means.

### 4.3.2 Pots of Gold: Bloated Market Projections

Forecasts of the commercial returns nanotechnologies will generate have played their part in opening state coffers. The first of these was the now land-mark projection made by the U.S. National Science Foundation in 2001 that the world market for nanoproducts would reach $1 trillion by 2015 (Roco and Bainbridge 2001). That figure has since been raised to $1.5 trillion (Cientifica 2008), and thoroughly outbid by visions of $3.1 trillion (Lux Research 2008a), since pared down to $2.5 trillion due to the recession (Lux Research 2009b).

Predicting market value of nanotechnologies is a creative exercise, not least because formal definitions on what constitutes nano are diverse and con-tested, and because the level of market activity is not fully known (Palmberg, Dernis, and Miguet 2009). Accordingly, accounts of the technologies' market share vary wildly even in hindsight; in 2009, the market value generated by nanotechnologies was either $11.7 billion (McWilliams 2010) or $254 billion (Lux Research, as cited in Roco et al. 2010), or somewhere in between.

Among the most widely cited forecasts are those that employ a value-chain approach in which the total market value for nanotechnologies is arrived at by adding the value of "raw" nanomaterials, the "nano-intermediates" these are incorporated into, and the final, "nano-enabled" product. The potential for bloated figures is considerable because, although the nanocomponent in either intermediate or end products classed as nano-enabled may indeed be tiny, the value of the end product is counted as nanotechnology's market value (Berger 2007). Just 0.2 percent ($1.8 billion) of the total market value for nanotechnologies in 2012, generated by consultant Lux Research's value-chain modeling, comes from raw nanomaterials, while end products account for 86 percent of the total. Looking forward to 2015, the value of raw nanoma-terials drops to 0.1 percent of the total value chain (see Table 4.1).

The sums for raw nanomaterials would tend to support Lux's assess-ment that big money is not to be made manufacturing nanoparticles but in "nano-intermediate" products (Lux Research 2009a). The data also confirms the OECD's caution that the value-chain approach is likely to generate "significant overstatements" (Palmberg, Dernis, and Miguet 2009, 22), "rather inflated" forecasts (OECD 2010a, 29), or as one industry

**TABLE 4.1**

Lux Research's Nano Value-Chain Predictions 2012–2015

| | 2012 | | 2013 | | 2014 | | 2015 | |
|---|---|---|---|---|---|---|---|---|
| Value-Chain Stage | Value ($ million) | Value chain share (%) | Value ($ million) | Value chain share (%) | Value ($ million) | Value chain share (%) | Value ($ million) | Value chain share (%) |
| *Nanomaterials* | 1,798 | 0.20 | 2,098 | 0.16 | 2,462 | 0.14 | 2,916 | 0.11 |
| *Nano-Intermediates* | 120,206 | 13.6 | 206,823 | 16.0 | 322,691 | 17.9 | 498,023 | 20.2 |
| *Nano-Enabled Products* | 762,204 | 86.2 | 1,081,025 | 83.8 | 1,480,928 | 81.9 | 1,962,950 | 79.7 |
| Total | 884,208 | 100 | 1,289,947 | 100 | 1,806,081 | 100 | 2,463,890 | 100 |

*Source:* Adapted from Lux Research, *The Recession's Impact on Nanotechnology*, 2009. With permission.

commentator puts it plainly, "terribly deceiving numbers" (Shalleck 2010). Nevertheless, the value-chain predictions have received considerable uncritical airtime and are regularly cited by governments and industry, typically without the more sobering breakdown of the value chain. Such wholesale repetition may arise, in part, because the analytical detail is found inside the covers of consultancy reports that are so expensive that, anecdotally, even governments are known to rely on the free summaries (for example, Lux Research has made the 2009–2015 nano value chain widely available but did not release accompanying analysis that shows how much of the value of the final products is due to nanocomponents). However, not all states can cry poor, and usual practice has been to cite the most compelling figures, in part to justify high levels of government investment in the technology.

Projections of nanotechnology's contribution as a job creator have similarly been aspirational and vary widely. Forecasts include the National Science Foundation's estimate of up to 7 million workers by 2015 (Roco 2003) and what the OECD labels "even more optimistic forecasts" of 10 million manufacturing jobs related to nanotechnology emerging by 2014 (Palmberg, Dernis, and Miguet 2009, 26).[2]

Attempts to put numbers to the current nanotechnology workforce are rare and, like market predictions, complicated by lack of government oversight of nanotechnology R&D and commercial activity, and a lack of consensus definition for nano. Rarer still are assessments of the jobs that nanotechnologies may take away or replace (such that net job creation is not known). In any case, an OECD review suggests there is "a large discrepancy" between the projections and the state of the workforce (Palmberg, Dernis, and Miguet 2009, 27).

### 4.3.3 The Partial Economics of Nanotechnologies

Also noticeably absent is analysis by governments to determine how much investment will be required before the technologies can begin to deliver on promises. Naturally, widely varying time frames across different areas of experimentation and application are likely. Furthermore, the scientific breakthroughs required to access the so-called revolutionary applications will, by their very nature, defy precise time frames, and research to achieve these is often being pursued in more than one country. Nevertheless, the apparent lack of attempts to characterize the investment required is another manifestation of the poor economic analysis and the opaqueness of government policy.

What does seem clear is that considerably more state funding will be required to make nanotechnologies roadworthy and able to occupy the industrial pole position forecast for them (Dosch and Van de Voorde 2009). Substantial technological and industrial hurdles remain. The nanomanufacturing industry, according to an OECD assessment, "is still in its infancy" and characterized by "lack of infrastructure equipment for

## THE SLUGGISH COMMERCIAL DAWN

Commercialization of nanotechnologies has not occurred to the extent expected (Shalleck 2009). That said, the full extent of the progress of nanotechnologies to market is unclear due to the lack of labeling requirements, premarket assessments, market monitoring by governments, and industry avoidance of voluntary reporting schemes. Most governments currently rely upon charity—a freely available online inventory of consumer products developed by the nongovernmental, U.S.-based Project on Emerging Nanotechnologies (PEN)—to provide examples of products that indicate progress in commercialization. (The consumer product inventory can be viewed at http://www.nanotech-project.org/inventories/.) According to that catalogue, there were over 1,300 products on the market as of March 2011. That is likely to be well below the number of commercialized products, as the inventory lists only those products claimed by the manufacturer to incorporate nano-technologies and does not cover intermediate products. (Note: Variable survey results reflect the lack of clarity around nanocommercializa-tion. A survey conducted by Health Canada and Environment Canada in 2009 identified around 1,600 nanoproducts on the Canadian market alone [OECD 2010b]. Conversely a Dutch government study also based on manufacturer claims found 858 nanoproducts on the European market in 2010 [Wijnhoven et al. 2011], while a survey by European consumer organizations BEUC and ANEC found 475 products on the European market [ANEC/BEUC 2010a, 2010b]. Of note, three of the sur-veys were based on product claims by manufacturers. Wijnhoven et al. attribute the range of results to different approaches to data collection [the method used to identify products on the Canadian market was not described in the OECD report].)

nanomanufacturing, and few efficient manufacturing methods especially in bottom-up approaches to nanoscale engineering" (Palmberg, Dernis, and Miguet 2009, 79; U.S. Department of Defense 2009). Mass production, scaling up, and quality control—fundamental for the cost-effective nanomaterials that the wider manufacturing industry will use—still present considerable challenges for most nanomaterials (OECD 2010a; Kelly 2011). Today, pro-duction is typically a low-volume affair, generating considerable waste and by-products, making some nanomaterials, at least, prohibitively expensive (Kiparissides 2009; Shalleck 2010).

Sustained, heavy investment will be required to advance basic science through to new industrial infrastructure, manufacturing techniques, and processing technologies (Australian Academy of Technological Sciences and Engineering 2008; Kiparissides 2009; Sargent 2011). As such, the technologies

Whatever the actual number of products, it is generally agreed that nanotechnology commercialization is at an early stage, and the products currently available reflect immature technological and manufacturing capabilities in the field. The sluggish pace of commercialization has made many governments anxious, and a common lament is the lack of commercial products to show for the considerable investment (see, for example, European Commission 2009b; Padma 2008; PCAST 2010). Certainly, many of the nano-enabled consumer products lay claim to somewhat trivial gains when set against the revolutionary achievements predicted. In 2008, at least, stain-resistant trousers had been for one industry consultant the best the technology had to offer in terms of "real-life" products for half a decade (Cientifica 2008), although other commentaries point to commercial success stories including semiconductor applications, lithium-ion batteries incorporating multiwalled carbon nanotubes (Eklund et al. 2007), and nanoscale lithium phosphate cathodes (PCAST 2010).

Technological hurdles are not the only cause of the slow progress to market. Wider business interest in nanotechnologies is reportedly low. According to Lux Research, nanotechnology is broadly seen as "a technology without a product" (Lux Research 2008b, 2). The lack of headlining market success and the prediction of marginal profitability for nanoproducts is not helping to generate interest (Holman 2007; Lux Research 2008b). Without governments, investors, and the like lining up to purchase early stage products, one industry analyst warns, "disruptive nanotechnologies will primarily remain as science projects and underfunded start-ups" (Gordon 2010). More difficult to overcome than disinterest, perhaps, is wariness. Wider manufacturing industry concerns about nanosafety, regulatory uncertainty, and public perceptions are impeding technology uptake (Liroff 2009; Lux Research 2008b; PCAST 2010).

are likely to remain heavily dependent on state subsidies for some time; the long haul, "high-risk, high-reward" nature of much of the R&D agenda sits well beyond the horizons of much of the private sector (Harper 2009; Kiparissides 2009). Meanwhile, nanotechnologies' entrance to the marketplace has proved slower than predicted.

### 4.3.4 Python Economics: Always Look on the Bright Side of Life

The considerable interest that governments have shown in the disruptive technological potential of nanotechnology is not matched by an interest in the distributive effects of a nanoeconomy and, in particular, negative economic impacts that could occur domestically or in countries not engaging with nanoinnovation.

The OECD, despite having economic development as its focus, has made little contribution to addressing the deficit in critical, economic analysis. The scope of socioeconomic impacts set out in its first statistical analysis framework proposes to document only the positive side of the ledger with respect to market forecasts in industrialized nations, job creation, and end products (Palmberg, Dernis, and Miguet 2009). Left outside its analysis is consideration of what nanotechnology products and production systems may leave in their wake by way of obsolete industries, job losses, and environmental degradation; which communities, populations, and countries might be most affected by these; and how such costs might weigh on the public purse.[3]

Eschewing critical evaluation on how the future nanoeconomy will affect vulnerable communities, nano nations instead tend to focus political debate on benefits that individual applications or areas of application (such as water purification) may offer. It is not only northern, industrialized countries that sidestep examination of the broader, distributive effects of the technologies; countries of the Global South that have committed to nanotechnologies apparently assume that the technologies will benefit marginalized communities by a combination of targeted individual applications and the projected contribution that the technologies will make to economic well-being through increased economic competitiveness.

This is of concern, not because there should be no enquiry into how specific products—properly evaluated and compared against competing systems, technologies, and approaches could benefit marginalized communities—but rather because assessing the merits of individual applications is not a substitute for a broader evaluation of the total, potential benevolence of a nanoeconomy. It is possible, in other words, that individual applications could prove to be technically viable, safe, superior, and accessible for the communities that typically exist outside the mainstream economy even while the broader workings of the nanoeconomy undermine livelihoods or introduce new forms of contaminants that disturb the resources upon which those same communities depend.

### 4.3.5 Technological Favoritism

Commitment to nanotechnologies to realize economic growth and competitiveness aspirations further skews public policy because, in order to justify the scale of investment, the technologies will need to be "the answer" somewhat irrespective of "the question."

Indicating that the cart has been hitched ahead of the horse, the Russian government's primary vehicle for nanotech development has mooted a pilot project "to stimulate market demand" for a selection of nanoproducts (Rusnano 2009), and an advisory group reviewing the EU's considerable R&D funding has emphasized the "need for clear market drivers, for example, industrial problems that can be solved by the application of nanotechnologies" in order to remove barriers to commercialization (Kiparissides 2009, xi). Across the Atlantic, an assessment by the U.S. Congress's World Technology Evaluation Center cites the "development of applications to create uses and demand" as the single greatest challenge for carbon nanotubes (CNT) commercialization (Eklund et al. 2007, 15).

Privileging nanotechnologies could cast a long shadow over a range of competing strategies and approaches that have the potential to deliver less risky alternatives and, indeed, bring about the necessary civilizational shifts so that humans can live sustainably on a planet with finite resources.

Nowhere is the political desire to buoy nanotechnologies more evident than in the ballooning area of "clean tech." While there may be differences of opinion within the industry about which technologies belong under the clean-tech banner, there does appear to be agreement that clean tech is, by definition, novel and high tech (for example, not bicycles) (Stack et al. 2007). Others say it presents "the largest economic opportunity of the 21st century" (Wesoff 2009) and offers a "natural fit" between economic growth and projected environmental gains (Rutt and Wu 2009). Governments, too, have preordained that the technology will be a primary vehicle of sustainability (Government of Korea 2010; Government of Malaysia 2010).

Nanotechnologies' recent inclusion, at least by some, in the clean-tech repertoire has provided a boost for an industry struggling with the recession and scant market success (Rutt and Wu 2009; Shalleck 2009). The clean-tech rhetoric is seductive. As Davies, Macnaghten, and Kearnes note, "who wouldn't want a technology that is 'safe by design,' that can deliver clean water to billions, or enable consumption without negative effects on ourselves or our environment?" (2009, 40). By 2008, nanotechnology applications in the energy and environment sectors reportedly accounted for 29 percent of U.S. federal nanotech spending, 13 percent of corporate spending, and 41 percent of venture capital (Lux Research 2009a).

Yet there is little to support the assertion that nanotechnologies are, by definition, clean. While certain applications such as supercapacitors could be beneficial (Friends of the Earth 2010), the fog around nanosafety and the absence of life-cycle analyses make the claims to ecological sensitivity and sustainability premature, at the very least. Certainly, the historically frequent inclusion under the clean-tech banner of nuclear power generation—the technology that was to provide electricity "too cheap to meter" but whose safety and waste-stream problems, decades on, remain too difficult to adequately address—should be sufficient to encourage critical evaluation of the nano clean-tech brand.

Furthermore, the environmental gains predicted from the use of nanotechnologies are speculative and contingent on overcoming a number of technological hurdles that, if they are overcome at all, may take years or decades. Within such time frames, a range of other technologies, systems, and approaches, given adequate funding and political consideration, could make equal or greater contributions to sustainability.[4]

## 4.4 Surface Chemistry

### 4.4.1 Global Spread and Share of Nanotechnologies

The promise that the nanotechnology revolution will include, not exclude, those who have been on the "wrong" side of the earlier technology divides has become a regular feature of nanotechnology's narrative and accompanies promises of a host of nanoproducts aimed at alleviating poverty.

The nanotechnology programs embarked upon by "emerging economies" such as China, Brazil, India, Russia, and South Africa would seem to have extended the technology race beyond the OECD and have led nanotechnologies to be labeled a "more international scientific project" (UNESCO 2006, 13). Yet, with few exceptions, countries of the Global South have made late and modestly funded bids to develop capacity in the technologies. For example, Thailand is reportedly investing US$9 million per annum, while Indonesia dedicated US$29 million in 2010 (You 2010). In 2007, the Indian government pledged $50 million per annum over five years (India PRWire 2007) and South Africa's per annum investment was reportedly between US$21 and $60 million over the 2006–2009 period (Government Communication and Information System 2010), with additional funding packages of US$74 million in laboratory equipment and university curricula (Claassens 2008; Tobin 2009).

Indicating the limited spread of nanotechnology enterprise internationally, a recent review by ICPC-Nanonet (an EU-funded, joint program with China, India, and Russia that aims "to provide wider access to published nanoscience research, and opportunities for collaboration between scientists" [Tobin 2010, 242]) reported that of nanotechnology activity in 140 non-OECD countries, no nanotechnology R&D activity was evident in more than half of these; in at least 20 other countries, isolated R&D was occurring at tertiary research institutions; around 20 countries make explicit reference to nanotechnologies in national science policies; fewer have dedicated national nanotechnology strategies coupled with dedicated government funding programs; and research facilities and equipment, human resources, and related infrastructure are still scarce in most countries.

Thus, in the wide world that lies beyond the circle drawn by OECD wagons, China—and to a much lesser extent, South Korea—represent exceptions. Looking at the standard metrics for measuring technological progress together—R&D funding, publications, patents, and education—suggests that, for the foreseeable future, the nanotechnology race will continue to be an OECD-plus-China event. A survey of publications authored by researchers in the EU, United States, and International Cooperation Partner Countries (ICPC)[5] provide some illustration (see Table 4.2).

Even if countries of the Global South were faring better in the conventional technology-advancement metrics, this would not, of itself, guarantee that

**TABLE 4.2**

Nanotechnology Publications 2008 by the EU,
United States, and ICPC Countries

| Country/Region | No. of Papers |
|---|---|
| EU | 36,822 |
| USA | 21,185 |
| China | 19,053 |
| East and West Asia (minus China) | 6,529 |
| Eastern Europe and Central Asia | 4,158 |
| Latin America | 3,302 |
| Mediterranean countries | 1,616 |
| Africa | 366 |
| Western Balkan countries | 364 |
| Caribbean countries | 70 |
| Pacific Island countries | 3 |

*Note:* In the analysis undertaken for publication activity cited here, publications with multiple authors from different countries are counted for each country. That means that the sum of all countries' nano-related publications reported in the ICPC-Nanonet reports will be greater than the actual number of worldwide publications for that year.

*Source:* Data from Wang, L. and Notten, A., *Observatory NANO Benchmark Report: Nano-Technology and Nano-Science, 1998– 2008.* 2nd Observatory NANO Benchmark Report, Maastricht, The Netherlands, UNU-MERIT, 2010; Tobin, L. and Dingwall, K., *Second Annual Report on Nanoscience and Nanotechnology in Africa*, ICPC-Nanonet, 2010, http://www.icpc-nanonet.org; You, Z., *Second Annual Report on Nanoscience and Nanotechnology in East-Asia*, ICPC-Nanonet, 2010, http://www.icpc-nanonet.org. With permission.

the nanoeconomy would make any significant inroads to reducing, let alone eliminating, inequity and poverty. Notably, most of the advances in technological capacity and prosperity made in recent decades have bypassed populations of the Global South and the "human development gap" between the North and South has been widening rather than narrowing (UNDP 2005). South Africa, Nigeria, and Sri Lanka are among the southern countries that have acknowledged this and have identified applications directed to poverty alleviation and meeting local needs as priorities in their respective national nanotechnology strategies. Under the South African program, a cluster of "social R&D" sits alongside industrial nanotech priorities, with "flagship" programs in water, energy, and primary health care (Department of Science and Technology, Republic of South Africa 2006). The impact on the quality of life of previously marginalized sectors of the community is also a key indicator of the strategy's success over the medium to long term. Subsequent to the formulation of its nanotechnology mission, the Indian government began to articulate the need to harness the technology for the benefit of rural India and the poor.[6] However, data on the share of government funding devoted to "social R&D" relative to industrial manufacturing are not readily available. It is unclear what level of priority social R&D actually enjoys or, indeed, what concrete gains can be achieved from the considerable public funds being invested.

There is considerable cooperation and collaboration across borders, in the form of bi- and multilateral science agreements and R&D funding programs between governments, joint research programs spanning science institutions in the North and South, as well as corporate financing of national and tertiary science facilities. Yet technical and financial support of southern state nanotechnology efforts by the north is double-edged. There is concern that northern government and private-sector financing of southern R&D may divert southern research agendas to market-led product development that will service affluent populations (Tobin and Dingwall 2010). Alternatively, south–south cooperation and the pooling of resources are seen as a way to place R&D efforts in the service of local needs and priorities (Hassan 2008; Sawahel 2008). A number of south–south initiatives have been instigated to pool resources and to set out southern-driven research agendas. Examples include several initiatives on the African continent, such as the iThemba LABS (regionally dedicated research facilities), the ECO Nanotechnology Network of Asian and Eurasian states, the Latin American Cooperation of Advanced Networks (CLARA), and the trilateral India–Brazil–South Africa Dialogue Forum (IBSA). In the latter, nanotechnologies are one arena of science collaboration, led by India, which has four priority areas and is expected to be resourced by a $3 million research pool. IBSA cooperation areas are biotech, HIV/AIDS, malaria, nano, oceanography, and tuberculosis. ICPC-Nanonet reports provide an overview of such collaborations, researchers, and programs in which each country is involved.

### 4.4.2 Geopolitics of Ownership and Control

To a large extent, access to and control over nanotechnologies will be determined in the intellectual property (IP) arena. There, the scope and speed by which nanomaterials, instrumentation, methods, and applications are being patented further suggests that nanotechnology divides will resemble existing technology divides.

Through nanotechnologies, the reach of IP extends to the biological and nonbiological fundaments of nature, and patent offices are proceeding apace to facilitate that expansion. Indeed, while scientific uncertainty has reportedly prevented government agencies from regulating nanomaterials, it does not appear to have caused the same inertia in patent offices, which have managed to negotiate their way around the absence of internationally standardized definitions of nanotechnologies and the characterization methodologies that would support them.

Proprietary capture of the technologies is also said to be occurring at a very early stage in their development: Basic concepts are being patented before end products have been developed. And, because of their cross-sectoral relevance, patents over basic concepts in nanotechnologies are expected "to cast a larger shadow" (Lemley 2005, 618; Watal and Faunce 2011). According to OECD analysis of nanopatenting between 1995 and 2005, nanomaterials account for the largest share of patent activity (38 percent), followed by nano-electronics (25 percent), nano-optics (11 percent) and instrumentation (9 percent). Nanobiotechnologies appear to be a late developer, at 13 percent (Palmberg, Dernis, and Miguet 2009).

By 2006, around 12,000 nanotechnology patents had been granted by the patent offices in the United States, Europe, and Japan. Over the 2000–2008 period, 20,000 nanotechnology patent applications were lodged with the U.S. Patent Office and around 18,500 at the Chinese Patent Office (Dang et al. 2009).[7] And while nanotechnology-related patenting may account for just 1 percent of all patent activity, nanopatenting publications have grown annually by 34.5 percent since 2000. In 2008 alone, 10,000 unique nanotechnology patent applications were filed globally (compared to 1,153 in 2000) (Dang et al. 2009). Nanotechnology, at least in the area of privatization, appears to be "recession-proof," according to World Intellectual Property Organization (WIPO) statistics. As overall patent activity in 2009 dropped 4.5 percent from the previous year, nanopatenting continued its upward trajectory, growing 10.2 percent.

In the United States and the EU, the extent of government investment in nanotechnologies is not reflected in their share of IP, with the private sector accounting for the bulk of patenting activity. The private sector reportedly holds 61 percent of all nanopatents awarded in the United States and 66 percent of all patents awarded in the EU over the period 1995–2005 (Palmberg, Dernis, and Miguet 2009). In 2010, eight of the top ten patent holders at the USPTO were companies (Graham and Iacopetta, in Roco et al. 2010).

The IP rankings shadow overall trends in the nanotechnologies league table. Here, an early lead by the traditional science and technology leaders is giving way to the rise of the new nanoscience power, China. An OECD review of the 1995–2005 period attributes 84 percent of all nanopatents to the United States, Japan, and the EU (Igami and Okazaki 2007). By one account, over 10,000 nanopatents are held by U.S. entities (a more modest tally than that offered by U.S.-based commentators; see Lux Research in PCAST 2010). Yet patent applications—a "forward indicator" of technology capture—show China has overtaken the United States (PCAST 2010). In 2008, applicants in China filed almost twice as many nanopatent applications (4,409) as applicants in the United States (2,228) at their respective patent offices; for the period 1991–2008, applicants in China had filed more applications in total (16,348) than U.S. applicants (12,696) (Dang et al. 2009). Based on filings at their domestic patent offices, Russia holds only 711 nanopatents, Brazil 116, and Mexico 28;[8] meanwhile, patent activity in countries such as India and South Africa also remains low (Tobin 2009; Dang et al. 2009).

## TUBES TIED: THE CARBON NANOTUBE THICKET

Allocations by the USPTO on carbon nanotube applications have been described by patent law commentators as "generous" (Harris 2009, 168). Review of early CNT patents suggests that fundamental issues, such as patentability, prior art, adequate disclosure, and nonobviousness, have not been properly addressed. The extent of the problem created by the patent office's openhandedness has yet to manifest, as CNTs are still, by and large, a technology without commercial applications. Nevertheless, the IP tangle is cited as one of the most acute challenges for those wishing to commercialize nanotechnology applications (Escoffier 2009; Harris 2009) and the expectation is that the courts will be called in to clear a path through the CNT patent jungle (Harris 2009). Anecdotally, these uncertainties have had a chilling effect on the uptake of CNT technologies, and elaborate fixes, such as nanotube patent forums, are now being concocted to navigate the confusion brought on by early IP awards and to enable commercial CNT applications to flow.

Meanwhile, patenting in the CNT area continues unabated. In 2008, CNTs were the primary nanomaterials in more than a quarter of the nanopatents awarded by the USPTO and were the primary nanomaterials in a third of all applications filed in the United States that year. (Data are based on a review of patents awarded and applications filed at the USPTO over the period December 31, 2007, through December 31, 2008. The database search was conducted on February 16, 2010, and drew a total of 429 patents awarded and 684 applications filed. It is likely that the number of applications filed will be higher, as there is a lag of some months before filings are published online. This tally does not include patents or applications in which CNT is one of a range of nanomaterials provided for in the descriptions, and the overall number that includes these may be rather higher number as the term *nanowire* may sometimes be used to denote *nanotubes*.)

Pledges that nanotechnologies will benefit the peoples of the Global South are difficult to reconcile with such vigorous patenting activity. While the World Intellectual Property Organization continues to work slowly toward a "development agenda" intended to assist the dissemination of useful technologies in the Global South (WIPO 2007), no such considerations have figured in the trilateral meetings of the USPTO, the European Patent Office (EPO), and the Japanese Patent Office (JPO). The resolve of such groups has been to agree on how to identify nano-inventions within the International Patent Classification system (IPC), which will further facilitate nanopatenting (Trilateral Cooperation 2008).

Evidence of the effects IP has had on access to medical treatments and health care in the Global South, for example, appears to have made little impression on governments navigating this latest IP frontier. In the area of health care and medical applications, nanopatent activity has risen sharply (particularly in drug delivery systems and diagnostics), providing patent holders with twenty-year monopolies "during a critical time window of innovation" (Tyshenko 2009, 12). The European Group on Ethics's warning that broad patents limit access to nanomedicines, as well as its calls for a thorough examination of the innovation/reward equation and a broader review of whether the patent system is appropriate for new technologies (European Group on Ethics 2007), have evidently not gained traction within EU policy circles.

Patenting of energy-generation technologies is another area where divides could arise or existing ones widen, particularly in light of the USPTO's resolution to confer "accelerated status" on technologies to combat climate change and foster job creation in the green technology sector (Kappos 2009). That fast-track policy is likely to exacerbate tensions between the Global South and North over energy-related IP, with the issue of accessibility to new energy technology generation having been an ongoing source of disagreement at Kyoto Protocol negotiations (Syam 2010).

However, not even the north, it appears, will be immune to the effects of the patent rush. As early as 2002, the fledgling nano industry lobby group in the United States, the NanoBusiness Alliance, warned that the breadth of patents being awarded could slow commercial development of the technologies (Modzelewski 2002). In the broad and complex area of nanobiotechnologies, meanwhile, patents awarded by the USPTO are reportedly paving the way for "patent thickets" (Berger 2008). Similar problems are looming due to IP awards on carbon nanotube technologies, materials for which a wide range of uses are speculated (Harris 2009).

---

## 4.5 The Emperor's Stain-Resistant Trousers: Governance of Nanotechnologies

### 4.5.1 The Age of Responsible

Accompanying the nanotechnology push by states and the private sector is a governance model that originates from Washington, D.C., and Brussels. As explored more fully in Chapters 10–12, the nanogovernance approach is a mix of traditional regulation, "soft law" (voluntary schemes and codes of conduct), foresight, and public deliberation (Kearnes 2009; NSET 2007). This blend, it is suggested, creates an anticipatory, participatory, and adaptable regime that is well suited for situations where scientific uncertainty reigns. It is further suggested that the emerging nanogovernance model will serve as a blueprint for other new technologies (Roco et al. 2010).

The turn of phrase recruited to convey the spirit of this approach is "responsible development"—which ascended with the launch of the International Dialogue for Responsible Research and Development of Nanotechnology in 2004—and has since become the byline of many state nanotechnology programs. Examples of other countries that have adopted responsible development as a central theme of national policy include Australia (Australian Government 2008), Korea (Government of Korea 2010), New Zealand (Government of New Zealand 2006), Taiwan (Roam, Wu, and Lien 2009) and the United Kingdom (HM Government 2010). While not all countries use the term *responsible*, many high-level state strategies plot a similar course of a determination to pursue nanotechnologies with a mind to the consequences, while tilting at similar thematic areas (economic competitiveness, commercialization, public awareness and participation), although how these are implemented may differ greatly (see, for example, Government of Malaysia 2010; Government of Sri Lanka 2010; Department of Science and Technology, Republic of South Africa 2006).

The pledge to be responsible is welcome but also prompts questions: To whom? For what? Who determines the nature of those responsibilities? And how are these to be enforced? On current course, the definition of responsibility in relation to nanotechnologies is being "drafted" largely by governments and the private sector and is adhering closely to the political pedigree of the term. Its most notable use in recent years has been "Responsible Care," the global self-regulation initiative developed by the chemical industry in the late 1980s to restore public confidence in the industry and to ward off government regulation (Moffet, Bregha, and Middelkoop 2004). According to a leading nanotechnology proponent, the responsible development governance system centers on the move away from top-down regulatory culture to a system where governments set the parameters within which industry self-regulates: a shift from "powers over" to "powers to" operators (Renn and Roco 2006).

Indeed, a decade into the official nanotechnology era, the governance of the technologies remains largely limited to promotional policies (see Chapters 11 and 12 for more detail). The political understanding of responsibility appears to be the product of a government–business partnership and construed as "the necessity to commercialise nanotechnologies for the benefit of human kind, and for this commercialisation not to be 'held back'" (Kearnes and Rip 2009, 113) (see also Chapter 11). The state–private-sector convergence is, in large part, forged by a common interest in commercializing technologies. In the case of governments, most have set product commercialization as a priority of their investment policies (Spinverse 2010), an emphasis that makes nanotechnologies, according to one commentator, "the first emerging technology in which the federal government's efforts included 'commercialization as a specific goal'" (Sargent 2008, 6). University participation in advancing corporate and state research agendas in nanotechnologies would also appear to be high. By way of indication, a Lux Research review (2008b) found that all thirty-one global corporations interviewed had recruited academic institutions to meet their nano R&D objectives; meanwhile, 70 percent of the nanotechnology patents to which U.S. federal military agencies have rights cover research conducted in universities using funds granted by the agencies.

Governments' dedication to nanotechnology commercialization further erodes the traditional boundaries between government and the commercial sector (already under strain with industry's often disproportionate influence on public policy). That has implications for the business of governing: States have commercial interests that conflict with their duties to set the conditions for the technology in the interests of the wider community and the environment, as we discuss in Sections 4.5.2 through 4.5.5

### 4.5.2 The Inclusive Revolution, or Short Engagements and Shotgun Weddings?

The nanoeconomy, the citizens of many countries have been promised, will be inclusive. Leaders in Brussels, Washington, London, and Johannesburg, among others, affirm that involvement of the wider community in plotting the technology's path is an integral feature of introducing nanotechnology "the right way," although the framing of that commitment and the level of participation it implies varies from country to country.

The apparently progressive policy commitment to an inclusive nanogovernance regime has yet to translate into a more enlightened and participatory innovation culture. There has been a considerable amount of engagement activity (particularly in the EU, which, in crude output terms, has been the most engaging). Yet, by early 2008, few of the roughly seventy government and nongovernment exercises reviewed by European researchers broke out of the "tokenism" category, and most were not conceived with a view to effectively incorporate the results in decision making (Baya Laffite and Joly 2008).[9]

By and large, government and industry efforts to engage the wider community have taken place after, not before, determining that a nanotechnology revolution is desirable (and inevitable). Even if the intention to create a more inclusive innovation culture around nanotechnologies was heartfelt at the time such commitments were first articulated, this has given way to more strategic motivations, such as removing potential obstacles to the rollout of the technologies (Baya Laffite and Joly 2008; Davies, Macnaghten, and Kearnes 2009).

For example, four years after the launch of the state nanotechnology program, the South African government initiated a public awareness program to educate South Africans so that negative perceptions arising from public ignorance do not "mar and prematurely suffocate what is otherwise a hugely promising development" (Government of South Africa 2009; also see South African National Research Foundation 2010). Sri Lanka's nanotech strategy plots a similarly top-down approach; acknowledging the influence public attitudes can have on market success and the regulatory environment, activities for "improving the public understanding, acceptance and debate on nanotechnology" are foreseen (Government of Sri Lanka 2010, 23). The European Commission's bold and commendably phrased pledges to "an open, traceable and verifiable development of nanotechnology, according to democratic principles" (EC 2004, 19), "a true dialogue with the stakeholders" (EC 2005, 9), and most recently, to an "audacious communication roadmap" (EC 2010, 5) sit relatively unexercised next to its balder determinations that public concerns need to be addressed "to avoid delays in introduction of new technologies in the EU" (EC 2009c, 6).

Nor does the U.K. government, which has made ambitious commitments to public engagement, prove an exception. Reviews of engagement exercises suggest that those run by the government, in particular, are seen by the policymaking establishment as "one-way forms of consultation or communication" (Gavelin, Wilson, and Doubleday 2007, 72), that genuine openness to public involvement in early decisions about technology and governance has been "elusive," and that enthusiasm has outstripped political commitment or capacity to do anything with the results where these might conflict with pre-ordained innovation plans (Royal Commission on Environmental Pollution [RCEP] 2008, 75).

The gulf between what dialogue, debate, and engagement appear to promise and how far governments are prepared to entertain participation in "upstream" policy and decision making has led to understandable skepticism in civil society about such exercises, most starkly perhaps in protests accompanying the series of public debates funded by the French government—a common concern being that genuine debate was not possible because the government had already committed to the technology (see McAlpine 2010). Meanwhile, public awareness of nanotechnologies is reportedly low in most countries (Satterfield et al. 2009; Macnaghten, Davies, and Kearnes 2010). Four years of public surveys in the United States consistently report that more than two thirds of respondents have heard little or nothing about nanotechnologies (Peter D. Hart Research Associates 2006, 2007, 2008; Hart Research Associates 2009).

### 4.5.3 Regulatory Holidays

Regulation of nanomaterials may be a more frequent topic within policy circles (Mantovani, Porcari, Morrison, et al. 2010; Roco et al. 2010), but this has yet to translate into meaningful action. In 2011, nanotechnology activity is occurring largely untouched by nano-specific regulatory requirements (Mantovani, Porcari, Morrison, et al. 2010), and fears expressed by practitioners that nanotechnologies could be regulated to a standstill (Drezek and Tour 2010) are, on current course, unfounded (see Chapter 12 for a more comprehensive overview of nanotechnology and regulation).

Tours around the OECD Working Party table report little regulatory activity by member-states (OECD 2009b, 2010b, 2010c). Among them, Japan has no nano-specific regulation in place, and high-ranking nano nation South Korea has only begun to consider what a regulatory framework engaging directly with nano could look like (OECD 2010b). Small regulatory steps taken in the EU (such as in the area of cosmetics) are in large part due to European parliamentarians, who have become impatient with what they consider the commission's laissez-faire approach (European Parliament 2009). Despite claims to having regulated a number of nanomaterials, the U.S. Environmental Protection Agency has been struggling to get basic rules on just two carbon nanotube products in place and provisions for other rules are reportedly not yet in force (Government Accountability Office 2010). Nor do OECD observer countries have much to announce. In the most recent reports to the Working Party tour de table, Russia had conducted no nanotechnology-related regulatory risk assessments (OECD 2010b), although its government cited at least one law scrutinizing nanoproducts and claimed commercial activity of around $700 million by 2008 (RosBusiness Consulting 2008; Rusnano 2008; Johnston 2008). South Africa has acknowledged that risk assessment research, and presumably risk assessment, has "yet to take root" while also acknowledging that worker exposure and commercialization are on the rise (OECD 2010b, 56). Meanwhile, China, India, and Taiwan reportedly have no nano-specific regulation in place and no imminent plans to legislate (Mantovani, Porcari, Morrison, et al. 2010; Mantovani, Porcari, and Azzolini 2010).[10]

Various grounds are offered for the regulatory holidays nanotechnologies are enjoying in most countries, including a lack of evidence of harm and insufficient data and information to support the development of nano-specific laws. The regulatory reticence is also likely due to oft-expressed fears that placing any limitations on nanotechnology activity will jeopardize that country's competitiveness. The South African government's pledge to create "the best possible climate—regulatory, politically, and economically—for [n]anotechnology investment" (Government of South Africa 2006) raises the specter of regulatory discounts and meek legislative requirements that will give industry the benefit of the doubt and externalize the unexpected. Similarly, recommendations by presidential advisors who want U.S. federal agencies, such as the Food and Drug Administration whose role is ensuring food safety, to "help accelerate technology transfer to the marketplace" represent an irreconcilable conflict of interest (PCAST 2010, 31).

### 4.5.4 Speeding around Blind Corners: Commercialization Racing Ahead of Nanosafety

Existing occupational, environmental, and public health protection laws are often said to be sufficient to manage nanotechnologies, but to the extent that this is true, they are largely immobilized by the lack of nano-specific risk assessment methodologies, characterization methods, and adequate safety data on individual nanomaterials. As such, existing law, as one member of the European Parliament put it, is as effective in regulating nanotechnology "as trying to catch plankton with a cod fishing net" (Schlyter 2009, 8).

The nanosafety challenge is daunting. The scale where quantum changes occur is a vast new territory of science, and assessments of the state of nanosafety research and technology-specific risk assessment methodologies confirm that considerable research effort is required to enable quantitative risk assessment of even the first generation of nanomaterials (see, for example, Royal Commission on Environmental Pollution 2008; SCENIHR 2009; EFSA 2009; National Research Council 2008; Aitken et al. 2009; Stone et al. 2010; Council of Canadian Academies 2008). The immaturity of nanosafety research is also due to its relative neglect in government funding.[11] This is of particular concern given governments' ambitions to have nanotechnologies pervade industrial production as quickly as possible. At any rate, the proposition that, for the first time in the emergence of a new technology, risk assessment will be in "on the ground floor" of the technology rollout (Weber 2009; also see Mantovani et al. 2009) is unconvincing. This is the case for nanomaterials already in commercial circulation; the prospects for more complex products in development are arguably bleaker. In 2008, the U.K. Royal Commission on Environmental Pollution put nanosafety at least a generation behind the development and commercialization of the technology and was pessimistic that international collaborations would deliver results "before irreparable harm is done to individuals or ecosystems" (RCEP 2008, 57). The much-heralded "anticipatory governance"—pegged to move risk governance from post- to pre-market assessment—appears fanciful under such conditions of ignorance.

The largely unregulated, unassessed, and unmonitored status of nanoscience and nanotechnology stands at considerable distance from political pledges to an "open, traceable and verifiable development," *ex ante* risk assessment and life-cycle analysis (EC 2004, 19). Multinational insurance company Lloyd's has likened the current practice—where nanoproducts are allowed into wide market and environmental circulation, and workers and citizens are exposed to nanomaterials without knowledge of their effects— to the financial crisis, with its origins in "blithe acceptance of complex products that many didn't understand" (Gray 2009). Risk rankings reflect the ignorance about the safety of nanomaterials and the potential for harm from their widespread deployment; the technologies are rated variously as one of three major technological risks facing the planet, a ranking conferred by the World Economic Forum for five years running (World Economic Forum 2006, 2007, 2008, 2009, 2010); the top emerging workplace risk in Europe (European Agency for Safety and Health at Work 2009); and a new environmental threat to child health, according to the WHO International Conference on Children's Environmental Health (WHO 2009).[12]

### 4.5.5 Nanotechnologies' Reluctant Volunteers

In the new technology governance model, voluntary and self-regulatory approaches are presented as state-of-the-art management, and a suite of nonbinding arrangements, such as self-regulation, codes of conduct, and voluntary reporting schemes, has been created. Among them are the European Commission's Code of Conduct for Responsible Nanosciences and Nanotechnologies Research, the Responsible NanoCode, and the voluntary reporting schemes run by regulators in the United Kingdom, the United States, and Australia. These mechanisms have been advanced by governments and the private sector as appropriate means of introducing accountability and transparency in this early stage of development.

Voluntary arrangements are controversial and have been broadly rejected by civil society because they offer "discount rates" to commercial operators, generally shielding them from full accountability and liability to the wider community and circumventing public participation in determining acceptable levels of risk and how risk is to be managed (Coalition of Civil Society Organisations 2007a, 2007b).

Thus far, however, the theory that voluntary mechanisms will introduce transparency to nanotechnology activity has been disproved, due to high levels of truancy on the part of nano developers. As such, voluntary schemes have done little to make nanotechnology activity traceable or to dispel skepticism within civil society that governments, in partnership with industry, can be transparent and accountable. Indeed, the failure of voluntary schemes has proven embarrassing for governments, which have been unable to marshal the private sector, even on such favorable terms. Industry response to government-run voluntary reporting initiatives has been so low that the U.K. government's two-year voluntary scheme was assessed to be "pathetic" by the chair of the Royal Commission on Environmental Pollution (ENDS 2008, 8),[13] while across the Atlantic, the Environmental Protection Agency was forced to concede that participation in its two-year Nanoscale Materials Stewardship Program suggested that "most companies are not inclined to voluntarily test their nanoscale materials" (U.S. Environmental Protection Agency 2009, 27).

Codes of conduct are faring little better. In the United Kingdom, a joint initiative by public science institutions and industry to develop a code of conduct encountered industry resistance when corporate attorneys suggested the structure of the code could create legal liabilities for members, and then faltered for lack of funding at the point that benchmarking and systems for monitoring adherence were in sight (Responsible Nano Code Initiative 2008; Sutcliffe 2009). Meanwhile, uptake of the European Commission's Code of Conduct, a document that contains some commendable provisions but little by way of implementation, has been labeled "tepid" (Mantovani, Porcari, Morrison, et al. 2010, 57), with only one EU country of seven reviewed having formally adopted the code (Grobe, Kreinberger, and Funda 2011).[14]

The invisibility cloak that surrounds nanotech activity—due to industry's reluctance to report its activities (typically concealed by claims of "confidential business information")—is hardly consistent with pledges of transparency or inclusiveness. This has not been lost on European members of Parliament, who laconically noted the contradiction between the voluble claims to wide-ranging future benefits and the silence on current uses (European Parliament 2009). Among the sectors to have disappeared from view is the food industry. Its failure to publicly account for its plans in the area of food-related nanotechnologies has been labeled "an almost inevitable communications disaster" (Grobe, Renn, and Jaeger 2008, 14), leading bodies such as a U.K. House of Lords select committee to call on industry to adopt a culture of transparency (House of Lords 2010).

There have been suggestions that incentives have not been properly crafted to ensure participation in voluntary schemes (Nanotechnology Industries Association 2006, 2009; Hansen and Tickner 2007). That, however, is a tactical question about how to ensure that, in the absence of the stick, the carrot is sufficiently appealing. At the higher level of principle, there is another interpretation: The industry's presumption of a "right to operate" in conditions that frustrate regulatory scrutiny and in the absence of legislated structures for accountability demonstrates an unwillingness to answer to the wider community that has, in large measure, underwritten its commercial activities. Given this, mandatory measures are required to bring commercial nanotech activities into view.

## 4.6 International Bodies: Governance Vacuum without Borders

The fainthearted governance of nanotechnologies (beyond promotional programs and policies) at the national level is mirrored at the international level. There, the nature of nanotechnologies has yet to be framed as economically and ecologically "transboundary," to be understood for their potential to intensify inequitable resource and economic relations and, further, as requiring governance by the international community rather than just those pursuing the technologies. For now, nanotechnology regulation is cast as the exclusive business of individual countries and a matter for dialogue at the international level. That political construction finds particular expression in the International Dialogue on Responsible Research and Development of Nanotechnology (a nonbinding, invitation-only forum that sits outside existing intergovernmental structures and agreements) and by the OECD Working Parties, with their emphasis on information exchange, cooperation on technology promotion, and nanosafety research collaboration.

By and large, UN institutions have sidestepped nanotechnologies, and in the vacancy they have left, governments and industry intent on pursuing nanotechnologies have created forums that facilitate the transition from laboratory to marketplace. In doing so, a rather spectacular divide in the governance of nanotechnologies is opening up, as forums that favor the resources, influence, and political interests of northern industrialized countries and the private sector dominate international discussion and assessment.

Of the international forums, the OECD hosts the most regular and ambitious platform through two working parties: the Working Party on Manufactured Nanomaterials (WPMN) and the Working Party on Nanotechnologies (WPN). Joining the thirty-four OECD member-countries at the working parties as observers are just seven non-OECD countries (one country from Latin America [Brazil], one from Sub-Saharan Africa [South Africa], three from Asia [China, Singapore, and Thailand], and the Russian Federation). Despite this meager representation and the emphasis on the preoccupations of industrialized economies, the OECD activities on nanomaterials are likely to be influential in shaping the scope and culture of domestic regimes in other countries, as well as of intergovernmental arrangements in the future (Bowman and Gilligan 2007), particularly if no other entity at the intergovernmental level develops a serious agenda on nanotechnologies.

The picture is similar in the development of international standards for nanotechnologies. Standards are considered fundamental to a global market in the technologies, and influence over their development is fiercely contested. A primary forum is the International Organization for Standardization (ISO), a "network" of national standards institutes from different countries. Of the thirty-three member-institutes to its Technical Committee on Nanotechnologies (TC 229), around two thirds are OECD countries, with low participation from the Global South (refer to Table 4.3). Technically, the ISO is open to participation by any country, but resourcing effective participation presents a challenge for many countries of the Global South. Furthermore, the ISO's insistence that the standards developed under its roof are voluntary is somewhat of a political fiction (Hatto 2009; OECD 2010b). The Agreement on Technical Barriers to Trade under the WTO, for example, requires signatories to participate (wherever possible) in international standards development, to avoid duplication with international activities, and to use existing standards as a basis for any national standards (ISO, UNIDO 2008). This is likely to make a persuasive case for countries to follow the course set at the ISO.

In this rather exclusive global governance landscape, the tripartite Strategic Approach to International Chemicals Management (SAICM) and the related International Forum on Chemical Safety (IFCS) have recently begun to open up new terrain. The IFCS is a forum that arose in 1992 and was initially convened by the United Nations Environmental Program, the International Labor Organisation, and the WHO, who describes it as "a global platform where governments, international, regional and national organizations, industry groups, public interest associations, labour organizations, scientific associations and representatives of civil society meet to build partnerships,

**TABLE 4.3**

Participation and Observation by Non-OECD Countries in ISO TC229

| | Africa | Asia | Caribbean | Central America | Latin America | Middle East | Pacific Island Nations | Eastern Europe/ Central Asia | Total non-OECD |
|---|---|---|---|---|---|---|---|---|---|
| Participants | 2 | 5 | – | – | 1 | 1 | – | 3 | 12 |
| Observers | 2 | 3 | – | – | 1 | – | – | 3 | 9 |

provide advice and guidance, make recommendations and monitor progress" (WHO 2010).

SAICM is a policy framework that arose from the IFCS and is dedicated to achieving the target agreed to at the 2002 World Summit on Sustainable Development: that, by 2020, significant adverse impacts on the environment and human health arising from chemical production are minimized. Explicit in SAICM's understandings is the recognition that a fundamental change in chemicals management is required, its awareness of the vulnerability of particular parts of communities to chemical pollution, and that inclusiveness is needed to achieve its mandate.

These forums—which are more representative in membership and are aligned, if somewhat weakly, to sustainability—have cast more inclusive political understandings about nanotechnologies and have determined that the broader implications of nanotechnologies are to be mapped and understood—particularly their effects on more vulnerable groups within communities and the countries of the Global South. For example, at the sixth session of the IFCS in Dakar, 2008, delegates unanimously adopted a resolution affirming the right of countries to accept or reject nanomaterials, emphasized the particular vulnerability of certain groups within communities (children, pregnant women, and the elderly) to the risks nanomaterials pose as well as the absence of a global policy framework, and urged application of the precautionary principle (IFCS 2008).[15] Parties to the SAICM have backed a report regarding the implications of nanotechnologies for the Global South—the first of its kind at the intergovernmental level—that will be presented in 2012. These developments indicate the interest and political will to subject nanotechnologies to the broader, critical scrutiny that can arise in a more representative forum. Although this is positive, these forums are less regular and arguably less resourced than the OECD working parties and thus more vulnerable to "capture" by the more affluent countries and less likely to influence the shape of international governance arrangements. The future of the IFCS is also under some doubt, due largely to the withdrawal of funds by the United States and Japan (International Institute for Sustainable Development 2009).

## 4.7 Conclusions

A decade since the official beginning of the global nanotechnology race, the picture is of a technology dominated by northern countries, captured by economic growth and competitive motivations, and continuing to gain leverage from hyped visions and often highly speculative future benefits. Meanwhile, governance of nanotechnologies is virtually nonexistent beyond promotional policies. The world that matters in this frame is narrow.

There is, then, "plenty of room at the top." At the international level, global coordination and national capacity-building for the monitoring and evaluation of rapidly emerging technologies are required. The ETC Group has identified the need for a permanent international forum wherein governments, scientists, civil society organizations, social movements, and industry can meet together in participatory and transparent processes that support societal understanding, encourage scientific discovery, and facilitate equitable benefit sharing from new technologies (ETC Group 2010).

## References

Adams, J. and Pendlebury, D. 2010. *Global Research Report. United States.* Evidence: Thomson Reuters. http://researchanalytics.thomsonreuters.com/m/pdfs/globalresearchreport-usa.pdf (accessed July 28, 2011).

Adewoye, O.O. 2008. *Status of Nanotechnology in Nigeria—Prospects, Options, and Challenges.* Presentation to the sixth meeting of the Intergovernmental Forum on Chemical Safety (IFCS), Dakar, Senegal, September 15–19. http://www.who.int/ifcs/documents/forums/forum6/ppt_nano_adewoye.pdf (accessed July 28, 2011).

African Regional Meeting, SAICM. 2010. *Resolution on Nanotechnologies and Manufactured Nanomaterials by Participants in the African Regional Meeting on Implementation of the Strategic Approach to International Chemicals Management.* Abidjan, Côte D'Ivoire, 25–29 January. http://www.sciencecorps.org/nano_resolution_from_African_region.doc (accessed April 26, 2011).

Aitken, R.J., Hankin, S.M., Ross, B., Tran, C.L., Stone, V., Fernandes, T.F., Donaldson, K., Duffin, R., Chaudhry, Q., Wilkins, T.A., Wilkins, S.A., Levy, L.S., Rocks, S.A. and Maynard, A. 2009. *EMERGNANO: A Review of Completed and Near Completed Environment, Health and Safety Research on Nanomaterials and Nanotechnology.* SAFENANO. http://www.iom-support.co.uk/Portals/3/SN_Content/Documents/EMERGNANO_CB0409_Full.pdf (accessed September 15, 2011)

Altmann, J. 2009. Military nanotechnology: Issues for ethical assessment. *Nanomagazine* 15. http://www.nanomagazine.co.uk.

ANEC/BEUC. 2010a. *How Much Nano Do We Buy? ANEC & BEUC Updated Inventory on Products Claiming to Contain Nanomaterials.* http://www.anec.org/attachments/ANEC%20BEUC%20leaflet%20on%20nano%20inventory_How%20much%20nano%20do%20we%20buy.pdf (accessed September 15, 2011).

ANEC/BEUC. 2010b. *ANEC & BEUC Updated Inventory on Products Claiming to Contain Nanomaterials.* http://www.anec.org/ATTACHMENTS/ANEC-PT-2010-Nano-017.xls (accessed July 28, 2011).

Associated Press. 2010. *UN Patent Filings Dropped for 1st Time since 1978.* February 8. http://www.businessinsider.com/un-patent-filings-dropped-for-1st-time-since-1978-2010-2 (accessed May, 2 2011).

Australian Academy of Technological Sciences and Engineering. 2008. *Energy and Nanotechnologies: Strategy for Australia's future.* Parkville, Vic.: Australian Academy of Technological Sciences and Engineering. http://www.atse.org.au/resource-centre/func-startdown/64/ (accessed September 15, 2011).

Australian Government. 2008. *National Nanotechnology Strategy*. Department of Innovation, Industry, Science and Research. http://www.innovation.gov.au/Industry/Nanotechnology/NationalEnablingTechnologiesStrategy/Documents/NNSImplementationPlan.pdf (accessed July 28, 2011).

Bai, C. 2008. Nano rising. *Nature* 456:36–37.

Baya Laffite, N. and Joly, P.-B. 2008. Nanotechnology and society: Where do we stand in the ladder of citizen participation? *CIPAST Newsletter*, March. http://www.cipast.org/download/CIPAST%20Newsletter%20Nano.pdf (accessed July 28, 2011).

BBC. 2007. Russia tests giant thermobaric/nanotechnology bomb. *BBC News*, September 18.

BCC Research. 2008. *Nanotechnology: A Realistic Market Assessment*. News release, May. http://www.bccresearch.com/report/NAN031C.html (accessed April 21, 2011).

Berger, M. 2007. Debunking the trillion dollar nanotechnology market size hype. *Nanowerk Spotlight*, April 18. http://www.nanowerk.com/spotlight/spotid=1792.php (accessed April 25, 2011).

Berger, M. 2008. Key patent strategies for nanotechnology inventors. *Nanowerk Spotlight*, September 15. http://www.nanowerk.com/spotlight/spotid=7238.php (accessed April 25, 2011).

Bowman, D. and Gilligan, G. 2007. How will the regulation of nanotechnology develop? Clues from other sectors. In *New Global Frontiers in Regulation. The Age of Nanotechnology*, edited by G. Hodge, D. Bowman, and K. Ludlow, 353–84. Cheltenham: Edward Elgar.

Canton, J. 2001. The strategic impact of nanotechnology on the future of business and economics. In *Social Implications of Nanoscience and Nanotechnology*, edited by M. Rocco and W. Bainbridge, 91–96. Arlington, Va.: National Science Foundation.

Chitty, A. 2008. Introduction. In *The Storm Breaking upon the University*. March 20. http://stormbreaking.blogspot.com/2008/03/introduction.html (accessed May 10, 2010).

Cientifica. 2007. *Nanotech: Cleantech*. White paper. http://cientifica.eu/blog/white-papers/nanotechcleantech/ (accessed July 28, 2011).

Cientifica. 2008. *Nanotechnology Opportunity Report*. Executive Summary, 3rd ed.

Cientifica. 2009. *Nanotechnology Takes a Deep Breath … and Prepares to Save the World! Global Nanotechnology Funding in 2009*. http://cientifica.eu/blog/white-papers/nanotechnologies-in-2009/ (accessed July 28, 2011).

Citizen Participation in Science and Technology. 2008. *Publishable Finale Activity Report and Final Recommendations*. http://www.cipast.org/cipast.php?section=35 (accessed July 28, 2011).

Claassens, C. 2008. Country profile: South Africa, *Nanomagazine* 7, June. http://www.nanomagazine.co.uk.

Coalition of Civil Society Organisations. 2007a. *An Open Letter to the International Nanotechnology Community at Large, Civil Society-Labor Coalition Rejects Fundamentally Flawed DuPont-ED Proposed Framework*. http://www.etcgroup.org/upload/publication/610/01/coalition_letter_april07.pdf (accessed September 19, 2011).

Coalition of Civil Society Organisations. 2007b. *Principles for the Oversight of Nanotechnologies and Nanomaterials*. http://www.nanoaction.org/nanoaction/page.cfm?id=223 (accessed April 26 2011).

Council for Science and Technology (U.K.). 2007. *Nanosciences and Nanotechnologies: A Review of Government's Progress on its Policy Commitments*. London: Council for Science and Technology. http://www.bis.gov.uk/cst/business/nanoreview (accessed April 26, 2011).

Council of Canadian Academies. 2008. *Small Is Different: A Science Perspective on the Regulatory Challenges of the Nanoscale*. Report of the Expert Panel on Nanotechnologies. Ottawa: Council of Canadian Academies.

Council of the European Union. 2007. *Council Conclusions on Nanosciences and Nanotechnologies*. 2832nd Competitiveness (Internal market, Industry and Research) Council meeting. Brussels, November 22–23. http://www.consilium.europa.eu/uedocs/cms_Data/docs/pressdata/en/intm/97238.pdf (accessed April 15, 2011).

Dang, Y., Zhang, Y., Fan, L., Chen, H., and Roco, M.C. 2009. Trends in worldwide nanotechnology patent applications: 1991 to 2008. *Journal of Nanoparticle Research* 12(3): 687–706.

Davies, S., Macnaghten, P., and Kearnes, M. (eds.). 2009. *Reconfiguring Responsibility. Deepening Debate on Nanotechnology*, Durham: Durham University. http://dro.dur.ac.uk/6399/1/6399.pdf?DDD14+dgg1mbk (accessed April 15, 2011).

Department of Science and Technology, Republic of South Africa. 2006. *The National Nanotechnology Strategy*. Pretoria: Department of Science and Technology. http://www.dst.gov.za/publications-policies/strategies-reports/reports/Nanotech.pdf (accessed April 15, 2011).

Dingwall, K. 2010. *Second Annual Report on Nanoscience and Nanotechnology in the Pacific*. ICPC-Nanonet. http://www.icpc-nanonet.org (accessed April 15, 2011).

Dosch, H. and Van de Voorde, M.H. (eds.). 2009. *GENNESYS White Paper*. Stuttgart: Max-Planck-Institut für Metallforschung. http://www.mf.mpg.de/mpg/websiteMetallforschung/english/veroeffentlichungen/GENNESYS/index.html (accessed April 15, 2011).

Drezek, R.A. and Tour, J.M. 2010. Is nanotechnology too broad to practise? *Nature Nanotechnology* 5:168–69.

Eklund, P., Ajayan, P., Blackmon, R., Hart, A.J., Kong, J., Pradhan, B., Rao, A., and Rinzler, A. 2007. *International Assessment of Carbon Nanotube Manufacturing and Applications*. World Technology Evaluation Center Panel Report. Baltimore: World Technology Evaluation Center.

Elder, M. 2007. Nanotechnology: Big bang aims to start boom. *Financial Times*, October 2. http://www.ft.com/reports/investrussia2007 (accessed April 26, 2011).

ENDS. 2008. RCEP calls for tougher nanotech measures. *ENDS Report* 406:7–9.

Escoffier, L. 2009. International IP and regulatory issues involved in CNT commercialization: Two case studies. *Nanotechnology Law and Business* 6(2): 311–19. http://www.nanolabweb.com/ (accessed April 15, 2011).

Esteban, M., Webersik, C., Leary, D., and Thompson-Pomeroy, D. 2008. *Innovation in Responding to Climate Change: Nanotechnology, Ocean Energy and Forestry*. Tokyo: United Nations University Institute of Advanced Studies. http://www.ias.unu.edu/sub_page.aspx?catID=111&ddlID=738 (accessed April 15, 2011).

ETC Group. 2010. Geopiracy: The case against geoengineering. *Communiqué #103*, November. http://www.etcgroup.org/upload/publication/pdf_file/ETC_geopiracy_4web.pdf (accessed April 15, 2011).

European Commission. 2004. *Towards a European Strategy for Nanotechnology*. Luxembourg: Office for Official Publications of the European Communities. http://ec.europa.eu/nanotechnology/pdf/nano_com_en_new.pdf (accessed April 15, 2011).

European Commission. 2005. *Nanosciences and Nanotechnologies: An Action Plan for Europe 2005–2009*, Luxembourg: Office for Official Publications of the European Communities. http://ec.europa.eu/research/industrial_technologies/pdf/nano_action_plan_en.pdf (accessed April 15, 2011).

European Commission. 2008. *Commission Recommendation of 7/02/2008 on a Code of Conduct for Responsible Nanosciences and Nanotechnologies Research.* http://ec.europa. eu/nanotechnology/pdf/nanocode-rec_pe0894c_en.pdf (accessed April 15, 2011).

European Commission. 2009a. *On the Progress Made under the Seventh European Framework Programme for Research.* http://eur-lex.europa.eu/LexUriServ/LexUriServ. do?uri=COM:2009:0209:FIN:EN:PDF (accessed April 15, 2011).

European Commission. 2009b. *Preparing for Our Future: Developing a Common Strategy for Key Enabling Technologies in the EU.* Current situation of key enabling technologies in Europe. Staff Working Document. {COM(2009) 512}. http:// ec.europa.eu/enterprise/sectors/ict/files/staff_working_document_sec512_ key_enabling_technologies_en.pdf (accessed April 15, 2011).

European Commission. 2009c. *Preparing for Our Future: Developing a Common Strategy for Key Enabling Technologies in the EU.* Communication from the Commission to the European Parliament, the Council, the European Economic Social Committee and the Committee of the Regions. {SEC(2009) 1257}. http://ec.europa.eu/ enterprise/seict/files/communication_key_enabling_technologies_sec1257_ en.pdf (accessed April 15, 2011).

European Commission. 2010. *Communicating Nanotechnology. Why, to Whom, Saying What and How?* Directorate-General for Research. Brussels: European Commission. ftp://ftp.cordis.europa.eu/pub/nanotechnology/docs/communicating-nanotechnology_en.pdf (accessed April 15, 2011).

European Food Safety Authority (EFSA). 2009. Scientific opinion of the Scientific Committee on a request from the European Commission on the potential risks arising from nanoscience and nanotechnologies on food and feed safety, *EFSA Journal* 958:1–39.

European Group on Ethics in Science and New Technologies to the European Commission. 2007. *Opinion on the Ethical Aspects of Nanomedicine.* Opinion No 21. Luxembourg: European Commission. http://ec.europa.eu/bepa/european-group-ethics/docs/publications/opinion_21_nano_en.pdf (accessed September 19, 2011).

European Parliament. 2009. *Resolution of 24 April 2009 on Regulatory Aspects of Nanomaterials (2008/2208(INI)).* http://www.europarl.europa.eu/sides/getDoc.do?type= TA&reference=P6-TA-2009-0328&language=EN (accessed September 19, 2011).

Federal Ministry of Education and Research. 2007. *Nano-Initiative—Action Plan 2010.* Berlin: Federal Ministry of Education and Research. http://www.research-in-germany.de/dachportal/en/downloads/download-files/28608/nano-initia-tive-action-plan-2010.pdf (accessed September 19, 2011).

Federal Ministry of Education and Research. 2009. *nano.DE-Report 2009. Status Quo of Nanotechnology in Germany.* Bonn: Federal Ministry of Education and Research. http://www.bmbf.de/pub/nanode_report_2009_en.pdf (accessed April 15, 2011).

Foladori, G. 2006. Nanoscience and nanotechnology in Latin America, *Nanowerk Spotlight.* www.nanowerk.com/spotlight/spotid=767.php (accessed April 15, 2011).

Friends of the Earth. 2010. *Nanotechnology, Climate and Energy: Over-Heated Promises and Hot Air?* Washington, D.C.: Friends of the Earth. http://www.foe.org/nano-climate (accessed April 15, 2011).

Gavelin, K., Wilson, R., and Doubleday, R. 2007. *Democratic Technologies? The Final Report of the Nanotechnology Engagement Group (NEG).* London: Involve. http:// www.involve.org.uk/democratic-technologies/ (accessed April 15, 2011).

Gordon, N. 2010. Will 2010 be a better year for nanotechnology? *Nanotechnology Now.* February 24. http://www.nanotech-now.com/columns/?article=416 (accessed April 15, 2011).

Gosling, T. 2007. Big things expected from nanotech. *Russia beyond the Headlines,* August 25. http://rbth.ru/articles/2007/08/25/big_things_expected_from_nanotech.html (accessed April 26, 2011).

Government Accountability Office. 2008. Nanotechnology: Accuracy of data on federally funded environmental, health, and safety research could be improved. Testimony before the Subcommittee on Science, Technology, and Innovation, *Committee on Commerce, Science, and Transportation, U.S. Senate,* April 24. http://www.gao.gov/new.items/d08709t.pdf (accessed April 15, 2011).

Government Accountability Office. 2010. *Nanotechnology. Nanomaterials Are Widely Used in Commerce, but EPA Faces Challenges in Regulating Risk.* Report to the Chairman, Committee on Environment and Public Works, U.S. Senate. http://www.gao.gov/new.items/d10549.pdf (accessed April 15, 2011).

Government Communication and Information System. 2010. *South Africa Yearbook 2009/10.* Republic of South Africa. http://www.gcis.gov.za/resource_centre/sa_info/yearbook/2009-10.htm (accessed September 19, 2011).

Government of India. 2008. *Vice President Inaugurates Bangalore Nano-2008.* Press Information Bureau. Press release, December 13 2008. http://www.pib.nic.in/release/release.asp?relid=45620 (accessed April 25 2011).

Government of Korea. 2006. *Nanotechnology Korea.* Ministry of Science and Technology. www.bioin.or.kr/upload.do?cmd=download&seq=2236&bid=tech (accessed April 15, 2011).

Government of Korea. 2010. *Nanotechnology for Dynamic Korea.* Ministry of Education, Science and Technology. http://www.kontrs.or.kr/data/pdf/Nanotechnology%20for%20Dynamic%20Korea.pdf (accessed April 15, 2011).

Government of Malaysia. 2010. *National Nanotechnology Statement.* Ministry of Science, Technology and Innovation Malaysia. http://www.nano.gov.my/ (accessed April 15, 2011).

Government of New Zealand. 2006. *Nanoscience and Nanotechnologies.* Roadmap for Science. Wellington: Ministry of Research, Science and Technology. http://www.morst.govt.nz/current-work/roadmaps/nanotech/ (accessed April 15, 2011).

Government of South Africa. 2006. *Keynote Address by Deputy Minister Derek Hanekom, at the Launch of the South African National Nanotechnology Strategy.* April 13. http://www.dst.gov.za/media-room/speeches/speech.2007–02–16.1265313839/ (accessed April 25, 2011).

Government of South Africa. 2009. *Keynote Address by Deputy Minister of Science and Technology, Derek Hanekom, at the Official Opening of South Africa's First International Nanoschool,* November 23 2009. http://www.dst.gov.za/keynote-address-by-deputy-minister-derek-hanekom-at-the-official-opening-of-south-africa2019s-first-international-nanoschool (accessed April 25, 2011).

Government of Sri Lanka. 2010. *National Nanotechnology Policy.* [Draft]. National Science Foundation and National Science and Technology Commission (Ministry of Technology and Research of Sri Lanka). http://www.nsf.ac.lk/images/stories/TD/version10.pdf (accessed September 19, 2011).

Gray, D. 2009. Nanotechnology: Balancing risk and opportunity. *Lloyd's 360 Risk Insight,* March 26. http://www.lloyds.com/News-and-Insight/News-and-Features/360-News/Emerging-Risk-360/Nanotechnology_balancing_risk_and_opportunity (accessed September 19, 2011).

Grobe, A., Kreinberger, N., and Funda, P. 2011. *Stakeholder's Attitudes towards the European Code of Conduct for Nanosciences and Nanotechnologies Research. Synthesis Report*. NanoCode Report. http://www.nanocode.eu/files/reports/nanocode/nanocode-consultation-synthesis-report.pdf (accessed April 15, 2011).

Grobe, A., Renn, O., and Jaeger, A. 2008. *Risk Governance of Nanotechnology Applications in Food and Cosmetics*. A report for the International Risk Governance Council. Geneva: International Risk Governance Council. http://www.irgc.org/IMG/pdf/IRGC_Report_FINAL_For_Web.pdf (accessed April 15, 2011).

GRULAC. 2010. *Resolution on Nanotechnologies and Manufactured Nanomaterials by Participants to the GRULAC Regional Meeting on the Implementation of the Strategic Approach to International Chemicals Management (SAICM)*. Kingston, Jamaica, March 8–9.

Guillermo, F. and Invernizzi, N. 2008. Introduction. In *Nanotechnologies in Latin America*, edited by F. Guillermo and N. Invernizzi, 7–10. Rosa Luxemburg Stiftung Manuskripte 81. Berlin: Karl Dietz Verlag.

Hansen, S.F. and Tickner, J.A.. 2007. The challenges of adopting voluntary health, safety and environment measures for manufactured nanomaterials: Lessons from the past for more effective adoption in the future. *Nanotechnology Law & Business* 4(3): 341–59.

Harris, D.L. 2009. Carbon nanotube patent thickets. *Nanotechnology and Society*, edited by F. Allhoff and P. Lin, 163–84. Boston: Springer.

Harris, D.L. and Bawa, R. 2007. The carbon nanotube patent landscape in nanomedicine: An expert opinion. *Expert Opinion on Therapeutic Patents* 17(9): 1–11.

Harper, T. 2009. *How to Make Money from Emerging Technologies. Rational Investing in an Age of Rampant Hype*. White Paper. Cientifica. http://cientifica.eu/blog/white-papers/how-to-make-money-from-emerging-technologies/ (accessed April 15, 2011).

Hart Research Associates. 2009. *Nanotechnology, Synthetic Biology and Public Opinion*. Based on a national survey among adults conducted on behalf of project on emerging nanotechnologies. Woodrow Wilson International Center for Scholars. http://www.nanotechproject.org/publications/archive/8286/ (accessed April 15, 2011).

Hassan, M.H.A. 2008. Small things and big changes in the developing world. *Nanomagazine* 7. http://www.nanomagazine.co.uk (accessed April 15, 2011).

Hatto, P. 2009. Supporting Stakeholders' Needs and Expectations through Standardization. In *"No Data, No Market?" Challenges to Nano-Information and Nano-Communication along the Value Chain Fifth International NanoRegulation Conference, November 25—26*. Conference report, edited by S. Knébel and C. Meili, 23–24. Rapperswil, Switzerland: Innovation Society. http://www.innovationsgesellschaft.ch/media/archive2/publikationen/NanoRegulation_5_Report_2009.pdf (accessed April 15, 2011).

Hellsten, E. 2008. *Environment and Health Aspects of Nanomaterials—An EU Policy Perspective*. Presentation to the Nanotech Northern Europe 2008, Copenhagen.

HM Government. 2005. *The Government's Outline Programme for Public Engagement on Nanotechnologies (OPPEN)*. http://www.bis.gov.uk/files/file27705.pdf (accessed April 15, 2011).

HM Government. 2010. *UK Nanotechnologies Strategy. Small Technologies, Great Opportunities*. http://www.bis.gov.uk/assets/BISPartners/GoScience/Docs/U/10-825-uk-nanotechnologies-strategy (accessed September 19, 2011).

Hobson, D.W. 2009. How the new regulatory environment will affect manufacturers in the U.S. and abroad. *Controlled Environments Magazine*, May. http://www.cemag.us/article/nanotech-safety (accessed April 15, 2011).

Holman, M. 2007. *Nanotechnology's Impact on Consumer Products*. Lux Research presentation, October 25. http://ec.europa.eu/health/archive/ph_risk/committees/documents/ev_20071025_co03_en.pdf (accessed April 26, 2011).

House of Lords Science and Technology Committee. 2010. *Nanotechnologies and Food*. Vol. 1: Report, *Summary*. First Report of Session 2009–10. London: Stationery Office. http://www.publications.parliament.uk/pa/ld200910/ldselect/ldsctech/22/22i.pdf (accessed April 15, 2011).

Hsinchun, C., Roco, M.C., Li, X., and Lin, Y. 2008. Trends in nanotechnology patents. *Nature Nanotechnology* 3:123–25.

Huang, Z., Hsinchun, C., Yip, Ng, G., Guo, F., Chen, Z.-K., and Roco, M.C. 2003. Longitudinal patent analysis for nanoscale science and engineering: Country, institution and technology field. *Journal of Nanoparticle Research* 5:333–63.

Ibrügger, L. 2005. *The Security Implications of Nanotechnology*. Report. NATO Parliamentary Assembly, Sub-Committee on Proliferation of Military Technology. 179 STCMT 05 E. http://www.nato-pa.int/default.asp?SHORTCUT=781 (accessed April 15, 2011).

ICPC-Nanonet. (N.d.). List of International Partner Co-operation Countries and Associated Consortium Partners. http://www.icpc-nanonet.org/images/stories/Regions/ListofICPCs.pdf (accessed April 20, 2011).

Igami, M. and Okazaki, T. 2007. Capturing nanotechnology's current state of development via analysis of patents. *OECD Science, Technology and Industry Working Papers,* 2007/4. OECD Publishing. http://www.oecd.org/dataoecd/6/9/38780655.pdf (accessed April 15, 2011).

India PRwire. 2007. National mission to make India global nanotechnology hub. *Nanowerk News*, November 5. http://www.nanowerk.com/news/newsid=3182.php (accessed April 18, 2011).

International Forum on Chemical Safety (IFCS). 2008. *Sixth Session of the Intergovernmental Forum on Chemical Safety. Dakar, Senegal, 15—19 September 2008*. Final Report. http://www.who.int/ifcs/documents/forums/forum6/f6_finalreport_en.pdf (accessed September 19, 2011).

International Institute for Sustainable Development. 2009. Summary of the second session of the International Conference on Chemicals Management, 11–15 May 2009. *Earth Negotiations Bulletin* 15(175). http://www.iisd.ca/vol15/enb15175e.html (accessed September 19, 2011).

Invernizzi N. 2008. Brazilian scientists embrace nanotechnologies. In *Nanotechnologies in Latin America*, edited by F. Guillermo and N. Invernizzi, 40–52. Rosa-Luxemburg-Stiftung Manuskripte 81. Berlin: Karl Dietz Verlag.

ISO, UNIDO. 2008. *Fast Forward. National Standards Bodies in Developing Countries*. Geneva: ISO. http://www.unido.org/fileadmin/user_media/Publications/documents/fast_forward.pdf (accessed September 19, 2011).

Ivanov, A. 2010a. *ICPC-Nanonet. Second Annual Report on Nanoscience and Nanotechnology in the Eastern European and Central Asia Countries—Russia and Ukraine (Provisional Version)*. ICPC-Nanonet.

Ivanov, A. 2010b. *Second Annual Report on Nanoscience and Nanotechnology in the Western Balkan Countries*. ICPCNanonet.

Jaspers, N. 2010. *International Nanotechnology Policy and Regulation. Case Study: India*. (Draft version). London School of Economics and Political Science and Lee Quan Yew School of Public Policy. http://ww2.lse.ac.uk/internationalRelations/centresandunits /regulatingnanotechnologies/nanopdfs/India2010.pdf (accessed September 19, 2011).

Johnston D. 2008. Nanotechnology Booming in Russia… Or is it? Tech talk. *IEEE Spectrum Blog*, March 19. http://spectrum.ieee.org/tech-talk/semiconductors/devices/nanotechnology_in_russia_is_bo (accessed April 25, 2011).

Kalyuzhnyi, S.V. 2010. "Rusnano": Nanoindustry is the way to innovation economics. In *Proceedings of the II International Conference, Nanotechnology for Green and Sustainable Construction*. March 14–17, 2010, Cairo, Egypt, edited by G. Yakovlev, p. 7. Izhevsk Publishing House of ISTU.

Kappos D. 2009. *Remarks at Press Conference Announcing Pilot to Accelerate Green Technology Applications*. U.S. Department of Commerce, December 7. http://www.uspto.gov/news/speeches/2009/2009nov07.jsp (accessed February 23, 2010).

Kearnes, M. 2009. The emerging governance landscape of nanotechnology: European and international comparisons. Presentation at the UK–China workshop *Nano: Regulation and Innovation: The Role of the Social Sciences and Humanities*, Beijing, January 14, 2009.

Kearnes, M. and Rip, A. 2009. The emerging governance landscape of nanotechnology. In *Jenseits von Regulierung: Zum politischen Umgang mit der Nanotechnologie*, edited by S. Gammel, A. Lösch, and A. Nordmann. Berlin: Akademische Verlagsgesellschaft.

Kelly, M.J. 2011. Intrinsic top-down unmanufacturability. *Nanotechnology* 22. http://stacks.iop.org/Nano/22/245303 (accessed May 2, 2011).

Khan, N. 2011. Experts call for more debate on pros, cons of Nanotechnology. *Kuwait News Agency*, April 28. http://menafn.com/qn_news_story_s.asp?storyid=1093410095 (accessed April 29, 2011).

Kiparissides, C. (ed.). 2009. *NMP Expert Advisory Group (EAG)*. Position paper on future RTD activities of NMP for the period 2010—2015. Brussels: European Commission. http://ec.europa.eu/research/industrial_technologies/pdf/nmp-expert-advisory-group-report_en.pdf (accessed April 15, 2011).

Krishnadas, K.C. 2007. India rolls out nanotechnology initiative. *EE Times* December 6. http://www.eetimes.com/electronics-news/4075544/India-rolls-out-nanotech-initiative (accessed April 25, 2011).

Kulkarni, G.U., Sundaresan, A., Eswarmoorthy, M., Angappane, S., Srinivas, S. and Vanitha, B. 2010. *Second Annual Report on Nanoscience and Nanotechnology in West Asia*. ICPC-Nanonet. http://www.icpc-nanonet.org (accessed April 15, 2011).

Langley, C., Parkinson, S., and Webber, P. 2008. *Behind Closed Doors. Military Influence, Commercial Pressures and the Compromised University*. Scientists for Global Responsibility. http://www.sgr.org.uk/ArmsControl/BehindClosed-Doors_jun08.pdf (accessed April 15, 2011).

Lemley, M.A. 2005. Patenting nanotechnology. *Stanford Law Review* 58:601–30. http://www.stanfordlawreview.org/system/files/articles/lemley.pdf (accessed April 15, 2011).

Liroff, R. 2009. Nanomaterials: Why your company should sweat the small stuff. *Green Biz*, September 9. http://www.greenbiz.com/blog/2009/09/09/nanomaterials-why-your-company-should-sweat-small-stuff (accessed April 15, 2011).

Liu, X., Zhang, P., Li, X., et al. 2009. Trends for nanotechnology development in China, Russia, and India. *Journal of Nanoparticle Research* 11:1845–66.

Lok, C. 2010. Small wonders. *Nature* 467(2).

Lux Research. 2004. *Revenue from Nanotechnology-Enabled Products to Equal IT and Telecom by 2014, Exceed Biotech by 10 Times*. News release, October 25.

Lux Research. 2008a. *Overhyped Technology Starts to Reach Potential: Nanotech to Impact $3.1 Trillion in Manufactured Goods in 2015.* Press release, July 22.

Lux Research. 2008b. Nanotechnology corporate strategies. *Lux Research Nanomaterial Intelligence.* http://www.printedelectronicsnow.com/whitepapers/download/10/Lux_Research_Nanomaterials_Intelligence_—_Nanotechnology_Corporate_Strategies.pdf (accessed April 15, 2011).

Lux Research. 2009a. *Profits in Nanotech Come from Intermediate Products, Not Raw Materials.* Press release, January 22.

Lux Research. 2009b. *Economy Blunts Nanotech's Growth.* Press release, June 24.

Maclurcan, D. 2005. Nanotechnology and developing countries—Part 2: What realities? *Online Journal of Nanotechnology.* http://www.azonano.com/article.aspx?ArticleID=1429 (accessed April 25, 2011).

Macnaghten, P., Davies, S. and Kearnes, M. 2010. Narrative and public engagement: Some findings from the DEEPEN project. In *Understanding Public Debate on Nanotechnologies. Options for Framing Public Policy*, edited by R.V. Schomberg and S. Davies, 13–20. Belgium: European Commission Directorate-General for Research, Science, Economy and Society.

Malsch, I. 2010. *Second Annual Report on Nanoscience and Nanotechnology in Latin America.* ICPC-Nanonet. http://www.icpc-nanonet.org (accessed April 15, 2011).

Mantovani, E., Porcari, A., and Azzolini, A. 2010. *Synthesis Report on Codes of Conduct, Voluntary Measures and Practices Towards a Responsible Development of N&N.* NanoCode Report. http://www.nanocode.eu/files/reports/nanocode/nanocode-project-synthesis-report.pdf (accessed September 19, 2011).

Mantovani, E., Porcari, A., Meili, C., and Widmer, M. 2009. *Mapping Study on Regulation and Governance of Nanotechnologies.* Framing Nano Report. http://www.framing-nano.eu/images/stories/FramingNanoMappingStudyFinal.pdf (accessed April 15, 2011).

Mantovani E., Porcari A., Morrison M.J., and Geertsma, R.E. 2010. *Developments in Nanotechnologies Regulation and Standards 2010.* Report of the Observatory Nano. http://www.observatorynano.eu.

McAlpine, K. 2010. Chaos at public nanotechnology debates in France. *Chemistry World*, posted on *Nanowerk News*, January 26. http://www.nanowerk.com/news/newsid=14522.php (accessed April 26, 2011).

McWilliams, A. 2010. *Nanotechnology: A Realistic Market Assessment.* Report Highlights. BCC Research. http://www.bccresearch.com/report/nanotechnology-realistic-market-assessment-nan031d.html (accessed April 23, 2011).

Milieu Ltd. and Risk and Policy Analysts. 2009. *Information from Industry on Applied Nanomaterials and Their Safety.* Background paper on options for an EU-wide reporting scheme for nanomaterials on the market prepared for European Commission DG Environment. www.nanomaterialsconf.eu/documents/Nanos-Task1.pdf (accessed April 15, 2011).

Modzelewski, M. 2002. *Prepared Statement of Mark Modzelewski, Executive Director, NanoBusiness Alliance.* Hearing before the Subcommittee on Science, Technology and Space of the Committee on Commerce, Science and Transportation. United States, 107th Congress, Second Session, September 17. Washington D.C.: U.S. Government Printing Office. http://www.gpo.gov/fdsys/pkg/CHRG-107shrg93633/html/CHRG-107shrg93633.htm (September 19, 2011).

Moffet, J., Bregha, F., and Middelkoop, M.J. 2004. Responsible care: A case study of a voluntary environmental initiative. In *Voluntary Codes: Private Governance, the Public Interest and Innovation*, edited by K. Webb, 177–208. Carleton Research Unit for Innovation, Science and Environment, Carleton University.

Musee, N., Brent, A.C., and Ashton, P.J. 2010. A South African research agenda to investigate the potential environmental, health and safety risks of nanomaterials. *South African Journal of Science* 106(3/4): 1–6. http://www.sajs.co.za/index.php/SAJS/article/view/159/273 (accessed April 15, 2011).

Nanotechnology Industries Association. 2006. *Response to the Consultation on a Proposed Voluntary Reporting Scheme for Engineered Nanoscale Materials from the Nanotechnology Industry Association.* http://www.nanotechia.org/nia-publications/consultation-responses (accessed April 15, 2011).

Nanotechnology Industries Association. 2009. *NIA Comment on the UK House of Lords Science and Technology Select Committee Call for Evidence: Nanotechnologies and Food.* http://www.nanotechia.org/nia-publications/consultation-responses (accessed April 15, 2011).

Nasu Nasu, H. and Faunce, T. 2010. Nanotechnology and the international law of weaponry: Towards international regulation of nano-weaponry: Towards international regulation of nano-weapons. *Journal of Law, Information and Science.* http://www.austlii.edu.au/au/journals/JlLawInfoSci/2010/ (accessed April 15, 2011).

National Research Council (U.S.). 2008. *Review of Federal Strategy for Nanotechnology-Related Environmental, Health and Safety Research.* Washington D.C.: National Academies Press. http://www.nap.edu/catalog.php?record_id=12559 (accessed April 15, 2011).

Newton, R. and Xie, S. 2010a. *Second Annual Report on Nanoscience and Nanotechnology in the Mediterranean Partner Countries.* ICPC-Nanonet. http://www.icpc-nanonet.org (accessed April 15, 2011).

Newton, R. and Xie, S. 2010b. *Second Annual Report on Nanoscience and Nanotechnology in the Caribbean.* ICPCNanonet. http://www.icpc-nanonet.org (accessed April 15, 2011).

Nordan, M.M. 2008. *Change Required for the National Nanotechnology Initiatives as Commercialization Eclipses Discovery.* Lux Research testimony before the Senate Committee on Commerce, Science, and Transportation, April 24, 2008. http://www.hsdl.org/?view&did=32278 (accessed September 19, 2011).

NSET—U.S. Subcommittee on Nanoscale Science, Engineering, and Technology Committee on Technology, National Science and Technology Council. 2007. *The National Nanotechnology Initiative. Strategic Plan.* http://www.sandia.gov/NINE/documents/NNI_Strategic_Plan_2007.pdf (accessed April 15, 2011).

NSET—U.S. Subcommittee on Nanoscale Science, Engineering, and Technology Committee on Technology, National Science and Technology Council. 2009. *The National Nanotechnology Initiative. Research and Development Leading to a Revolution in Technology and Industry.* Supplement to the President's FY2010 Budget. http://www.nano.gov/sites/default/files/pub_resource/nni_2010_budget_supplement.pdf (accessed April 15, 2011).

NSET—U.S. Subcommittee on Nanoscale Science, Engineering, and Technology Committee on Technology, National Science and Technology Council. 2010. *The National Nanotechnology Initiative. Research and Development Leading to a Revolution in Technology and Industry.* Supplement to the President's FY2011 Budget. http://www.nano.gov/node/219 (accessed September 19, 2011).

NSET—U.S. Subcommittee on Nanoscale Science, Engineering, and Technology Committee on Technology, National Science and Technology Council. 2011. *The National Nanotechnology Initiative. Research and Development Leading to a Revolution in Technology and Industry Supplement to the President's FY 2012 Budget*. http://www.nano.gov/about-nni/what/funding (accessed September 19, 2011).

OECD. 2009a. *Inventory of National Science, Technology and Innovation Policies for Nanotechnology 2008*. OECD Working Party on Nanotechnology. http://www.oecd.org/dataoecd/38/32/43348394.pdf (accessed April 15, 2011).

OECD. 2009b. *Current Developments in Delegations and Other International Organisations on the Safety of Manufactured Nanomaterials—Tour de Table. Fifth Meeting of the Working Party on Manufactured Nanomaterials*. Paris, France 4–6 March 2009. Series of Safety of Manufactured Nanomaterials, No 17. http://www.oecd.org/document/53/0,3746,en_2649_37015404_37760309_1_1_1_1,00.html (accessed April 15, 2011).

OECD. 2010a. *The Impacts of Nanotechnology on Companies: Policy Insights from Case Studies*. OECD Publishing. http://dx.doi.org/10.1787/9789264094635-en (accessed April 15, 2011).

OECD. 2010b. *Current Development/Activities on the Safety of Manufactured Nanomaterials- Tour de Table. Sixth Meeting of the Working Party on Manufactured Nanomaterials*. Paris, France 28–30 October 2009. Series of Safety of Manufactured Nanomaterials, No 20. http://www.oecd.org/document/53/0,3746,en_2649_37015404_37760309_1_1_1_1,00.html (accessed April 15, 2011).

OECD. 2010c. *Current Developments/Activities on the Safety of Manufactured Nanomaterials—Tour de Table*. Seventh Meeting of the Working Party on Manufactured Nanomaterials. Paris, France 7–9 July 2010. Series of Safety of Manufactured Nanomaterials, No 26. http://www.oecd.org/document/53/0,3746,en_2649_37015404_37760309_1_1_1_1,00.html (accessed April 15, 2011).

OECD. 2011. *Fostering Nanotechnology to Address Global Challenges*. http://www.oecd.org/dataoecd/22/58/47601818.pdf (accessed April 15, 2011).

Padma, T.V. 2007. India "must regulate nanotechnology" urgently. *SciDevNet*, October 11. http://www.scidev.net/en/news/india-must-regulate-nanotechnology-urgently.html (accessed April 15, 2011).

Padma, T.V. 2008. Lack of industry links "keeping Indian nanotech small." *SciDevNet*, September 8. http://www.scidev.net/en/science-and-innovation-policy/tuberculosis/news/lack-of-industry-links-keeping-indian-nanotech-sma.html (accessed September 19, 2011).

Padma, T.V. 2010a. Safety ignored in nanotech rush, warn experts. *SciDevNet*, January 12. http://www.scidev.org/en/china/news/safety-ignored-in-nanotech-rush-warn-experts.html (accessed April 15, 2011).

Padma, T.V. 2010b. Regulation must match advances in nanotechnology. *SciDevNet*, December 15. http://www.scidev.net/en/news/regulation-must-match-advances-in-nanotechnology-1.html (accessed April 15, 2011).

Palmberg, C., Dernis, H., and Miguet, C. 2009. *Nanotechnology: An Overview Based on Indicators and Statistics*. OECD STI Working Paper 2009/7 Statistical Analysis of Science, Technology and Industry. http://www.oecd.org/dataoecd/59/9/43179651.pdf (accessed April 15, 2011).

Peter D. Hart Research Associates. 2006. *Report Findings Based on a National Survey of Adults Conducted on Behalf of the Woodrow Wilson International Center for Scholars Project on Emerging Nanotechnologies*. http://www.nanotechproject.org/news/archive/public_awareness_nano_grows_-/ (accessed April 15, 2011).

Peter D. Hart Research Associates. 2007. *Awareness of and Attitudes towards Nanotechnology and Federal Regulatory Agencies Based on a National Survey of Adults Conducted on Behalf of the Woodrow Wilson International Center for Scholars Project on Emerging Nanotechnologies*. http://www.nanotechproject.org/news/archive/poll_reveals_public_awareness_nanotech/ (accessed April 15, 2011).

Peter D. Hart Research Associates. 2008. *Awareness of and Attitude towards Nanotechnology and Synthetic Biology Based on a National Survey of Adults Conducted on Behalf of the Woodrow Wilson International Center for Scholars Project on Emerging Nanotechnologies*. http://www.nanotechproject.org/publications/archive/synbio/ (accessed April 15, 2011).

President's Council of Advisors on Science and Technology. 2010. *Report to the President and Congress on the Third Assessment of the National Nanotechnology Initiative*. Washington, D.C.: Executive Office of the President. http://www.whitehouse.gov/sites/default/files/microsites/ostp/pcast-nano-report.pdf (accessed April 15, 2011).

Press Trust of India. 2010. India to have Nanotechnology Regulatory Board soon. *Business Standard*, February 18. http://www.business-standard.com/india/news/india-to-have-nanotechnology-regulatory-board-soon/86186/on (accessed April 15, 2011).

Project on Emerging Nanotechnologies. 2011. *Nanotech-Enabled Consumer Products Continue to Rise Oversight Challenges Still Exist*. Press release, March 10 2011. http://www.nanotechproject.org/news/archive/9231/ (accessed April 15, 2011).

Ray, S.G. 2009. Nanotech, a big way to beat recession. *Sakaal Times* (online), March 11.

Renn, O. and Roco, M.C. 2006. Nanotechnology and the need for risk governance. *Journal of Nanoparticle Research* 8(2): 153–91.

Responsible Nano Code Initiative. 2008. *Working Group Meeting Seven, Record of Deliberations*, May 13. http://www.responsiblenanocode.org/pages/progress/index.html (accessed April 15, 2011).

RIA Novosti. 2007. Putin vows to bankroll nanotechnology, stresses payoff. *RIA Novosti*, April 4. http://en.rian.ru/russia/20070418/63882148.html (accessed April 15, 2011).

Rickett, S.E. 2009. Taking the NanoPulse—Nano-nomics 101: What drives growth in 2009? *Industry Week*, January 7. http://www.industryweek.com/readarticle.aspx?ArticleID=18143 (accessed April 15, 2011).

Roam, G.-W., Wu, W.-Y., and Lien, Y.-U. 2009. *Taiwan's Strategic Plan for Responsible Nanotechnology (2009–2014)*. International Symposium on Nano Science and Technology (ISNST). Tainan, Taiwan, November 20–21 2009. Office of Sustainable Development, Environmental Protection Administration. http://www.epa.gov.tw/FileLink/FileHandler.ashx?file=13491 (accessed April 15, 2011).

Roco, M.C. 2003. Converging science and technology at the nanoscale: Opportunities for education and training. *Nature Biotechnology* 21(10): 1247–49.

Roco, M.C. 2005. International perspective on government nanotechnology funding in 2005. *Journal of Nanoparticle Research* 7(6): 707–12.

Roco, M. and Bainbridge, W. (eds.). 2001. *Social Implications of Nanoscience and Nanotechnology*. NSET Workshop Report. Arlington: National Science Foundation. http://www.wtec.org/loyola/nano/societalimpact/nanosi.pdf (accessed April 15, 2011).

Roco, M.C., Harthorn, B., Gutson, D., and Shapira, P. 2010. Innovative and Responsible Governance on Nanotechnology for Societal Development. In *Nanotechnology Research Directions for Societal Needs in 2020. Retrospective and Outlook,* edited by M.C. Roco, C.A. Mirkin, and M.C. Hersam, 441–88. World Technology Evaluation Center Panel Report. Boston: Springer. http://www.wtec.org/nano2/ (accessed April 15, 2011).

RosBusiness Consulting. 2008. *Russia Estimates 2008 Nanotechnology Sales.* March 19. http://www.rbcnews.com/free/20080319123649.shtml (accessed April 15, 2011).

Royal Commission on Environmental Pollution (RCEP). 2008. *Novel Materials Report.* http://webarchive.nationalarchives.gov.uk/20110118100717/http://www.rcep.org.uk/reports/27-novel%20materials/documents/NovelMaterialsreport_rcep.pdf (accessed April 15, 2011).

Royal Society and Royal Academy of Engineering. 2004. *Nanoscience and Nanotechnologies: Opportunities and Uncertainties.* Royal Society Policy Document 19/04. London: Royal Society. http://royalsociety.org/Nanoscience-and-nanotechnologies-opportunities-and-uncertainties-/.

Royal Society and Royal Academy of Engineering. 2006. *Nanoscience and Nanotechnologies: Opportunities and Uncertainties. Two-Year Review of Progress on Government Actions: Joint Academies' Response to the Council for Science and Technology's Call for Evidence.* Royal Society Policy Document 35/06. http://royalsociety.org/Report_WF.aspx?pageid=8241&terms=nanotechnology (accessed April 15, 2011).

Rusnano. 2008. *Business Strategy of State Corporation "Russian Corporation of Nanotechnologies" until the Year 2020.* http://www.rusnano.com/Document.aspx/Download/17905 (accessed April 15, 2011).

Rusnano. 2009. *RUSNANO CEO Anatoly Chubais Outlined the Main Tasks for Corporation for 2009.* Press release, March 19. http://www.rusnano.com/Post.aspx/Show/18087 (accessed April 15, 2011).

Rutt, J.S. and Wu, B. 2009. A "nanotech" and/or a "clean tech" revolution?! *Foley and Lardner Nano and Cleantech Blog,* March 17. http://www.nanocleantechblog.com/2009/03/articles/clean-tech/a-nanotech-andor-a-clean-tech-revolution/ (accessed April 25, 2011).

Sargent, J.F. 2008. *Nanotechnology and U.S. Competitiveness: Issues and Options.* Congressional Research Service. http://handle.dtic.mil/100.2/ADA483318 (accessed April 15, 2011).

Sargent, J.F. 2011. *Nanotechnology: A Policy Primer.* Congressional Research Service. http://assets.opencrs.com/rpts/RL34511_20110119.pdf (accessed April 15, 2011).

Sastry, R.K., Rao, N.H., Cahoon, R., and Tucker, T. 2007. *Can Nanotechnology Provide the Innovations for a Second Green Revolution in Indian Agriculture?* U.S. National Science Foundation Nanoscale Science and Engineering Grantees Conference, December 3–6, 2007. www.nseresearch.org/2007/presentations/0000_sastry.ppt (accessed April 25, 2011).

Satterfield, T., Kandlikar, M., Beaudrie, C.E.H., Conti, J., and Herr Harthorn, B. 2009. Anticipating the perceived risk of nanotechnologies. *Nature Nanotechnology* 4:752–58.

Sawahel, W. 2008. IP model proposed for North-South nanotechnology divide. *Intellectual Property Watch,* November 19. http://www.ip-watch.org/weblog/2008/11/19/ip-model-proposed-for-north-south-nanotechnology-divide/ (accessed April 25, 2011).

Saxl, O. 2009. Can nanotechnology save the planet? *Nanomagazine* 10. http://www.nanomagazine.co.uk (accessed April 15, 2011).

Schlyter, C. 2009. *Draft Report on Regulatory Aspects of Nanomaterials* (2008/2208[INI]). European Parliament Committee on the Environment, Public Health and Food Safety. http://www.europarl.europa.eu/meetdocs/2004_2009/documents/pr/763/763225/763225en.pdf (accessed September 19, 2011)

Scientific Committee on Emerging and Newly-Identified Health Risks (SCENIHR). 2009. *Risk Assessment of Products of Nanotechnologies*. European Commission. http:// ec.europa.eu/health/ph_risk/committees/04_scenihr/docs/scenihr_o_023. pdf (accessed April 15, 2011).

Service, R.F. 2010. Obama nano budget not small. *ScienceInsider blog*, February 1. http://news.sciencemag.org/scienceinsider/2010/02/obama-nano-budg.html (accessed April 25, 2011).

Shalleck, A. 2009. The solution is the government. Get stimulus money to survive. *Nanotechnology Now*, May 18. http://www.nanotech-now.com/columns/?article=305 (accessed April 25, 2011).

Shalleck, A. 2010. 2010 outlook–2009 recap. *Nanotechnology Now*, January 5. http:// www.nanotech-now.com/columns/?article=396 (accessed April 25, 2011).

South African National Research Foundation, South African Agency for Science and Technology Advancement. 2010. *Call for Proposals: Public Engagement with Nanotechnology*. March 18. www.saasta.ac.za/pdf/SAASTA_Nanotechnology_ Call_2010–04.pdf (accessed April 15, 2011).

Spinverse Capital and Consulting. 2010. *Public Funding of Nanotechnology*. Observatory Nano Economic Report. http://www.observatorynano.eu/project/filesystem/ files/Economics_PublicFundingAnalysis_2010.pdf (accessed September 19, 2011).

Stack, J., Balbach, J., Epstein, B., and Hanggi, T. 2007. *Cleantech Venture Capital: How Public Policy Has Stimulated Private Investment*. Environmental Entrepreneurs and Cleanttech Venture Network LLC. http://www.e2.org/ext/doc/CleantechReport2007.pdf (accessed April 15, 2011).

Stavrianakis, A. 2009. In arms' way: Arms company and military involvement in education in the UK. *ACME* 8(3): 505–20. http://www.acme-journal.org/vol8/ Stavrianakis09.pdf (accessed September 19, 2011).

Stone, V., Hankin, S., Aitken, R., Aschberger, K., Baun, A., Christensen, F., Fernandes, T., Hansen, S.F., Bloch Hartmann, N., Hutchison, G., Johnston, H., Micheletti, C., Peters, S., Ross, B., Sokull-Kluettgen, B., Stark, D., and Tran, L. 2010. *Engineered Nanoparticles: Review of Health and Environmental Safety*. Final Report. http:// ihcp.jrc.ec.europa.eu/whats-new/enhres-final-report (accessed April 15, 2011).

Sutcliffe, H. 2009. *Submission to the House of Lords Science and Technology Select Committee Call for Evidence on Nanotechnologies and Food*. http://www.parliament.uk/ documents/lords-committees/science-technology/st129hilarysutclifferesponsiblenanoforum.pdf (accessed April 25, 2011).

Syam, N. 2010. Rush for patents may hinder transfer of new climate-related technologies. *Policy Innovations Briefings*, March 12. http://www.policyinnovations.org/ ideas/briefings/data/000162 (accessed March 20, 2010).

TERI. 2009. *Nanotechnology Development in India: The Need for Building Capability and Governing the Technology*. Briefing paper. The Energy Resources Institute (TERI). http://www.teriin.org/div/ST_BriefingPap.pdf (accessed April 15, 2011).

Times of India. 2010. Soon, a national regulatory framework for nanotechnology. *Times of India*, October 15. http://articles.timesofindia.indiatimes.com/2010-10-15/ pune/28257523_1_nanotechnology-r-d-prithviraj-chavan (accessed September 19, 2011).

Tobin, L. 2009. *First Annual Report on Nanosciences and Nanotechnology in Africa*, ICPC-Nanonet. http://www.icpc-nanonet.org (accessed April 15, 2011).

Tobin, L. 2010. Bridging the nano knowledge divide through the ICPC Nanonet project. *African Journal of Science, Technology, Innovation and Development* 2(3): 241–47.

Tobin, L. and Dingwall, K. 2010. *Second Annual Report on Nanoscience and Nanotechnology in Africa*. ICPC-Nanonet. http://www.icpc-nanonet.org (accessed April 15, 2011).

Tour, J.M. 2011. *Nanotechnology: Oversight of the National Nanotechnology Initiative and Priorities for the Future*. Testimony of James M. Tour, PhD. Richard E. Smalley Institute for Nanoscale Science and Technology Rice University before House Science, Space and Technology Subcommittee on Research and Science Education, April 14. http://science.house.gov/sites/republicans.science.house.gov/files/documents/hearings/Tour%20Testimony.pdf (accessed April 15, 2011).

Trilateral Cooperation. 2008. *Summary of the 26th Trilateral Conference*. The Hague, The Netherlands, 14 November 2008. European Patent Office, Japan Patent Office, U.S. Patent and Trademark Office. http://www.trilateral.net/conferences/2008.pdf (accessed April 15, 2011).

Tyshenko M.J. 2009. The impact of nanomedicine development on North–South equity and equal opportunities in healthcare. Studies in Ethics, Law, and Technology. *Nanotechnology and Health* 3(3). http://www.bepress.com/selt/vol3/iss3/art2/?sending=10920 (accessed April 15, 2011).

United Nations Development Program (UNDP). 2005. *Human Development Report 2005. International at a Crossroads. Aid, Trade and Security in an Unequal World*. New York: UNDP. http://hdr.undp.org/en/reports/global/hdr2005/ (accessed April 15, 2011).

UNESCO. 2006. *The Ethics and Politics of Nanotechnology*. Paris: UNESCO.

U.S. Department of Defense. 2007. *Defense Nanotechnology Research and Development Program*. http://www.fas.org/irp/agency/dod/nano2007.pdf (accessed April 27, 2011).

U.S. Department of Defense. 2009. *Defense Nanotechnology Research and Development Program*. Department of Defense Director, Defense Research and Engineering. www.nano.gov/html/res/NanoReporttoCongressFINAL1MAR10.pdf (accessed April 20, 2011).

U.S. Environmental Protection Agency. 2009. *Nanoscale Materials Stewardship Program*. Interim Report. Office of Pollution Prevention and Toxics. http://www.epa.gov/opptintr/nano/nmsp-interim-report-final.pdf (accessed April 15, 2011).

U.S. Government. 2010. *National Nanotechnology Initiative. Investments by Agency FY2001–2010*. http://www.whitehouse.gov/files/documents/ostp/opengov/Copy%20of%20NNI%20Investments%20by%20Agency%20and%20PCA%202001–2010%20(2).pdf (accessed April 21, 2011).

U.S. National Science and Technology Council, Committee on Technology. 1999. *Nanotechnology. Shaping the World Atom by Atom*. Washington, D.C.: Interagency Working Group on Nanoscience, Engineering and Technology. http://www.wtec.org/loyola/nano/IWGN.Public.Brochure/IWGN.Nanotechnology.Brochure.pdf (accessed April 15, 2011).

Vahtra, P. 2010. *The Rusnano Corporation and Internationalisation of Russia's Nanotech Industry*. Electronic Publications of Pan-European Institute 11/2010. http://www.tse.fi/FI/yksikot/erillislaitokset/pei/Documents/Julkaisut/Vahtra2.pdf (accessed April 15, 2011).

Walsh, B. 2007. *Environmentally Beneficial Nanotechnologies. Barriers and Opportunities*. A report for the Department for Environment, Food and Rural Affairs. Oakdene Hollins. http://archive.defra.gov.uk/environment/quality/nanotech/documents/envbeneficial-report.pdf (accessed April 15, 2011).

Wang, L. and Notten, A. 2010. *Observatory NANO Benchmark Report: Nano-Technology and Nano-Science, 1998–2008.* 2nd Observatory NANO Benchmark Report. Maastricht, The Netherlands: UNU-MERIT.

Watal, A. and Faunce, T. 2011. Patenting nanotechnology: Exploring the challenges. *WIPO Magazine,* April. http://www.wipo.int/wipo_magazine/en/2011/02/article_0009.html (accessed September 19, 2011).

Weber, B. 2009. *Unknowns Raise Environmental Concerns.* Canadian Press, January 9. http://www.theglobeandmail.com/news/technology/science/unknowns-raise-environmental-concerns/article965439/ (accessed April 27, 2011).

Wesoff, E. 2009. KP's John Doerr on greentech: 'The largest economic opportunity of the 21st century.' *GreentechMedia,* November 19. http://www.greentechmedia.com/green-light/post/kps-john-doerr-greentech-the-largest-economic-opportunity-of-the-21st-cen/ (accessed March 31, 2010).

WHO. 2009. *Busan Pledge for Action on Children's Health and Environment,* 3rd WHO International Conference on Children's Environmental Health. Busan, Republic of Korea, June 2009. www.who.int/phe/busan_pledge.pdf (accessed April 15, 2011).

WHO. 2010. *International Forum on Chemical Safety.* http://www.who.int/ifcs/page2/en/index.html (accessed April 12, 2010).

Wijnhoven, S.W.P., Dekkers, S., Kooi, M., Jongeneel, W.P., and de Jong, W.H. 2011. *Nanomaterials in Consumer Products. Update of Products on the European Market in 2010.* National Institute for Public Health and the Environment. Ministry of Health, Welfare and Sport. RIVM Report 340370003/2010. http://www.rivm.nl/bibliotheek/rapporten/340370003.pdf (accessed April 15, 2011).

Wilsdon, J. and Keeley, J. 2007. *China: The Next Science Superpower?* The Atlas of Ideas: Mapping the new geography of science. DEMOS. London: Good News Press. http://www.demos.co.uk/files/China_Final.pdf?1240939425 (accessed April 15, 2011).

WIPO. 2007. *The 45 Adopted Recommendations under the WIPO Development Agenda.* http://www.wipo.int/export/sites/www/ip-development/en/agenda/recommendations.pdf (accessed April 15, 2011).

WIPO. 2010. *International Patent Filings Dip in 2009 amid Global Economic Downturn.* WIPO Press release, February 8. PR/2010/632. http://www.wipo.int/pressroom/en/articles/2010/article_0003.html (accessed April 15, 2011).

World Economic Forum. 2006. *Global Risks 2006.* https://members.weforum.org/pdf/CSI/Global_Risk_Report.pdf (accessed September 19, 2011).

World Economic Forum. 2007. *Global Risks 2007.* https://members.weforum.org/pdf/CSI/Global_Risks_2007.pdf (accessed September 19, 2011).

World Economic Forum. 2008. *Global Risks 2008.* https://members.weforum.org/pdf/globalrisk/report2008.pdf (accessed September 19, 2011).

World Economic Forum. 2009. *Global Risks 2009.* https://members.weforum.org/pdf/globalrisk/globalrisks09/global_risks_2009.pdf (accessed September 19, 2011).

World Economic Forum. 2010. *Global Risks 2010.* http://www3.weforum.org/docs/WEF_GlobalRisks_Report_2010.pdf (accessed April 15, 2011).

World Economic Forum. 2011. *Global Risks 2011.* http://www.weforum.org/issues/global-risks (accessed April 15, 2011).

WTO. *Agreement on Technical Barriers to Trade.* Annex 3. Code of Food Practice for the Preparation, Adoption and Application of Standards. WTO.

You, Z. 2010. *Second Annual Report on Nanoscience and Nanotechnology in East-Asia.* ICPC-Nanonet. http://www.icpc-nanonet.org.

Záyago Lau, E. and Rushton, M. 2008. Nanotechnologies for development in Latin America. In *Nanotechnologies in Latin America*, edited by F. Guillermo and N. Invernizzi, 11–26. Rosa Luxemburg Stiftung Manuskripte 81. Berlin: Karl Dietz Verlag.

Zhenzhen, L., Jiuchun, Z., Ke, W., Thorsteinsdóttir, H., Quach, U., Singer, P.A., and Daar, A.S. 2004. Health biotechnology in China—Reawakening of a giant. *Nature Biotechnology* 22. Supplement, DC13–18.

## Endnotes

1. For example, for country-specific statements, see Germany (Federal Ministry of Education and Research 2007); Europe (Kiparissides 2009); Russia (Gosling 2007); Malaysia (Government of Malaysia 2010); South Korea (Government of Korea 2006); Latin America (Záyago Lau and Rushton 2008); NSET-US 2009; Department of Science and Technology, Republic of South Africa 2006; Council of the European Union 2007; and Government of Sri Lanka 2010. For the proposition that, from an industrial perspective, the twenty-first century will be a nano-economy, see Canton 2001.

2. Data relating to funding and orientation of state military nano R&D programs are generally not readily available (Nasu Nasu and Faunce 2010). For an outline of the U.S. federally funded R&D program and strategy and for some discussion of other countries' activities, see U.S. Department of Defense (2009).

3. The projection of seven million workers by 2015 (Roco 2003) was arrived at by extrapolating future workers required from then users of nanotechnology instrumentation such as atomic force microscopes and scanning tunneling microscopes. By this calculation, two million workers would be needed to support global nanotech activity by 2015 (up to 45 percent of these in the United States alone; 30 percent in Japan). Using information technology as a model, it was determined that a further five million jobs would be generated around nanotech activity (2.5 additional jobs for every nanotech position).

4. With UN institutions broadly passive in this area, a multiparty forum—the Global Dialogue on Nanotechnology and the Poor—was a solitary international exercise to examine the potential for nanotechnologies to undermine commodity-dependent economies in the Global South. As part of its consideration of the impact of nanotech on commodities and commodity-dependent countries, the dialogue hosted a workshop in Brazil and prepared background analysis and a final report (see http://www.merid.org/nano/commoditiesworkshop/ [accessed February 16, 2010]). The program was subsequently suspended due to lack of funding. Initial funding for the dialogue was provided by the Rockefeller Foundation, the U.K. Department for International Development, and the Canadian International Development Research Council.

5. Consultants to the U.K. government advised that nano-applications in energy efficiency and generation that are considered years away from market might offer significant gains but that these might not necessarily outperform

competing technologies and that forecasts "probably underestimate techno-logical advances in non-nanotechnological innovations" (Walsh 2007, 95). As we were finalizing this chapter, the OECD published a review of nanotechnol-ogy for water purification (OECD 2011). The report acknowledged that nano-technologies offered "complementary" approaches for water treatment and underscored that the potential contribution of nano-based systems needed to be considered in context: Access to clean water is not merely a technological issue but has significant economic and social dimensions. How far this thinking will permeate OECD and OECD member-country policy is, of course, another matter.

6. The ICP countries, by region, where there is no government policy, government funding, or nano R&D activity: six countries of Latin America; eleven Caribbean countries; ten Eastern European and Central Asian countries; fourteen Pacific Island countries (although a handful of scientific papers were identified); eleven Asian countries; twenty-three Sub-Saharan countries, and five Western Balkan countries. While the second round of ICPC-Nanonet reports suggests increased activity in some countries from the previous year, the dominant pattern for each of the nine regions canvassed is of a minority of countries with significant activity. A notable exception is the East Asia region, with China significantly out in front.

7. There is a considerable range in assessments of patent activity, including how many nanotechnology patents have been awarded or applications filed. Definitions are one cause of this. For example, the USPTO has adopted the "1–100 nm" guideline, a framing that is likely to leave nanobiotechnologies out-side the nanotechnologies class (Palmberg, Dernis, and Miguet 2009).

8. Patents awarded to U.S. entities since 1995, as identified by Lux Research, cited in PCAST 2010. In 2003, Huang and colleagues identified 70,000 patents related to nanotechnology at the USPTO for the period 1976–2003, and attributed around 57,000 of these to U.S. entities (Huang et al. 2003).

9. Based on filings at their respective patent offices (Dang et al. 2009). Applications tend to be filed domestically (at least first). This is the so-called home advantage and would suggest that patents held at the domestic offices will indicate fairly closely the extent of patent holdings.

10. It should be noted that the inventory includes what might be termed "traditional policy consultation processes." Europeans are talking the most, according to the French researchers, with forty-seven dialogue exercises; the North Americans apparently less (twelve events); with Latin America and Australasia trailing. Asian countries were not profiled, but it would appear that little has happened in the countries of the region investing most heavily in nano (China, Japan, India, and South Korea). See, for example, reports on activities to the OECD Working Party on Manufactured Nanomaterials (OECD 2009b, 2010b, 2010c). The Council of Canadians considers that Canada and the United States have largely side-stepped such exercises (Council of Canadian Academies 2008), while media commentators suggest that the emerging economies of India and South Africa are not engaging in public nanotechnology discourse (Padma 2010a).

11. There have been intermittent calls from civil society and scientists for the Indian government to regulate nanotechnologies (see, for example, Padma 2007, 2010a, 2010b; TERI 2009). In early 2010, the head of India's state-sponsored NanoMission announced the imminent establishment of a Nanotechnology Regulatory Board, which was to be charged with developing a regulatory agenda (Press Trust of

India 2010). However, as of mid-2010, no regulation was in place (Mantovani, Porcari, Morrison, et al. 2010; Jaspers 2010), although the government has made further pledges to introduce regulation (Times of India 2010).

12. In the 2011 Global Risk report, threats from new technologies such as nanotech, synthetic biology, and genetic engineering are rated as "outliers" because their risk profile is difficult to determine and is only coming into view, although it was acknowledged that they "could move rapidly to the centre of the risk landscape" (World Economic Forum 2011, 47).

13. The Royal Society and the Royal Academy of Engineering (2006) and the Council for Science and Technology (2007) also called for mandatory reporting if industry did not participate. The U.K. reporting scheme was run by the Department for Environment, Food and Rural Affairs and ran for two years (2006–2008) for the purpose of gathering information on the risks associated with nanomaterials produced there. During that time, the scheme received thirteen submissions, which was, as the government observed, a small proportion of the nano-activity believed to be occurring (HM Government 2010).

14. A two-year EU-funded multi-stakeholder program began in 2010 with the purpose of developing a framework for promoting uptake and implementation of the Code of Conduct. See http://www.nanocode.eu. Issue of a revised code had been signaled for mid-2010 but had not occurred at the time of writing (May 2011).

15. See also subsequent resolutions made by African, Caribbean, and Latin American countries during regional awareness-raising workshops held under the umbrella of the SAICM (African Regional Meeting, SAICM 2010; GRULAC 2010). The event was one in a series of regional awareness-raising workshops in response to the International Conference on Chemicals Management-2 plan of action, and was organized by UN Institute for Training and Research and the OECD, with funding from Switzerland, the United Kingdom, and the United States.

# 5

# Nanotechnology, Agriculture, and Food

**Kristen Lyons, Gyorgy Scrinis, and James Whelan**

## CONTENTS

## 5.1 Introduction

There is a crisis in global agri-food systems. Environmental degradation of agricultural landscapes, skyrocketing food prices, and growing rates of hunger, with close to one billion of the world's population recognized as "food insecure," are demonstrative of this crisis (Holt-Gimenez and Patel 2009; Lawrence, Lyons, and Wallington 2010). Pat Mooney (2010), one of the world's leading critics of agri-food nanotechnologies, has linked the contemporary crisis in food, along with the crises in "fuel, finance, and Fahrenheit," with the expansion of technological (and capital- and resource-intensive) approaches to farming and food, including chemical, genetic, and nanotechnological applications. Yet others argue industrial and high-tech agriculture and food production systems—including nanotechnologies—will be central to addressing the range of ecological, public health, and socioeconomic concerns that define the contemporary farming and food crisis.

In this chapter we examine the extent to which the agricultural and food industries have embraced nanotechnologies and the contributions such technologies may play in addressing the agri-food crisis. From the development of nano-seeds with in-built pesticides, to food able to alter its nutritional composition based on perceived consumer deficiencies, and "intelligent packaging" that can change color when spoilage is detected, nanotechnologies may soon permeate the entire agri-food system. Nano-agriculture and food proponents hold out great hopes with promises that nanotechnologies may reduce pesticide use on farms, increase agricultural productivity, reduce food wastage, and improve the nutritional value of food. More generally, some proclaim that nanotechnologies—often in combination with other technologies, such as genetic engineering (GE) and synthetic biology—offer magic bullet solutions to the current crisis in agriculture and food (Dane 2005). In Australia, for example, the federal government describes nanotechnologies as part of a suite of "enabling technologies" that will be capable of addressing problems as diverse as climate change, hunger, and the global financial crisis (AGDIISR 2009). Nanotechnologies are also argued to be central to a future green economy—claims that will be scrutinized at the forthcoming United Nations Rio +20 Summit in 2012 (Thomas 2011). As highlighted in Chapter 4, such potential innovations and the techno-fix paradigm that underpins them have captured the imagination of the agriculture and food industries, including companies such as Kraft Foods, H. J. Heinz, Syngenta, and Monsanto, as well as many governments around the world, who are investing heavily in nano research and development (R&D) (ETC Group 2004; Friends of the Earth 2008; Kuzma and VerHage 2006; FAO/WHO 2010).

Despite the promises, we argue that nanotechnological innovations are set to further entrench the technological treadmill of agricultural production, to extend the reach and influence of corporate and industrial science across the entire agri-food system, and to perpetuate the underlying structural causes of many of the social, economic, and environmental problems that define agriculture and food production. In contrast to this modernist pathway of agri-food and rural development, many civil society groups, farm organizations, union groups, some scientists, and others oppose the spread of nanotechnologies across the agriculture and food sectors. We draw upon the work of the ETC Group and Friends of the Earth as a means of giving voice to advocates who have expressed their concerns with nanotechnological innovations through actions such as advocating moratoria on nanotechnology research and commercialization until comprehensive regulatory frameworks are developed that will ensure workplace safety (ETC Group 2009) and proposing regulatory arrangements to ensure public confidence and a precautionary approach (Friends of the Earth 2009). We conclude our chapter by examining civil society concerns related to nanotechnologies in the context of broader food sovereignty and food justice movements, and the opportunities for strengthening the cause of nano social movements by extending alliances with the broader food sovereignty movements.

## 5.2 Crisis in Global Agriculture and Food Systems

The scope of the current crisis in food and agriculture is plainly evident: between 2006 and 2008, global food prices rose by 83 percent (Loewenberg

2008) and peaked again in 2010/2011 (Henn 2011), which has propelled rates of chronic hunger among almost one billion of the world's 6.8 billion people (Cresswell 2009). For many of the world's poor, this food crisis was the "straw that broke the camel's back" (Bello and Baviera 2010, 62), driving riots over food prices in at least forty countries, including Bangladesh, Burkina Faso, Cameroon, Cote d'Ivoire, Egypt, Guinea, India, Indonesia, Mauritania, Mexico, Morocco, Mozambique, Senegal, Somalia, Uzbekistan, and Yemen. The police and military backlash to these riots in some countries, including in Mozambique and Haiti, resulted in many dozens of deaths (Holt-Gimenez and Patel 2009). The causes of these recent events have emerged over many decades, indeed centuries, and include the displacement of peasant farming with capitalist agriculture, structural adjustment, import substitution, and a range of other neoliberal economic and fiscal politics that reflect the dominance of a modernist rural development agenda, as well as a range of biophysical factors. The global food crisis is also connected to a technological treadmill, a paradigm that facilitates corporate concentration, enclosure of the biological commons, and the privileging of profits above environmental and social protection (Gould 2005; Patel 2009; see also Chapter 2).

Modern agricultural development demonstrates these goals and trajectories. It has been underpinned by a "productivist" logic that assumes, firstly, that the primary goal of agricultural development is to increase rates of productivity, and secondly, that such productivity increases will only be achieved via the uptake of higher levels of emerging technology (see Scrinis and Lyons 2010). This technological treadmill comes at great environmental cost by placing increasing demands on the use of natural resources and producing larger quantities of (often more toxic) pollution (Schnaiberg and Gould 1994; Shiva 2000).

Industrial agricultural development is founded on the input of fossil fuels—from production (including the manufacture of pesticides, herbicides, and seeds) as well as harvesting, manufacturing, processing, packaging, transportation, and refrigeration (Pfeiffer 2006; Dodson et al. 2010). The fossil fuel energy requirements of modern agri-food systems make a substantial contribution to global greenhouse gas (GHG) emissions, with agriculture alone responsible for between 17 and 32 percent of all global GHG emissions (Bellarby et al. 2008). The unsustainable dependence of agri-food systems on nonrenewable energy reserves is matched by dependence on the unsustainable use of water, soil, plant, genetic, and other resources. For example, agriculture consumes at least 70 percent of the world's fresh water reserves (Millstone and Lang 2003). Meanwhile, poor land management and the intensification of farming practices have resulted in severe land degradation and salinity problems (Roberts 1995; Gray and Lawrence 2001). The expansion of productivist agriculture also coincides with a rapid decline in the world's agricultural biodiversity (Fowler and Mooney 1990; Lockie and Carpenter 2009; Maina 2010). Most of the world's agricultural production comprises just thirty species of crops, and with concentration in ownership of the seeds for these to a few agrichemical corporations; the top ten seed companies control 67 percent of the global seed market (Sharife 2010). These conditions leave farming landscapes with little resilience to cope with the vagaries of climate change—including fluctuations in temperature, as well as drought and flood events—that are expected to intensify over the coming years (Stern 2007; IAASTD 2008).

Alongside these environmental problems, farming families and rural communities face a number of social and economic challenges. The globalization of agriculture and food exposes farmers to international markets where subsidies, inequitable trade rules, and declining terms of trade have reduced both commodity purchase prices and farm incomes—circumstances that threaten the very basis of farmers' livelihoods (Bello 2009). As a strategy to compete in the global marketplace, farmers have frequently been encouraged to adopt high-tech inputs, including "green revolution" technologies— many of which have contributed to the environmental problems outlined above. These technologies have also delivered mixed outcomes in terms of productivity (Holt-Gimenez and Patel 2009). Many farmers have gone into unmanageable debt to purchase these technologies, and the resulting bankruptcies are linked with an increase in rural suicides (an estimated 100,000 farmers committed suicide between 1993 and 2003 in India alone) (Sharma 2006). The social and economic pressures facing rural communities are felt acutely by women, due to their high levels of involvement in farming, especially in countries in the Global South (Millstone and Lang 2003).

Despite the diversity and complexity of ecological, social, and economic challenges that underpin contemporary agriculture and food production, the policy and commercial responses to these problems are frequently couched in narrow technological terms (Busch et al. 1991; IAASTD 2008). This technological fix approach—fueled by linear understandings of development pathways (see, for example, Rostow 1960)—assumes that a narrow set of technological innovations will provide the basis for improving agricultural productivity, and that productivity gains will in turn produce positive environmental, social, and economic outcomes. The current responses to the global food crisis—including, for example, activities by the Alliance for a Green Revolution in Africa (AGRA) to introduce genetically engineered seed (and other technologies) to increase food production on the African continent—demonstrate the hegemonic position of the technological fix paradigm. Similarly, the expanding agro-fuels industry is frequently touted as a possible solution to peak oil. Rather than substantially reducing the energy requirements of industrial agri-food systems, agro-fuels will simply replace one compromised energy source with another, introducing a new range of adverse ecological and social impacts (Murphy 2010). Other examples include the recent activity by multinational companies (including Monsanto, DuPont, and Bayer), filing patents on "climate ready" genes that may be able to adapt to climatic variations, rather than ebbing the increase in carbon dioxide emissions by developing strategies to reduce agriculture's dependence on fossil fuels (McMichael 2009).

Technological innovation is also a feature of the food manufacturing, packaging, and service sectors and has facilitated growth in the provision of processed and convenience food, and the establishment of extensive global distribution chains. Whether it's tinkering with the nutritional profiles of highly processed foods to address the perceived deficiencies of relatively well-off consumers in the North or genetically modifying the nutrient profiles of crops to address deficiency diseases of the world's poorest people—such as the genetically engineered, beta-carotene-enhanced Golden Rice—these represent narrowly framed technological solutions to systemic problems in dietary patterns, food quality, poverty, and socioeconomic structures (Bowring 2003).

Despite the promises of these technological fixes being applied across the entire agri-food system, in many instances their adoption directly or indirectly exacerbates social inequities, environmental degradation, and food insecurity (Kloppenburg 1991; Heinemann 2009). It is in this context that a growing number of researchers, policymakers, activists, and others argue that more diverse approaches will be required to address the complex array of ecological, social, and economic challenges we face (Bassey 2009).

The recent report by the International Assessment of Agricultural Science and Technology Development (IAASTD) has identified the limits of narrow technological approaches as they have been applied to agriculture, such as the simplification of complex agro-environments: "Over the last century, the agricultural sector has typically simplified systems to maximise the harvest of a single component.... This has often led to degradation of environmental and natural resources (e.g. deforestation, introduction of invasive species, increased pollution and greenhouse gas emissions)" (IAASTD 2008, 21).

While the IAASTD report only briefly addresses the implications of the new nanotechnologies, it draws attention to the limits of technological innovations in addressing global agriculture and food-related challenges. It identified the inability of technologies themselves to address the fundamental challenges of rising food prices, hunger, and poverty—challenges that are increasingly recognized as political and economic—"that cannot and will not be solved through science and technology" (Gould 2005, 6). In response to their evaluation, the IAASTD—and with the endorsement of many non-government organizations (NGOs)—has called for greater support in the development of agro-ecological farming systems (see, for example, Miller and Scrinis 2010). It is in this context of competing visions over the role of technological innovations in shaping the future of agriculture and food that nanotechnologies are being developed and applied.

## 5.3 Nanotechnology and Agri-Food

The manipulation of materials and living organisms at the nanoscale has a wide variety of applications, and these are being researched and commercialized by scientists and companies across all sectors of the agri-food system. Most of the world's largest agri-food corporations are reportedly investing in research on nanotechnology, including Nestlé, Kraft Foods, Unilever, Cargill, Pepsi-Cola, Syngenta, and Monsanto (Friends of the Earth 2008). Many applications are still in the relatively early stages of R&D, such as nano-enabled plant and animal breeding, and nanosensors for agricultural applications (FAO/WHO 2010). Other applications are being commercialized and are likely to be in use, such as nano-packaging materials (including adhesives for use in McDonald's hamburger containers), food processing equipment (including in the manufacture of Corona beer), nutritional supplements (including in infant formula), diet-related products, and kitchen equipment (such as cookware and refrigerators), as well as nanopesticides (Project on Emerging Technologies 2011).

We now turn to a review of these applications and their likely impacts for the future of agriculture and food. Our review starts on the farm, where research and development of nanotechnologies is underway.

### 5.3.1 Nano Farming

Nanotechnologies are being applied in a diversity of ways across the farming landscape, including reformulating on-farm inputs that are widely utilized in conventional farming systems—such as pesticides, fungicides, plant, soil, and seed treatments, as well as veterinary medicines (Friends of the Earth 2008; FAO/WHO 2010). A recent meeting by the FAO/WHO concluded that while many of these applications are currently in the R&D stages, "it is likely that the agriculture sector will see some large-scale applications of nanotechnologies in the future" (FAO/WHO 2010, 18). In the case of pesticides, particle size may be reduced to the nanoscale to harness the specific properties of nanoscale materials—such as increased toxicity, the ability to dissolve in water, or increased stability—with the intention of maximizing the effectiveness and targeted delivery of pesticides (ETC Group 2004; Kuzma and VerHage 2006; FAO/WHO 2010). So-called nano-encapsulation techniques are also being utilized to enclose pesticides for release in certain conditions (Syngenta 2007; Zhang et al. 2006). Seeds are also being atomically reformulated to express different plant characteristics, including color, growth season, and yield (FAO/WHO 2010).

Leading agrichemical companies, including BASF, Bayer Crop Science, Cargill, Monsanto, and Syngenta, are engaged in research in these areas, and it is likely that a number of these applications have already been commercialized (Friends of the Earth 2008). Syngenta, for example, has retailed chemicals with emulsions that contain what they label *micro-emulsions* for a number of years (ETC Group 2004) and have obtained patent protection for their *gutbuster* microcapsules containing pesticides that break open in alkaline environments, including the stomachs of certain insects (ETC Group 2004). Nano- (and micro-) encapsulation techniques not only provide in-built pesticides for crops—similar to genetically modified (GM) *Bt* insecticidal crops—but also in-built switches to control the release and subsequent availability of pesticides.

Proponents of these applications argue that nano-encapsulated pesticides facilitate greater control over the circumstances in which pesticides are applied, in turn reducing farm chemical use and the risks of chemical pollution in agricultural environments (Kuzma and VerHage 2006). Yet despite these promises, nanopesticides further normalize chemical farming systems, entrenching both agricultural chemical use and a paradigm of technological dependence. In a landmark study commissioned by the U.K. government to assess current and future developments in nanotechnologies and their impacts, the Royal Society and Royal Academy of Engineering also questioned the claim that nanotechnologies would reduce chemical use, amid similar unfulfilled promises by many of the same companies in relation to genetically engineered (GE) crops (RS-RAE 2004). Reflecting such concerns, the FAO/WHO recently suggested that the uptake of agriculture-related nanotechnologies was set to increase potential exposure to agricultural chemicals (FAO/WHO 2010). The use of nano-formulated agricultural chemicals raises specific safety concerns for both chemical manufacturers and agricultural workers, who are likely to be directly exposed to these new materials—concerns that have led the International Union of Food, Farm, and Hotel Workers to call for a moratorium on the use of nanotechnologies in agriculture and food (Foladori 2007; FAO/WHO 2010).

The application of nano-reformulated farm inputs is also likely to introduce a new order of environmental problems (RS-RAE 2004; Scrinis and Lyons 2007). While these remain poorly understood, there is already evidence that nanomaterials have greater potency, reactivity, and bioavailability than their conventional counterparts—these are the character traits they are designed to express (Moore 2006). As a result of the size, dissolvability, and other novel characteristics of nanopesticides, they may more readily contaminate soils, waterways, and food chains, directly affecting nontarget ecosystems and living organisms (Moore 2006). It is likely, for example, that nanomaterials (similar to genetically modified organisms—see, for example, "Organic farmer to sue over GM contamination" 2011) will trespass onto nano-free farms—including organic farms—resulting in contamination, and in the case of organics, result in decertification and loss of organic and high-value niche markets.[1] The United Kingdom's Royal Commission on Environmental Pollution (2008) also notes that most of the environment-related research on nanotechnology has focused narrowly on the cytotoxic (that is, toxic to cells) effects of nanomaterials. We have only limited insights into broader ecological and population health impacts, including reproduction, population dynamics, and impacts for biodiversity. Further research is urgently required to increase understandings in these fields where nanotechnologies are likely to have significant impacts. Gould also argues that nanotechnologies will further threaten the enclosure and subsequent destruction of natural habitats, "by facilitating the integration of new ecosystems into the global production treadmill" (2005, 5). The convergence of nanotechnology, genetic engineering, and synthetic biology may also introduce novel ecological hazards, including the introduction of new toxic, genetically engineered, or nano-engineered organisms that are able to rapidly self-replicate, thereby threatening to replace indigenous species and colonize natural environments (Schmidt 2010).[2] As a result of these environmental and health concerns, civil society organizations such as the ETC Group have repeatedly called for a moratorium on the development of nanomaterials until adequate regulations are in place (ETC Group 2004, 2010).

Farm families, farmworkers, and rural residents are likely to carry a disproportionate burden of health risks and costs associated with the introduction of nanomaterials across the farming landscape, risks that are magnified in the absence of any required safety testing or regulation of nanoscale formulations of already approved chemical pesticides (Lyons and Scrinis 2009). Studies are already showing that nanoparticles gain ready access to the bloodstream after being inhaled, while some can directly penetrate the skin (see, for example, Friends of the Earth 2008). Without legal requirements to label agricultural chemicals that contain nanomaterials or that are derived from nanotechniques, farmers will be unable to identify or seek to avoid these nano-reformulations. At the same time, nanochemical and nano-seed R&D is concentrated among a small number of agrichemical companies, including those that already dominate the agrichemical and seed market (ETC Group 2004). The expansion in the use of nano-reformulated farm inputs will further concentrate the sales of farm inputs among these few corporate actors, enabling them to further extend their reach across farming landscapes, as well as extending their control of the agricultural inputs market (Lyons 2006). Reminiscent of other technological developments, the economic benefits of nanotechnologies are likely to accrue to the corporate owners of these technologies, while the social and ecological costs are borne by farmers, farmworkers, and citizens (Gould 2005).

Nanotechnologies are also being applied in farming to develop new forms of surveillance and monitoring technologies. Nanosensors—nanoscale, wireless sensors—are products of the intersection of nanotechnologies and information technologies. Alongside geographical positioning systems and other information technologies, nanosensors could be scattered across farmers' fields to enable the real-time monitoring of crops and soils (ETC Group 2004). In Australia, for example, researchers have developed nanosensors that could be utilized in paddocks to monitor crop growth, as well as for use in animal breeding for disease diagnosis (Clifford 2007). Automated farm management systems might not only monitor farm conditions—including soil moisture, temperature, pH, nitrogen availability, pest attacks, the presence of weeds, disease, and the vigor of crops or animals, and other factors—but also respond with the automated delivery of water, minerals, fertilizers, herbicides, and other farm inputs. The potential capabilities of nanosensors will extend the logic of precision farming in new and novel ways. The U.S. Department of Agriculture, for example, is reported to be developing a Smart Field System that "automatically detects, locates, reports and applies water, fertilisers and pesticides—going beyond sensing to automatic application" (ETC Group 2004, 17). Such applications (further) disconnect agriculture from its biological and social bases, causing a metabolic rift that will replace farmers' local knowledge, narratives, and histories with a technological gaze and an automated laboratory response (McMichael 2009).

Nanosensors offer so-called on-farm efficiencies; backed by claims they will reduce crop losses and/or crop damage, mitigate animal diseases, as well as reduce human labor requirements (Joseph and Morrison 2007), any such outcomes are likely to be realized only by large farms with the economies of scale to purchase the new technologies. Similarly, the technology itself is most readily suited to large farm operations, where scale restricts farmers from those intimate, tactile, and sensory monitoring practices traditionally utilized by small-scale farmers. As a result, it can be expected that any improvements in production efficiencies associated with the use of nanosensors (or other nano-farming applications) will support improvements in production efficiencies on large-scale farms, further concentrating agricultural production to large farming operations (ETC Group 2003). At the same time, the automated delivery of agrichemicals and other high-tech inputs—many of which will likely contain nanoparticles and be manufactured via nanotechniques—will continue to entrench chemical-intensive agricultural practices.

Nanotechnology also has a range of potential applications for animal production systems, including new tools for application in animal breeding, targeted disease-treatment delivery systems, new materials for pathogen detection, and identity preservation systems (ETC Group 2004; FAO/WHO 2010). Examples include the use of micro- and nanofluidic systems for the mass production of embryos for breeding; drug delivery systems able to penetrate previously inaccessible parts of the body; more biologically active drug compounds; and sensors for monitoring livestock locations. In livestock production and handling, for example, the Queensland Department of Primary Industries in Australia is developing a needle-free vaccine delivery system to target bovine viral diarrhea virus, which affects large numbers of factory farmed and intensively reared livestock (Mittar 2008). Scientists claim the development of this nanoparticle vaccine delivery will enable more targeted delivery, thereby reducing rates of infection and, by default, improving rates of cattle production. The nanovaccines will also have magnetic properties, enabling scientists and industry to track animals treated with the nanovaccine. These nanotech animal-production technologies demonstrate ways of achieving new efficiency and productivity gains within capital- and input-intensive industrial production operations, including close confinement factory production. They largely involve reengineering and further adapting animals to the requirements of this mode of animal production.

Nanotechnologies are also being combined with genetic engineering to provide scientists and industry with a suite of new tools to enable the modification of animal and plant genes, with the intended outcome of greater control in delivering new character traits across animals and crops (ETC Group 2004; Friends of the Earth 2008). These new nanobiotechnologies include the use of nanoparticles, nanofibers, and nanocapsules to carry foreign DNA and chemicals that may be able to modify genes. For example, silica nanoparticles have been used to deliver DNA and chemicals into plant and animal cells and tissues (Torney et al. 2007). Additionally, researchers in this field have also already succeeded in drilling holes through the membranes of rice cells to enable the insertion of a nitrogen atom to stimulate rearrangement of the rice DNA (ETC Group 2004). This technique has been successful in altering the color of rice, and researchers aim to use this technique to extend the growing season for rice, enabling year-round production (Friends of the Earth 2008).

In addition to the reengineering of existing plants, novel plant varieties may be developed using the techniques of synthetic biology—a new branch of technoscience that draws on the techniques of genetic engineering, nanotechnology, and informatics. In a recent breakthrough in this area, entrepreneurial geneticist Craig Venter, along with a team of researchers, was successful in building the world's first synthetic life form (Gibson et al. 2010). The implications of this research are profound, providing the basis for building biological systems from scratch.

Advocates of nanotechnologies argue that agriculture-related nano-applications will be particularly useful in the Global South, where rates of food insecurity are especially acute and where many recent food riots have occurred. Nanotechnologies are widely framed as a one-size-fits-all technological solution to the agri-food crisis. This is despite the centralization of nanotechnology R&D activities and patent rights among corporate actors from the developed world,[3] and the notable lack of R&D activity in Africa and other parts of the Global South (see Maclurcan 2010).

The promised outcomes associated with the uptake of agriculture-related nanotechnologies are also reminiscent of those associated with previous technological revolutions. For example, the green revolution promised that a technological package of hybrid seeds, fertilizers, credit, and irrigation could increase agricultural productivity in the Global South. Yet this technological revolution delivered mixed outcomes and resulted in increasing rates of hunger and food insecurity in many parts of the world—including an 18 percent increase in rates of hunger in South America between 1970 and 1990, as well as a 9 percent increase in hunger in Asia over the same period (Holt-Gimenez and Patel 2009).

The mixed outcomes of the green revolution demonstrate the limits of magic bullets to address the complexity of socioeconomic and political structures that are at the basis of the food crisis. It is in this context that the ETC Group, a leading NGO in international nanotechnology debates, has declared that nanotechnologies are set to continue to marginalize farming communities in the Global South, particularly smaller-scale local market and subsistence-oriented farmers, by failing to address the underlying causes of hunger and food insecurity (ETC Group 2010; see also Invernizzi, Guillermo, and Maclurcan 2008). The ETC Group, alongside other NGOs such as La Via Campesina (the international peasant farmer organization), Friends of the Earth, and the Network of Farmers' and Agricultural Producers' Organizations of West Africa, advocates socioeconomic and political trans-formation to address the food crisis. These transformations are at the core of the food sovereignty movement, which acknowledges the right of nations and peoples to take control of their food systems (Whittman, Desmarais, and Wiebe 2010).

The recent growth of food sovereignty movements in the Global South demonstrates the extent to which many see high-tech crops as a threat to rural livelihoods (Holt-Gimenez and Patel 2009). The movements represent mounting opposition to efforts by the Alliance for a Green Revolution in Africa (AGRA) to introduce GE crops across Africa, including the "We Are the Solution: Celebrating African Family Agriculture" campaign, coordi-nated by the Alliance for Food Sovereignty in Africa. In contrast to high-tech solutions to the food crisis, including nanotechnologies, "We Are the Solution" advocates the rights of smallholder farmers, especially women, to define agriculture and food systems, as well as supporting biodiversity and agro-ecology, thereby challenging industrial methods of agriculture (Aziz 2011). Similarly, and in a powerful conclusion to the 2011 World Social Forum in Senegal, the Final Declaration of World Assembly of Social Movements (2011)—comprising large representation from the Global South—opposed genetically engineered foods, declaring to fight for food sovereignty so as to ensure the rights and dignity for the world's peasant farmers and agriculture workers. In addition, the recent decision to place a moratorium on the culti-vation of GE eggplant in India provides further evidence of this opposition, as well as pointing to the concerns high-tech transformations of agri-food systems present for both agriculture as well as food.

## 5.3.2 Nano-Processed Foods

In the GM foods debate, one of the explanations commonly put forward for public opposition to GM crops was that they offered too few direct and obvious benefits to consumers, such as cheaper or nutritionally enhanced foods. Instead, the two main commercial applications—herbicide-tolerant crops and insecticidal *Bt* crops—were intended to modify pesticide man-agement practices and were seen as beneficial primarily for farmers and the agri-biotech corporations that owned the seeds. Such explanations of public opposition tend to portray citizens as predominantly concerned with per-sonal well-being. It has therefore been assumed that offering individualized benefits of GE crops might override the range of concerns expressed about these crops, such as potential health and ecological hazards, the erosion of seed diversity, or corporate control of the food supply, as well as the range of adverse impacts for farmers in the South. Although the promise of genetically engineered, nutritionally enhanced crops has been repeatedly invoked over the past decade, to date no such crops have been commercialized. The most celebrated case of a nutritionally enhanced GM crop, still under development, is the beta-carotene-enhanced Golden Rice. Golden Rice has been promoted as capable of ameliorating the causes of blindness in the South as well as "feeding the world," and in so doing, offers a technical and narrowly framed approach to the challenge of world hunger (Scrinis 2005).

It is not surprising then that the potential applications of nanotechnology most commonly promoted are those that may have direct appeal to consumers, particularly nano-processed foods with modified nutrient traits that claim to offer individualized and targeted health benefits for individuals. Other novel applications being touted are the development of "smart" or interactive foods that would give consumers the ability to modify the nutritional characteristics or flavor of the food product after purchase (Friends of the Earth 2008; FAO/WHO 2010). Although much of the emphasis has been on applications that claim to deliver nutritional and health benefits, nanotechnology will also be used to achieve a range of other processing functionalities, such as modifications in the flavor, texture, speed of processing, heat tolerance, shelf life, and bioavailability of nutrients (Gardener 2002).

A range of nano-processing techniques and nano-engineered materials are being developed in an attempt to assert greater control over the properties and traits of foods. This includes the production of nanoparticle-sized formulations of existing food components and additives, the production of nano-emulsions containing nanoscale droplets of liquids, and nano-encapsulation techniques as delivery systems for nanoscale ingredients and additives. One of the broad aims of this research is likely to be the ability to achieve minor improvements or cost savings in the production of cheap, shelf-stable, diverse, and appealing convenience foods. For example, the German company Aquanova has developed nanosized food additives that may speed up the processing of industrial sausage and cured meat production (Friends of the Earth 2008). Other nanoparticle-sized formulations of existing additives and ingredients may increase the functionality or bioavailability of ingredients and nutrients, thereby minimizing the concentrations needed in the food product as well as associated costs (Weiss, Takhistov, and McClements 2006).

Nanostructured food ingredients and nanoparticles in emulsions are being developed in an attempt to control the material properties of foodstuffs, such as in the manufacture of ice cream to increase texture uniformity (Rowan 2004). For example, the development of food ingredients able to reproduce the creamy taste and texture of full-fat dairy products would enable the production of very low-fat ice cream, mayonnaise, and spreads (Chaudhry et al. 2008). The Unilever company has also reported breakthroughs in the development of stable liquid foams that may improve the physical and sensory properties of food products, as well as the ability to aerate products that currently do not contain air. This aeration is also seen as a means of reducing the caloric density of foods (Daniells 2008). Food company Blue Pacific Flavors has developed its Taste Nanology process for engineering ingredients with more concentrated flavors by targeting specific taste receptors, making it possible to remove the bitter taste of some additives and to reduce the required quantities of flavor additives (Food Navigator 2006).

A major growth area in the food manufacturing sector has been in the development of so-called functional foods—nutritionally engineered foods that are marketed with nutrient or health claims (Scrinis 2008). Nanotechnology provides a range of approaches for producing foods with modified nutrient profiles and novel traits. Nano-encapsulation techniques—the same as those utilized in the manufacture of nanopesticides—are being developed as part of a strategy to harness the controlled delivery of nutrients and other components in processed foods. For example, nanocapsules have been produced through the development of self-assembled nanotubes using hydrolyzed milk proteins (Chaudhry et al. 2008). Food companies are already utilizing microcapsules for delivering food components, such as omega 3-rich fish oil. The release of the fish oil in the stomach is intended to deliver the claimed health benefits of the oil while masking its fishy taste. The ability to engineer nano-encapsulated food components would potentially enable the enhancement and control of a number of functionalities, such as to "provide protective barriers, flavour and taste masking, controlled release, and better dispersability for water-insoluble food ingredients and additives" (Chaudhry et al. 2008, 244).

The nano industries and their advocates also anticipate nutritionally interactive foods that will be able to change the nutritional profile in response to an individual's allergies, dietary needs, or food preferences. Such applications assume not only the ability to precisely manipulate the nutrient properties of foods but also the precision with which they are claimed to be able to target and address the precise needs of individual bodies.

Yet there is reason to question the claimed individual and public health benefits of these nutritional modifications. Like all functional foods, the claimed health benefits of nutritionally engineered nanofoods are based on a reductionist understanding of food and the body. This ideology or paradigm of "nutritionism" typically involves the reduction of our understanding of food into its nutrient components and also a further reductive focus on single nutrients (Scrinis 2008). This reductive understanding of food is then translated directly into the nutritional engineering of food products whereby single nutrients that are considered *good* or *bad* are added to or removed from foods depending on the nutritional trends of the day. The efficacy of these nutritional modifications assumes that these single nutrients can be manipulated individually and that they can deliver their health benefits in isolation from the foods and the nutrient matrix within which they are contained. The marketing of the nutritional content and claimed health benefits of these foods also typically focuses on the single nutrients added or subtracted, thereby distracting from the overall food nutrient profile and quality (Lawrence 2010).

Through the ability to produce processed foods that are cheap, convenient, and palatable or nutritionally engineered processed foods with claimed health benefits, these nano-processing applications will be used to promote the consumption of highly processed foods. The enhanced ability to mimic the sensory experiences associated with natural food components—and in a sense to deceive the senses—is also an example of the way nanotechnologies may facilitate the increased production and consumption of highly processed yet sensually appealing foods. The use of nanotechnology to nutritionally engineer foods also extends the ability of food companies to commodify nutrients and nutritional knowledge by embedding them in value-added food products.

While the health benefits of these nano-engineered foods are put forward as addressing important individual and public health challenges, these novel nano-processing techniques and nanosized ingredients also introduce a new set of potential and novel health and safety hazards. The behavior and potential toxicity of ingested nanosized ingredients in food and drinks are not well understood. As Chaudhry et al. note, there is a "growing body of scientific evidence which indicates that free engineered nanoparticles can cross cellular barriers and that exposure to some forms can lead to increased production of oxyradicals, and consequently, oxidative damage to the cell" (2008, 248). This includes concerns arising from the greater ability of nanosized ingredients to cross the gut wall, as well as their enhanced absorption and bioavailability—the latter are the very same mechanisms for delivering the claimed nutritional benefits. Even the increased uptake of supposedly beneficial nutrients is a source of concern, as little is known about the health effects of such increased nutrient absorption and their interactions with other nutrients (Parry 2006).

Similarly, as Pustzai and Bardocz (2006) note in their review of the health risks of nanoscale food components, nanoparticle versions of the food additives titanium oxide and silicon dioxide are already used in foods and have been approved as GRAS (generally recognized as safe) by the U.S. Food and Drug Administration. Yet they argue that there is already sufficient scientific evidence that these nanoparticles are cytotoxic and that they have been incorporated into foods without appropriate safety testing (Pustzai and Bardocz 2006). Such results are backed by the U.K. House of Lords Report (2010), which criticized industry secrecy around the development of nanofoods, as well as warning that the health risks of nanofoods remain poorly understood.

Health and safety issues are also emerging alongside the application of nanotechnology to food packaging materials, food storage containers, and kitchen appliances. It is in these areas that the greatest commercialization of nanotechnologies has occurred and where current health, safety, and ecological issues are likely to be most acute, at least in the immediate term. It is to these applications that we now turn.

### 5.3.3 Nanofood Packaging

The nanofood packaging sector has experienced some of the most significant commercial development to date (Chaudhry et al. 2008), with an estimated 400 to 500 nanopackaging products currently on the market, and with estimates that 25 percent of all food packaging materials will utilize nanotechnology in the next ten years (Helmut Kaiser Consultancy Group 2007). Manufacturers claim nanopackaging will improve food quality by enhancing its shelf life, durability, and freshness, the outcomes of which, proponents argue, may assist to address the current food crisis by keeping food fresher longer—or at least slowing the rotting process—thereby reducing food waste.

There is a range of nanopackaging techniques, including those that promise to reduce gas and moisture exchange and ultraviolet (UV) light exposure, as well as emitting antimicrobials, antioxidants, and other inputs. Commercial examples include the use of nanocomposite barrier technology by Miller Brewing to create plastic beer bottles. The plastic contains nanoparticles that provide a barrier between carbon dioxide and oxygen, enabling beer to retain its effervescence and shelf life longer (ETC Group 2004). DuPont has also produced a nano titanium dioxide plastic additive, DuPont Light Stabilizer 210. By reducing UV exposure, DuPont claims its barrier technology will minimize the damage to food contained in transparent packaging (ElAmin 2007).

Nanopackaging is also designed to enable packaging material to interact with the food it contains. Chemical release packaging, or "smart" or "active" packaging—as manufacturers brand it—is being developed to respond to specific trigger events where nanosensors could change color if a food is beginning to spoil or if it has been contaminated by pathogens. To do this, electronic noses and tongues will be designed to mimic human sensory capacities, enabling them to "smell" or "taste" scents and flavors (ETC Group 2004). In Scotland, UV-activated nano titanium dioxide is utilized to develop tamper-proof packaging materials, while in the United States, carbon nanotubes are incorporated into packaging materials to detect microorganisms, toxic proteins, and food spoilage (ElAmin 2007).

On the one hand, nanopackaging applications may provide a number of benefits for manufacturers, including the capacity to keep packaged food edible for longer (Tarver 2008). This may enable food to travel longer distances and to sit for extended periods in storage. These characteristics are expected to support the expansion of food transport and those complex agri-food systems that rely on global distribution networks. By increasing the shelf life of food, nanopackaging will enable food manufacturers to sell "old" food, including that which would otherwise have decomposed. While this may reduce food waste—a practical outcome, especially in times of food insecurity—the nutrient value of nanopackaged food is likely to be greatly reduced, due to the extended distance and time between harvesting and eating. Nanopackaging is also likely to appeal to manufacturers due to the promises of surveillance technologies in reducing the risks of food-borne illnesses.

However, despite these claims, nanopackaging materials introduce new health- and environment-related food risks by introducing new and potentially toxic materials into the food chain. For example, nanopackaging is likely to increase consumer exposure to nanomaterials, as nanomaterials migrate from packaging materials into food (Friends of the Earth 2008). Nanopackaging materials with antimicrobial characteristics, including nano silver, nano zinc oxide, and nano chlorine oxide, demonstrate specific adverse health impacts. Research on nano silver[4]—used as an antimicrobial in food packaging and storage containers—demonstrated its high toxicity to rat liver and brain cells, as well as its further increasing bacterial resistance (Senjen and Illuminato 2009). Tests with nano zinc oxide also produced damaging health impacts in mice and rats, as well as being toxic to human cells, even at very low concentrations (Friends of the Earth 2008). Meanwhile, carbon nanotubes—widely utilized in packaging materials—have been likened to asbestos, and concerns have been raised that human exposure may lead to mesothelioma and lung cancer (Poland et al. 2008).

In terms of environmental impacts, the release and disposal of nanosilver raises a number of issues. Research shows nanosilver has the potential to contaminate water, interfere with beneficial bacteria, and cause further contamination downstream on agricultural land, landfill sites, and other locations where contaminated water is distributed (Aitken et al., in Senjen and Illuminato 2009).

Antimicrobial nanomaterials are not only utilized in food packaging materials, they are also increasingly incorporated in home appliances, cleaning cloths, and food preparation surfaces. For example, a number of companies— including LG Electricals, Samsung, and Daewoo—have designed smart refrigerators. The so-called intelligence of these refrigerators is attributed to the addition of silver nanoparticles, intended to inhibit bacterial growth and eliminate odors. Nanosilver is also being applied to chopping boards, baby bottles,[5] kitchen utensils, cups, bowls, and other food equipment for its antibacterial qualities. These and other applications can be expected to magnify exposure to nanomaterials and their health risks, as well as the extent of adverse environmental impacts.

## 5.4 The Role of Advocacy

The current global food crisis, described at the beginning of the chapter, demonstrates the urgent need to rethink our food systems. Nano-agriculture and food proponents promise these can be addressed through narrowly framed technological fixes that otherwise leave dominant economic structures and technological systems in place. While these technological innovations may provide short-term Band-Aid solutions for some of these challenges, they also threaten to extend corporate concentration and control and exacerbate socioeconomic inequalities and power imbalances (see also Scrinis and Lyons 2010).

Civil society groups, such as Friends of the Earth and the ETC Group, reflect expressions of concern that are echoed by a growing number of actors, including biological and organic farmers, employees, trade unions, environmentalists, and scientists. Together, these civil society groups constitute an emerging social movement capable of coordinated action and resistance, such as the 2007 campaign for nanomaterials to be classified and regulated as new substances, launched by an international coalition of forty groups (ICTA 2008), as well as the release of a declaration by nearly seventy civil society, public interest, environmental, and labor organizations on "Principles for the Oversight of Nanotechnologies and Nanomaterials" the same year (NanoAction 2007; see also Miller and Scrinis 2010).

The movement against nano converges around the social, ecological, and economic problems associated with agri-food nanotechnologies that we have discussed and offers a critique of techno-fix approaches for addressing the global food crisis. The movement is mobilized through a shared concern about the emerging tension between nanofoods and broader principles of food sovereignty and social justice. Activists espouse and advocate ideals including democratic governance, worker and consumer safety, and the precautionary principle, seeking to shape the ways in which nanotechnologies are developed, applied, and regulated. In their review of NGO campaigning related to governing nanotechnologies, Miller and Scrinis (2010) identify three demands of NGOs: (1) inclusion of the broader social, economic, ecological, ethical, and public policy dimensions related to nanotechnology's governance; (2) adoption of the precautionary principle; and (3) public involvement in decision making relating to nanotechnologies. Yet despite their commonalities in aspirations for the future of food and agriculture, to date there has been limited specific and targeted alliance building between nano campaigns and the broader food sovereignty movement.

Like the international social movement that has effectively resisted genetically modified organisms (GMOs) in many countries, the nano social movement presents a promising counterpoint to the accelerating advancement of the nano industries. Ultimately, the nano industries and their advocates likely require the informed endorsement of consumers and of voters. Yet, as long as community understanding of nanotechnology and its expansion into everyday consumer products remains very low internationally, the nano movement's capacity to mobilize consumers and other allies remains limited (see also Miller and Scrinis 2010). In Australia, for instance, just 8 percent of respondents to an annual survey said they know what nanotechnology is and how it works (MARS 2008).

Determined and creative communication by nano activists through the mainstream and electronic media promises to shift this dynamic. So too may increasing connections with the broader food sovereignty movements, which have arguably already achieved much success in reaching a broader public, demonstrated, for example, in the growing support for food local-ization, transition towns, and other alternative, food-related movements. As one source of information about nanotechnology, civil society groups ben-efit from their status as a trustworthy informant and from the widespread consumer support for transparent labeling to inform choice—labeling that remains rare in nanoproducts.

By increasing community awareness, civil society groups also create the potential for accountable and democratic modes of governance. Through protest and other expressions of community opposition to nondeliberative and industry-dominated governance arrangements (Lyons and Whelan 2010), the nano movement may bring about more inclusive policy dialogue—before technological trajectories have been "locked in."

## References

Australian Government Department of Innovation, Industry, Science and Research (AGDIISR). 2009. *National Enabling Technologies Strategy*. Canberra: Author.

Aziz, N. 2011. We are the solution. Celebrating African family agriculture! *Grassroots International*. http://www.grassrootsonline.org/news/blog/we-are-solution-celebrating-african-family-agriculture (accessed April, 8, 2011).

Baby Dream Co. Ltd. 2011. *Baby Dream Co. Ltd.* http://babydream.en.ec21.com/ (accessed April 14, 2011).

Bassey, N. 2009. AGRA—a blunt philanthropic arrow. In *Voices from Africa: African Farmers and Environmentalists Speak Out against a New Green Revolution in Africa*, edited by A. Mittal and M. Moore. Oakland, Calif.: Oakland Institute.

Bellarby, J., Foereid, B., Hastings, A., and Smith, P. 2008. *Campaigning for Sustainable Agriculture*. Amsterdam: Greenpeace International.

Bello, W. 2009. *The Food Wars*. London: Verso.

Bello, W. and Baviera, M. 2010. Capitalist agriculture, the food price crisis and peasant resistance. In *Food Sovereignty: Reconnecting Food, Nature and Community*, edited by H. Whittman, A. Desmarais, and N. Wiebe. Halifax: Fernwood Publishing, Food First and Pambazuka News.

Bowman, D. and Graham, H. 2007. A small matter of regulation: An international review of nanotechnology regulation. *Columbia Science and Technology Law Review* 8:1–32.

Bowring, F. 2003. *Science, Seeds and Cyborgs: Biotechnology and the Appropriation of Life*. London: Verso.

Brown, A. 2009. Standards under scrutiny as food giants explore nanotechnology. *ABC News*. http://www.abc.net.au/news/stories/2009/03/28/2528759.htm (accessed April 15, 2011).

Busch, L., Lacy, W., Burkhardt, J. and Lacy, L. 1991. *Plants, Power, and Profit: Social, Economic, and Ethical Consequences of the New Biotechnologies*. Cambridge: Basil Blackwell.

Chaudhry, Qasim, et al. 2008. Applications and implications of nanotechnologies for the food sector. *Food Additives and Contaminants* 25(3): 241–58.

Clifford, C. 2007. Nanotechnology finds its way onto farms. *ABC Rural Western Australia*. http://www.abc.net.au/rural/wa/content/2006/s1885825.htm (accessed September 12, 2007).

Cresswell, A. 2009. Soaring food prices, global warming and natural disasters have experts worried that the world is facing a food crunch. *The Australian*. www.theaustralian.news.com.au/story/0,25197,25041144-28737,00.html (February 19, 2009).

Dane, S. 2005. The magic bullet criticism of agricultural biotechnology. *Journal of Agricultural and Environmental Ethics* 18:259–67.

Daniells, S. 2008. *Unilever Breakthrough Could Take Food Foams to New Level*. www.foodnavigator.com (accessed March 12, 2008).

Dodson, J., Sipe, N., Rickson, R. and Sloan, S. 2010. Energy security, agriculture and food. In *Food Security, Nutrition and Sustainability* edited by G. Lawrence, K. Lyons, and T. Wallington, 97–114. London: Earthscan.

ElAmin, A. 2007. Carbon nanotubes could be new pathogen weapon. *FoodProductionDaily.com Europe*. http://www.foodproductiondaily.com/news/ng.asp?id=79393-nanotechnologypathogens-e-coli (accessed November 13, 2008).

ETC Group. 2003. *No Small Matter II: The Case for a Global Moratorium. Size Matters!* Ottawa: ETC Group.

ETC Group. 2004. *Down on the Farm: The Impacts of Nano-Scale Technologies on Food and Agriculture*. Canada: ETC Group.

ETC Group. 2007. *Extreme Genetic Engineering: An Introduction to Synthetic Biology*. Canada: ETC Group.

ETC Group. 2009. *Civil Society/Labor Coalition Rejects Fundamentally Flawed DuPont-Ed Proposed Nanotechnology Framework*. http://www.etcgroup.org/en/issues/nanotechnology.html (accessed December 22, 2009).

ETC Group. 2010. *The Big Downturn? Nanogeopolitics*. Ottawa: ETC Group.

FAO/WHO. 2010. *FAO/WHO Expert Meeting on the Application of Nanotechnologies in the Food and Agriculture Sectors: Potential Food Safety Implications*. Meeting report. Rome: Food and Agriculture Organisation of the United Nations and World Health Organisation

*Final Declaration of World Assembly of Social Movements*. 2011. http://www.fahamu.org/fellowship/?q=node/18 (accessed April 12, 2011).

Foladori, G. 2007. *Agriculture and Food Workers Challenge Nanotechnology*. http://scitizen.com/nanoscience/agriculture-and-food-workers-challenge-nanotechnologies_a-5-937.html (accessed April 10, 2011).

Food Navigator. 2006. *Flavor Firm Uses Nanotechnology for New Ingredient Solutions*. www.foodnavigator-usa.com (accessed July 10, 2006).

Fowler, C. and Mooney, P. 1990. *Shattering: Food, Politics, and the Loss of Genetic Diversity*. Tucson: University of Arizona Press.

Friends of the Earth. 2008. *Out of the Laboratory and on to Our Plates: Nanotechnology in Food and Agriculture*. Australia.

Friends of the Earth. 2009. *Nanotechnology*. http://www.foe.org.au/nano-tech/ (accessed November 12, 2009).

Gardener, E. 2002. Brainy food: Academia, industry sink their teeth into edible nano. *Small Times*. June 21.

Gibson, D., Glass, J., Lartigue, C., et al., 2010. Creation of a bacterial cell controlled by a chemically synthesized genome. *Science* 329(5987):52–56.

Gould, K. 2005. *Small, Not Beautiful: Nanotechnology and the Treadmill of Production.* Paper presented at the annual meeting of the American Sociological Association, Philadelphia.

Gray, I. and Lawrence, G. 2001. *A Future for Regional Australia: Escaping Global Misfortune.* Cambridge: Cambridge University Press.

Heinemann, Jack. 2009. *Hope Not Hype: The Future of Agriculture Guided by the International Assessment of Agricultural Knowledge, Science and Technology for Development.* Malaysia: Third World Network.

Helmut Kaiser Consultancy Group. 2007. *Nanopackaging Is Intelligent, Smart and Safe Life.* New world study by Hkc22.com Beijing office. Press Release 14.05.07. http://www.prlog.org/10006688nanopackaging-isintelligent-smart-and-safe-life-newworld-study-by-hke22-com-beiging-office.pdf (accessed February 15, 2008).

Henn, M. 2011. The speculator's break: What is behind rising food prices? *EMBO Reports.* http://www.makefinancework.org/IMG/pdf/henn_2011_speculators_bread_embo.pdf (accessed July 28, 2011).

Holt-Gimenez, E., Patel, R., and Shattuck, A. 2009. *Food Rebellions: Crisis and the Hunger for Justice.* Oxford: Food First Books, Fahamu Books, and Grassroots International.

House of Lords. 2010. *Nanotechnologies and Food.* House of Lords Science and Technology Committee. First report of session 2009–2010. Volume 1: Report. http://www.publications.parliament.uk/pa/ld200910/ldselect/ldsctech/22/22i.pdf (accessed April 10, 2011).

International Assessment of Agricultural Knowledge, Science and Technology for Development (IAASTD). 2008. *Global Summary for Decision Makers.* http://www.agassessment.org/index.cfm?Page=doc_library&ItemID=14 (accessed November 12, 2008).

International Centre for Technology Assessment (ICTA). 2008. *Principles for the Oversight of Nanotechnologies and Nanomaterials.* http://www.icta.org/global/actions.cfm?page=15&type=366&topic=8 (accessed January 25, 2008).

Invernizzi, N., Guillermo, F., and Maclurcan, D. 2008. Nanotechnology's controversial role for the South. *Science, Technology and Society* 13(1): 123–48.

Joseph, T. and Morrison, M. 2007. Nanotechnology in agriculture and food. *Nanoforum Report.* http://www.nanoforum.org/dateien/temp/nanotechnologhy%20in20agriculture%20and%food.pdf?08122006200524 (accessed July 15, 2007).

Kloppenburg, J. 1991. Social theory and the de-reconstruction of agricultural science: Local knowledge for an alternative agriculture. *Rural Sociology* 56(4): 519–48.

Kuzma, J. and VerHage, P. 2006. *Nanotechnology in Agriculture and Food Production: Anticipated Applications.* Woodrow Wilson International Center for Scholars. September.

Lawrence, G., Lyons, K., and Wallington, T. Introduction: Food security, nutrition and sustainability in a globalized world. In *Food Security, Nutrition and Sustainability*, edited by G. Lawrence, K. Lyons, and T. Wallington. London: Earthscan.

Lawrence, M. 2010. The food regulatory system—is it protecting public health and safety?, In *Food Security, Nutrition and Sustainability*, edited by G. Lawrence, K. Lyons and T. Wallington, 162–74. London: Earthscan.

Lockie, S. and Carpenter, D. 2009. *Agriculture, Biodiversity and Markets: Livelihoods and Agroecology in Comparative Perspective.* London: Earthscan.

Loewenberg, S. 2008. Global food crisis looks set to continue. *Lancet* 327(9645): 1209–10.

Lyons, K. 2006. Nanotech food futures? *Chain Reaction* (June): 38–39.

Lyons. K. and Scrinis, G. 2009. Under the regulatory radar? Nanotechnologies and their impacts for rural Australia. In *Tracking Rural Change: Community, Policy and Technology in Australia, New Zealand and Europe*, edited by F. Merlan and D. Raftery. Canberra: ANU E-Press.

Lyons, K. and Whelan, J. 2010. Community engagement to facilitate, legitimise and accelerate the advancement of nanotechnologies in Australia. *Nanoethics* 4: 53–66.

Maclurcan, D. 2010. *Nanotechnology and the Hope for a More Equitable World: A Mixed Methods Study*. PhD diss., University of Technology, Sydney.

Maina, A. 2010. Is seed recuperation possible? In *New Technologies and the Threat to Sovereignty in Africa*. Pambazuka News, 54–55.

MARS. 2008. *Australian Community Attitudes Held About Nanotechnology: Trends 2005–2008*. Presented to the Department of Innovation, Industry, Science and Research, Market Attitude Research Services Pty Ltd.

McMichael, P. 2009. Banking on agriculture: A review of the world development report 2008. *Journal of Agrarian Change* 9(2): 235–46.

Meridian Institute. 2007. *Global Dialogue on Nanotechnology and the Poor: Opportunities and Risks*. Background paper presented at the international workshop on Nanotechnology, Commodities and Development, Rio de Janeiro, Brazil.

Miller, G. and Scrinis, G. 2010. The role of NGOs in governing nanotechnologies: Challenging the "benefits versus risks" framing of nanotech innovation. In *International Handbook on Regulating Nanotechnologies*, edited by G. Hodge, D. Bowman, and A. Maynard. Cheltenham: Edward Elgar.

Millstone, E. and Lang, T. 2003. *The Atlas of Food: Who Eats What, Where and Why*. London: Earthscan.

Mittar, N. 2008. *Nanotechnology, Vaccine Delivery Systems and Animal Husbandry*. Paper presented at the Science in Parliament briefing on Nanotechnology—Ultra-Small Particles, Mega Impacts: Queensland's Future with Nanotechnology, Brisbane.

Mooney, P. 2010. The big squeeze: Geopirating and the remaining commons. In *New Technologies and the Threat to Sovereignty in Africa*, Pambazuka News, 5–9.

Moore, M. 2006. Do nanoparticles present ecotoxicological risks for the health of the aquatic environment? *Environ International* 32:967–976.

Murphy, S. 2010. Biofuels: Finding a sustainable balance for food and energy. In *Food Security, Nutrition and Sustainability*, edited by G. Lawrence, K. Lyons and T. Wallington. London: Earthscan.

NanoAction. 2007. *Principles for the Oversight of Nanotechnologies and Nanomaterials* www.nanoaction.org/nanoaction/page.cfm?id=223 (accessed April 15, 2011).

Organic farmers to sue over GM contamination. 2011. *ABC News*. http://www.abc.net.au/news/stories/2011/01/13/3112367.htm (accessed April 15, 2011).

Parry, V. 2006. Food fight on a tiny scale. *The Times*, October 21.

Patel, R. (2009). *The Value of Nothing*. Melbourne: Black Ink Press.

Pfeiffer, D.A. 2006. *Eating Fossil Fuels*. Gabriola Island, BC: New Society Publishers.

Poland, C., Duffin, R., Kinloch, I., et al. 2008. Carbon nanotubes introduced into the abdominal cavity of mice show asbestos-like pathogenicity in a pilot study. *Nature Nanotechnology* 3(May): 423–28.

*Project on Emerging Nanotechnologies*. 2011. http://www.nanotechproject.org/inventories/consumer/analysis_draft/ (accessed April 15, 2011).

Pustzai, A. and Bardocz, S. 2006. The future of nanotechnology in food science and nutrition: Can science predict its safety? In *Nanotechnology Risk, Ethics and Law*, edited by G. Hunt and M. Mehta. London: Earthscan.

Roberts, B. 1995. *The Quest for Sustainable Land Use*. Sydney: University of NSW Press.

Rostow, W. 1960. *The Stages of Economic Growth: A Non-Communist Manifesto*. Cambridge: Cambridge University Press.

Rowan, D. 2004. How technology is changing our food. *Observer*, Sunday, 16 May 2009.

Royal Commission on Environmental Pollution, 2008. *Novel Materials in the Environment: The Case of Nanotechnology*. London: Royal Commission on Environmental Pollution.

Royal Society and Royal Academy of Engineering (RS-RAE). 2004. *Nanoscience and Nanotechnologies: Opportunities and Uncertainties*. London: RS-RAE.

Schmidt, C. 2010. Synthetic biology: Environmental health implications of a new field. *Environmental Health Perspectives* 118:118–23.

Schnaiberg, A. and Gould, K. 1994. *Environment and Society: The Enduring Conflict*. New York: St. Martin's Press.

Scrinis, G. 2005. Engineering the food chain. *Arena Magazine* 77:37–39.

Scrinis, G. 2008. On the ideology of nutritionism. *Gastronomica* 8(1): 39–48.

Scrinis, G. and Lyons, K. 2007. The emerging nano-corporate paradigm: Nanotechnology and the transformation of nature, food and agri-food systems. *International Journal for the Sociology of Food and Agriculture* 15(2): 22–44.

Scrinis, G. and Lyons, K. 2010. Nanotechnology and the techno-corporate agri-food paradigm. In *Food Security, Nutrition and Sustainability*, edited by G. Lawrence, K. Lyons, and T. Wallington. London: Earthscan.

Senjen, R. and Illuminato, I. 2009. *Nano and Biocidal Silver*. Friends of the Earth Australia and Friends of the Earth United States.

Sharife, K. 2010. Biotechnology and dispossessions in Kenya. In *New Technologies and the Threat to Sovereignty in Africa*, Pambazuka News, 62–68.

Sharma, D. 2006. Farmer suicides. *Third World Resurgence* 191 (July).

Shiva, V. 2000. *Stolen Harvest: The Hijacking of the Global Food Supply*. Cambridge: South End Press.

Stern, N. 2007. *The Economics of Climate Change: The Stern Review*. Cambridge: Cambridge University Press.

Syngenta. 2007. *Primo MAXX Plant Growth Regulator*. Syngenta. http://www.syngentapp.com/prodrender/index.asp?nav=CHEMISTRY&ProdID=747 (accessed July 12, 2007).

Tarver, T. 2008. Novel ideas in food packaging. *Food Technology* 62(10): 54–59.

Thomas, J. 2011. Will Rio+20 squander green legacy of the original Earth Summit? *Guardian Environment Network*. http://www.guardian.co.uk/environment/2011/mar/31/rio-20-earth-summit (accessed April 15, 2011).

Torney, F., Trewyn, B., Lin, V., and Wang, K. 2007. Mesoporous silica nanoparticles deliver DNA and chemicals into plants. *Nature Nanotechnology* 2:295–300.

Weiss, J., Takhistov, P., and McClements, J. 2006. Functional materials in food nanotechnology. *Journal of Food Science* 71(9): 107–16.

Whittman, H., Desmarais, A., and Wiebe, N. 2010. *Food Sovereignty: Reconnecting Food, Nature and Community*. Halifax: Fernwood Publishing, Food First and Pambazuka News.

Zhang, Y., Zhang, Y., Chen, J., et al. 2006. A novel gene delivery system: Chitosan-carbon nanoparticles. *Nanoscience* 11(1): 1–8.

## Endnotes

1. In early 2008, the U.K. Soil Association banned the use of nanomaterials as a part of organic systems. Following this, the Australian Organic Standard was also amended to exclude nanomaterials and processes, and it is likely other organic certifiers will follow. See Kristen Lyons, "Nanotech Food and Farming and Impacts for Organics," *Australian Certified Organics* (Winter 2008): 30–31.

2. Synthetic biology reflects the convergence of a number of technologies, including biotechnology, nanotechnology, and information technology. See, for example, ETC Group, *Extreme Genetic Engineering: An Introduction to Synthetic Biology* (Canada: Action Group on Erosion, Technology and Concentration, 2007).

3. There are also a number of developing and transitional countries that are investing in nanotechnologies, including China, Russia, Nepal, Sri Lanka, and Pakistan, but the United States, the EU, and Japan still lead in terms of expertise, infrastructure, and capacity (ETC Group 2010).

4. These results are particularly alarming given a product inventory by the Project on Emerging Technologies (2011) identifies nanosilver to be the most common nanomaterial in commercial circulation, comprising around one quarter of all available nanoproducts.

5. Children's health concerns associated with exposure to nanosilver (and other nanoparticles) is especially acute given its use in a number of products especially oriented to babies and small children. Among these products includes the use of nanosilver in Baby Dream baby mug (Baby Dream Co. Ltd., 2011) alongside a range of bottles and cutlery, as well as the inclusion of nano ion particles in the infant nutrition supplement Toddler Health, alongside claims of increased bioavailability of nutrients (see, for example, Friends of the Earth 2008).

# 6

# Poor Man's Nanotechnology— From the Bottom Up (Thailand)

**Sunandan Baruah, Joydeep Dutta, and Gabor L. Hornyak**

## CONTENTS

## 6.1 Introduction

The phrase "poor man's nanotechnology" is an oxymoron. Nanotechnology is for the well-equipped, well-funded, well-PhD'd, and well-connected. It is for the risk taker, the visionary, and the entrepreneur. Regardless, we present a point of view from the poor man's end of this spectrum—the end where nanotechnology is accomplished by the seat of one's pants—chemically and financially speaking, from the bottom up, in our laboratory in Thailand.

In this chapter we do not seek to present a conventional view but what we call the poor man's nanotechnology (PMNT) point of view. We write from the perspective of the laboratory, presenting our most pressing concerns, issues, and findings, while interweaving discussion about the impact on end users and broader views regarding infrastructural needs in Thailand. Our perspective is presented from the point of view of scientists and engineers at the Center of Excellence in Nanotechnology (CoEN), located at the Asian Institute of Technology (AIT), north of Bangkok in Pathumthani province. PMNT's research and development (R&D) takes place at CoEN. The center sprang into existence by the force of sheer will, sweat equity, and the skillful placement and coordination of requisite partnerships. PMNT is nanotechnology produced by whatever means are available—whether building analytical instruments and experimental apparatus from scratch or synthesizing from the bottom up with inexpensive chemicals.

Poor man's nanotechnology is nanotechnology conducted with minimal resources. This interpretation applies to research, development, and education programs that lack a steady stream of funding or which, at best, is sparse or unreliable. The phrase can also be applied to start-up commercial enterprises that do not have the luxury of significant capital investment, government grants (such as Small Business Innovative Research grants in the United States), or other well-known mechanisms for support. Therefore, from our perspective as a laboratory, nanotechnology in this way is accomplished solely by the seat of one's pants to achieve experimental, pedagogical, or commercial objectives. At CoEN, partnering with "those who have" allows us access to equipment otherwise impossible for us to buy and maintain. For example, we pay an hourly fee for scanning electron (SEM) and transmission electron microscope (TEM) imaging services—this is PMNT with a big friend in the form of the Thai government. Without these tools and local access, CoEN would be blind to the nanoworld.

At CoEN we are interested in developing applications that address the needs of those who are without—without clean drinking water, cheap energy, unspoiled food, and the other necessities required to provide for decent living. We are dedicated to R&D that addresses these issues, but only by means that are affordable in the long term. Although we do not have direct contact with our end users—direct distribution of what we make for consumption or, at least at this time, other forms of outreach—thinking about end-user needs exists at the core of our mission. We are scientists and we are engineers, and we do our best to present the facts as they are with regard to the science and the technology and then develop practical options for potential applications. We may develop prototypes, file for patents, and then start companies. However, when all is said and done, we still require the assistance of infrastructure, government, and industry. It is up to these bodies to realize the means of human resource development, to find alternative pathways for southern innovation, and to build local capacity while keeping in mind the needs of the users—the poor man, woman, and child—and linking them back to the laboratory.

Therefore, an additional aspect to consider is how we ensure the equitable distribution of new technology. In a world of "everyone trying to make the big buck and the bottom line" as objectives, how can resources be distributed to those who are most in need? One can easily find examples of inequitable distributions of technology and technology-based services anywhere, especially—and ironically—in the "greatest" of industrialized nations. However, ensuring equitable distribution is not in our capacity here at CoEN. The best we can do in that regard is to promote efficient pedagogy—for example, teaching students about possible impacts of nanotechnology on end users as well as its relation to government and infrastructure.

The implied thrust of our chapter is a focus on what nanotechnology can do and, in particular, the type of nanotechnology that adds capability without adding cost and brings quality applications to underdeveloped regions. We commence this discussion by explaining how our research institute has grown so rapidly with few resources. We then describe the difference between bottom-up and top-down nanomaterial synthesis, detailing how our bottom-up methodologies, shaped and inspired by the natural world, present significant promise for scientific innovation in the Global South. As our expertise lies in developing technology in a lab environment, we then turn to a presentation of several key examples of bottom-up technologies that we have developed, before tracing future directions that can see nanotechnology further realize its significant potential. We conclude the chapter by returning to the key questions that face us as scientists and engineers, and the larger government and industrial context in which our work is located.

## 6.2 Thailand and the Foundations of CoEN

### 6.2.1 Scientific Innovation in Thailand

Thailand's economic and technological development has many strengths. It is geographically, demographically, and economically well positioned to play a major role in the region and is certainly one of the up-and-coming Tigers of Asia—this is, at least, our humble and nonexpert point of view. For example, it takes about three hours to fly from Bangkok to India, the Philippines, Taipei, Pakistan, Nepal, Bhutan, and Bangladesh, and less to fly to Malaysia, Indonesia, Myanmar, Hong Kong, Vietnam, Laos, and Cambodia. It is a few hours more to Australia, New Zealand, Japan, Korea, and the Gulf States to the West, as well as countries deeper into Central Asia. Thailand's airport (Suvarnabhumi), highways, and transportation systems are all world class.

**FIGURE 6.1**
Centre of Excellence.

The "Land of Smiles" is quickly becoming the "Land of Opportunity." Everything is in place for a technological explosion. The National Science and Technology Development Agency (NSDTA) of Thailand has spun out several laboratories devoted to high nanotechnology. The national centers, including NANOTEC (nanotechnology), MTEC (metals and materials), BIOTEC (genetic engineering and biotechnology), and NECTEC (electronics and computer technology), are all members of the community known as Science Park, conveniently a mere kilometer or so from AIT. The collective mission of these centers is to develop marketable technology and to support R&D in the region. The Nanotechnology Association of Thailand was launched in 2010—another key ingredient of Thai nano infrastructure.

Nanotechnology at AIT, soon to become the Center of Excellence in Nanotechnology (CoEN), laid foundations in Thailand in 2003 in a province called Pathumthani—the Land of the Lotus (ironically appropriate as we shall later explain). It started very small, on a start-up budget of 50,000 Thai Baht (equivalent to US$1,500), with no students. Driving the project was one of our authors, Professor Joydeep Dutta, a faculty member in the Department of Materials and Powders at the Swiss Federal Institute of Technology (EPFL). EPFL was and still is, without question, a world-class laboratory. In other words, there was no shortage of equipment, no shortage of funding, and no shortage of students or faculty; the funding of research projects was well established in Switzerland. In every way, EPFL and CoEN were at opposite ends of the spectrum.

### 6.2.2 Center of Excellence in Nanotechnology (CoEN)

On 25 May 2006, CoEN formally became the dedicated eighth Center of Excellence in Thailand under the auspices of the NSDTA and National Nanotechnology Center (NANOTEC) located in Science Park (Figure 6.1). The Thai government deserves commendation for such forward thinking because nanotechnology was seen, internationally, to have a high barrier for entry. Currently, about thirty students populate CoEN and more equipment is increasingly being made available for research. Since 2003, more than a hundred research papers have been published from the lab, along with two well-received pedagogical texts (*Introduction to Nanoscience*, CRC Press, 2009, and *Fundamentals of Nanotechnology*, CRC Press, 2009). The center offers a degree in nanotechnology for master's- and PhD-level students and boasts a well-developed curriculum.

From the outset, the CoEN laboratory's strategy was to produce prototypes with technology grounded in reality and practicality. It would only consider R&D that had the potential to be transformed into application—and transformed inexpensively at that. An assessment of what was needed in the South East Asia region yielded findings that Thailand, as the number-one exporter of rice in the world, has a great proportion of its workforce based in agriculture (LCFRD 2007). As an associated result, many rural communities do not have the infrastructure required to purify drinking water, provide local energy to power computers, or detect gas leaks and environmental toxins. The fundamental mission was established. The lab would dedicate its efforts to develop inexpensive products that could address these critical areas of need. Research would proceed from the bottom up—both from the perspective of the test tube where the synthesis occurs and from that of the lab where the work takes place.

The students at CoEN herald from all around the region, and indeed, even the world. Countries include Myanmar, Bangladesh, Indonesia, the Philippines, India, Pakistan, Nepal, Iran, Germany, and South Korea. We are expanding, with additional students expected from Malaysia, Vietnam, Cambodia, Laos, and Central Asian countries. Our students also have highly diverse backgrounds. Disciplines include chemistry, materials science, electronics, computer science, biology, physics, engineering, chemical engineering, and agriculture. We have core faculty from India and the United States and visiting faculty from Sweden, India, and other parts of Thailand. This diversity is, in no uncertain terms, the strength of our center. Each of these students comes prepared with a mission—a mission to take something back to his or her homeland—a technology, perhaps, that would purify local water supplies or deliver energy to power a computer.

Our job is to think of something useful and to project its potential application. Our job is to reflect, investigate, make, and test—and, of course, to publish and present. Our job is to patent, develop prototypes, and start up small companies. We make these things for the benefit of humankind and the environment—and, not to be naive, for the benefit of the lab and its living components as well. Nanotechnology is vast—even PMNT. The potential is unlimited. Countries are in a global competition to commercialize nano. At CoEN we believe that PMNT is playing an increasingly important role in circumventing traditional competition dynamics, challenging the established top-down field of nanotechnology.

## 6.3 Making Poor Man's Nanotechnology: Following Nature from the Bottom Up

As detailed in other chapters in this volume, the competition to commercialize products utilizing nanotechnology is fierce, with one source claiming that the value of nanotechnology by 2015 will be close to US$2.4 trillion (SASTR 2010). Information, bio, nano, and other emerging technologies are piling onto and into each other. Add to this the rapidity with which information is disseminated and assimilated on a global scale, and change can come upon us rather quickly. Nanotechnology is a horizontal technology, cutting across vertical industrial sectors and falling under the auspices of multiple government departments. Yet it also seemingly presents high entry barriers for less-resourced countries in the Global South, both in terms of R&D and market entry. However, nanotechnology cannot be overlooked as a field of potential in the Global South for, like any other technology, nanotechnology has the potential to change society. The change may be for the better or for the worse, depending on the orientation of its output and who wields its power. In our lab, we take a more modest approach. We believe that nanotech can enhance products and address challenges without a lot of fanfare and hullabaloo.

This becomes possible if we think about nanotechnology as the application of nanoscience. Although nanoscience is often associated with industry and commerce, it can also be thought of as the study of nature's nanotechnology. Nature is the master of nanotechnology. Everything in our bodies originates from atoms and molecules first and then from nanomaterials. For example, the nanometer domain encompasses the visible spectrum— an important coincidence with living things. The blue color of the *Morpho* genus's butterfly wings is due to interference created by nanostructures. The gecko sticks to ceilings due to millions of nanoscale setae on its feet that are approximately 200 nanometers in diameter. The setae take advantage of very weak intermolecular forces called van der Waals forces that collectively bear the weight of this remarkable creature—a force important at the nanoscale. Chromosomes, of course, are also nanomaterials. In our lab, in order to make PMNT a reality, we try to copy nature whenever we can. We recognize that we must listen to nature and learn how to make nanomaterial devices from the bottom up with minimal raw materials and input of energy (for examples of recent work, see Baruah and Dutta 2009a, 2009b, 2009c; Ullah and Dutta 2008; Baruah et al. 2008; Baruah et al. 2009; Baruah, Rafique, and Dutta 2008).

Based on fundamental thermodynamic arguments, breaking bonds and converting a big material into smaller materials requires energy. Conversely, energy is released when small materials combine to form big materials—also one of the primary reasons that nanomaterials are metastable: nanomaterials inherently want to become larger materials. In nature, biological materials are made exclusively from the bottom up. It seems to make logical sense that we would copy nature and make our materials in similar ways. We therefore argue that a bottom-up chemical synthesis strategy is the best way to develop nanotechnology in our lab environment and development context, for the simple fact that top-down fabrication is costly. By definition, top down requires significant input of energy and millions of dollars' worth of equipment. Top-down methods require ultra-high-vacuum and high-energy electron beams for micro- and nanolithography. In all cases, parts are checked for quality control by scanning electron microscopes, surface topography scanners like AFM (atomic force microscopy), and nanomanipulators. Thin film deposition methods also take place in ultraclean rooms that are generally expensive to maintain.

For the most part, bottom-up synthesis does not require highly complex and expensive equipment. However, such equipment can be needed when having to explore the characteristics of the materials formed during research and subsequent quality control steps. Expensive transmission electron and scanning electron microscopes (TEM and SEM respectively) are sometimes also needed, although only occasionally. However, while bottom-up methods are cheaper, with a higher throughput of product, the long-range order of nanostructures is not always apparent—a factor that affects repeatability, reproducibility, and perhaps overall reliability. Both manufacturing philosophies are therefore necessary to make nanotechnology continue its evolution successfully, and both methods are often mixed. During a manufacturing cycle, there may be a top-down application followed by a bottom-up application. Therefore in the short term, purely bottom-up synthesis of electronic products is currently not practical. So, while we claim to practice PMNT, we cannot yet claim to have done away with expensive equipment altogether. Thus, the goal of PMNT is to ultimately lower the overall cost of a product. Dollar for dollar, it must be less expensive to manufacture a device from the bottom up than it is from the top down.

## 6.4 The Development of Bottom-Up Technologies

### 6.4.1 The Beginning: Porous Anodic Aluminum Oxide Templates

Anodizing is a very old technology that has successfully made the transition into the nanoworld to form one of the most amazing template materials. First developed around the turn of the nineteenth century, the purpose of anodizing was to protect aluminum from corrosion. Later it was discovered that anodically formed alumina membranes were porous, and quite remarkably so. In anodizing, the aluminum metal is made and the anode then undergoes oxidation to form a thick aluminum oxide layer. Nearly any kind of material—whether it is a metal, semiconductor, insulator, or combinations

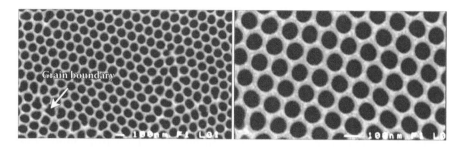

**FIGURE 6.2**
Porous alumina, left and right.

thereof—can be synthesized within the pore channels of the membrane. In this way, the new material takes on the size and shape of the pore channel; anodic aluminum oxide (AAO) is the perfect template.

Anodically formed alumina films are highly porous. The size of the pore channel is directly proportional to the applied voltage during anodizing. Of all the cheap technologies out there, this one is one of the cheapest—it could actually be classified as low tech. All that is required is aluminum (the substrate), a polyprotic acid (phosphoric, oxalic, chromic, or sulfuric), a DC power supply, and a chiller. However, from this low-technology PMNT we get a valuable nanomaterial. By way of the process of anodic oxidation of the aluminum metal, porous alumina, with a hexagonally distributed pore structure, is formed (Figure 6.2).

PMNT takes advantage of nature—specifically in the form of *Aspergillus niger*, a fungus that results in a common mold, through directed self-organization. We found that *Aspergillus niger* can act in the capacity of a template. To begin with, laboratory supplies at AIT consisted of five shot glasses, rudimentary equipment, and some spare parts. One of the first projects was to make gold colloids and stabilize them with a ligand (an organic material that is capable of binding to metal surfaces). Chemicals at the time were few and far between because we couldn't afford them, but some monosodium glutamate (MSG) was acquired at a local supermarket. During a holiday later that year, a gold colloidal solution was accidentally left out on a shelf for several weeks. A new student noted that a black scummy substance had formed at the bottom of the shot glass. Before disposing of the messy mire, he consulted Professor Dutta who noted that some shiny-looking material had precipitated underneath the scum. Scanning electron microscopy revealed that gold microwires had formed with assistance from the *Aspergillus niger* mold (see Figure 6.3).

Apparently *Aspergillus niger* was able to feast on the supermarket glutamate that was bound to the gold nanoparticles and during that process also served as a template to guide microwire growth of the gold. Once the glutamate was consumed, the nanoparticles were rendered unstable and hence, agglomerated into wires. Thus it began–the seeds for PMNT, with nature's assistance, had been sown.

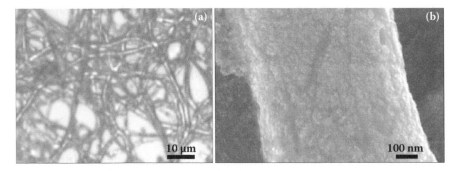

**FIGURE 6.3**
Porous alumina (a) and (b).

## 6.4.2 Applications of Nanoparticles

### 6.4.2.1 Detection: Fingerprints, Heavy Metals, and Gases

A unique method was developed for detecting latent fingerprints using gold nanoparticles (Ul Islam 2007). Apart from salts, a fingerprint also contains fatty acids. Chitosan, a lipophilic (fat-loving) natural polymer, was used to help the attachment of gold nanoparticles onto latent fingerprints to enhance optical contrast. The obtained contrast clearly distinguished the ridges, and this method stood a good chance of being applied to forensic identification. Further innovative examples include using chitosan-capped gold nanoparticles to detect heavy metal ions in water (Sugunan et al. 2005). We also fabricated zinc oxide (ZnO) nanostructure-based gas sensors using a simple and efficient hydrothermal method at low temperatures (below 100°C). When gas molecules come in contact with the surface of the nanorods, the resistivity of the nanorods changes. This change in resistivity, which gives an estimate of the number of gas molecules present, can be detected using a simple electrical circuit. The major focus was on enhancing the sensor performance for detection of harmful gases, like liquefied petroleum gas and ammonia. Ethanol sensors with very high sensitivity (to detect miniscule amounts) and stability (for long-lasting performance) are urgently needed in the field of chemical and food industries, breath analyzers, and quality monitoring, and therefore can find applications in areas like personal environmental monitoring, household and personal safety, and industry.

### 6.4.2.2 Dye-Sensitized Solar Cells

Dye-sensitized solar cells (DSSCs) are receiving a lot of attention from researchers because of their potential to achieve high efficiency at low manufacturing cost. The fabrication procedure is simple and environmentally friendly as compared to silicon solar cells. In a conventional photovoltaic system, the semiconductor serves a dual role of light absorption and charge carrier transport. In a DSSC, light is absorbed by a sensitizer dye onto the surface of a wide band semiconductor. Charge separation takes place at the interface through the injection of photoexcited electrons from the dye into the semiconductor that, from there, are channeled into the external circuit. The semiconductor layer in a DSSC can be grown using a simple hydrothermal process at low temperatures without the need of expensive equipment. Silicon solar cells, on the other hand, require sophisticated and expensive microfabrication units.

### 6.4.2.3 *ZnO-Based Super-Hydrophobic Surfaces*

A super-hydrophobic surface is able to totally deflect water, and therefore can be regarded to have a wide variety of significant applications such as self-cleaning windows, stain- and stink-free clothing, and so forth. We went by way of biomimetics and copied nature's own marvelous model—the lotus leaf. The water contact angle for a drop forming on the surface of the leaf is greater than 160 degrees. The maximum attainable is 180 degrees, resulting in a perfectly spherical drop of water. In other words, the drop does not spread out on the surface, as the surface is super-hydrophobic. This property is due to an array of micro- and nanostructures protruding from the surface of the leaf that minimize the energetic interaction between the water drop and the surface. An analogy can be made with a bed of nails, where, due to the strategic numbering and spacing of the nails, a person can actually lie upon them without risk of injury.

A super-hydrophobic surface is fabricated by placing arrays of ZnO nanorods on a surface consisting of microbump patterns. The microbumps were formed with the aid of a rebuilt inkjet printer (which we recovered from a junk heap and rebuilt for the purpose of patterning microstructures). The ink used was composed of ZnO seed particles. Depending on the spacing of the bumps and the height of the ZnO nanorods, varying degrees of hydrophobicity were obtained. In this way, we were able to tune the surface chemistry to suit specified design criteria—and very cheaply (Figure 6.4) (Myint et al. 2010).

## 6.5 Future Bottom-Up Directions

Our established projects have been presented and discussed, but there are several projects currently in progress that are worth mentioning, as they point toward our strategy for dealing with key issues facing Thailand, a developing economy, and more generally the rest of the world.

### 6.5.1 Hydrogen Production

The "hydrogen economy" is based on hydrogen as a primary fuel source—and could be regarded as impending, given current fuel crises. The implications

**FIGURE 6.4**
Super-hydrophobic cotton.

are immense. Rather than carbon dioxide, the only by-product yielded by the combustion is water. If hydrogen was used as a fuel source, there would be no need to chop trees, mine coal, or import oil. Rural or urban disadvantaged farmers could benefit from having access to an environmentally benign, cheap source of fuel to propel tractors and the like. The unwanted side effects of industrial agriculture (such as soil compaction) aside, this could potentially mitigate many of today's woes, especially in the Global South. In Thailand, for example, rice fields are annually burned for the sake of removal. Instead, if the biomass were applied to hydrogen production (as can be aided by nanotechnology), then not only would the farmers benefit but there also would be a concomitant reduction in the production of carbon dioxide and other pollutants.

We are involved in finding bottom-up ways to generate hydrogen cheaply from catalyst systems based on our nanomaterials. The major technological hurdle is actually with the safe and efficient storage of hydrogen, which we believe can be resolved over time.[1] The scale-up of the hydrocarbon reforming process is also expected to be economically viable. Similar to any catalytic process, the aim is to convert chemical reactions at lower temperatures and pressures. This can be achieved with nanoscale catalyst materials.

### 6.5.2 Capacitive Desalinization

The desalinization of brackish and salt waters is a new project for us that shows promise. Current methods to desalinate water are energy intensive. These include distillation, reverse osmosis, and filter methods. By using nanomaterials made by the layer-by-layer (L-B-L) method, which successively dips a substrate in different colloidal dispersions of nanoparticles, specialized desalination processes using charged electrodes (acting like a capacitor) can cut the cost of water purification by half or better. Clean drinking water can therefore be produced with low-energy devices that can be transported to and installed in remote regions. Filters remove the detritus and microbes, and capacitive desalination removes the salts.

### 6.5.3 Structures: Carbon Nanotubes

Another new venture for us involves the synthesis of carbon nanotubes (CNTs). Our goal is to make "extremely long" CNTs for use as reinforcement in polymers and building materials. Currently, the process is highly expensive. However, in the near future, simple geometric scale-up of the chemical vapor deposition process will reduce the cost of synthesis and as in the semiconductor industry and its expensive top-down equipment, the price for carbon nanotubes will eventually come down if there is demand.

Another remarkable and incredible property of carbon nanotubes is their ability to conduct electricity in a ballistic fashion—for example, without electrical resistance at room temperature. This is better than superconductivity as conduction of electricity without resistance across long distances would conserve energy. This saves money, preserves resources, and delivers cheaper power to those in need. There are, however, significant hurdles to overcome before this becomes reality any time soon, such as the synthesis of long nanotubes directly or development of a means to splice nanotubes of a reasonable length.

### 6.5.4 Antimicrobial Systems

Food preservation and medical salves are just a few applications of antimicrobial systems using nanotechnology. Compared to their bulk material form, silver nanoparticles have high antimicrobial activity due to high reactivity. Therefore, enhanced release of metal ions from metal/metal oxide nanostructures can be fatal to microbes as they can penetrate the cell envelope and disorganize cell membranes. Metal oxide nanostructures, in the presence of light, can generate reactive oxygen species that disrupt the microbial cell walls. ZnO nanorods grown on different substrates were successfully used for decontaminating water from bacteria like *E. coli, S. aureus,* and *B. subtilis* (Baruah et al. 2010; Sapkota et al. 2011). This field is already showing incredible growth worldwide. Antimicrobial systems would be able to fight infections in wounds encountered in agricultural regions.

### 6.6 Conclusion

We have presented and discussed ways to purify water, make and transport energy, sense gases, deflect water, and several other technologies with assorted potential applications—all made from bottom-up technology. Some of the materials and devices may have an impact on the way we collect energy and purify water. Others may have uses that we cannot yet begin to imagine. The bottom line is that we can make sophisticated nanomaterials and devices from relatively cheap stock, with equipment that was built from what is already available. As stated earlier, nanotechnology can be an expensive technology, but we have found a way to work with it as cheaply as possible. As engineers and scientists, we feel it is our responsibility to convince the powers that be that our technology is marketable. We seek to act as a mouthpiece to sell the idea to scientists, engineers, and nonscientists alike in order to take our technological innovations to the next level, for example, as a sustainable start-up company capable of employing people from a local workforce.

The important relation between the end user, the laboratory, and broader (associated) infrastructure is cultivated by networks comprising individuals and organizations within government agencies, industry, and universities at local, regional, national, and international levels. As nanotechnology commonly has high entry barriers, partnerships between and among business, government, and academia are necessities—especially for PMNT. The strength of collaborative networks will determine how well nanotechnology is delivered from the laboratory to the end user. This means considering the key issues and impediments to providing and implementing nanotechnology in the southern and southeastern regions of Asia, asking first if there are practical approaches to commercializing nanotechnology, and then developing the requisite human resources.

If global sustainability is to become a reality in our lifetimes, it is the job of governments and authorities to make equity the rule and not the exception. Nanotechnology's upside is immense—it offers hope for addressing many of the technical issues and dilemmas facing humankind in the twenty-first century. Some of the broader benefits of a PMNT lie in job creation, the provision of basic services, and the creation of resources necessary for a high quality of life. However, a question also begs to be asked as to whether the technology will only be available to those who have the means to afford it. We view our job as making logical overtures to government and industry and to engage them as often as possible, throughout the process of developing the technology, seeing it made into a product and then sold. As CoEN publishes more, attracts more funding, drives industry, and adds more faculty, visiting scientists, and students, we become more and more like a "rich man's nanotechnology" laboratory. Although this is something we cannot deny is desirable for us, we will never lose touch with our fundamental mission—to make materials cheaply in order to provide essential services to those in need.

## References

Baruah, S. and Dutta, J. 2009a. Hydrothermal growth of ZnO nanostructures. *Science and Technology of Advanced Materials* 10(1): 3001.

Baruah, S. and Dutta, J. 2009b. Effect of seeded substrates on the hydrothermally grown ZnO nanorods. *Journal of Sol-Gel Science and Technology* 50(3): 456–64.

Baruah, S. and Dutta, J. 2009c. pH dependent growth of ZnO nanorods. *Journal of Crystal Growth* 311(8): 2549–54.

Baruah, S., Jaisai, M., Imani, R., Nazhad, M.M., and Dutta, J. 2010. Photocatalytic paper using zinc oxide nanorods. *Science and Technology of Advanced Materials* 11(5): 5002.

Baruah, S., Rafique, R.F., and Dutta, J. 2008. Visible light photocatalysis by tailoring crystal defects in zinc oxide nanostructures. *Nano* 3(5): 399–407.

Baruah, S., Sinha, S.S., Ghosh, B., Pal, S.K.K., Raychaudhuri, A.K., and Dutta, J. 2009. Photo-reactivity of ZnO nanoparticles in visible light: Effect of surface states on electron transfer reaction. *Journal of Applied Physics* 105(7): 074308–6 pp.

Baruah, S., Warad, H.C., Chindaduang, A., Tumcharern, G., and Dutta, J. 2008. Studies on chitosan stabilised ZnS:Mn$^{2+}$ nanoparticles. *Journal of Bionanoscience* 2(1): 42–48.

Library of Congress—Federal Research Division (LCFRD). 2007. *Country Profile: Thailand*. July. Washington.

Myint, M.T., Kitsomboonloha, R., Baruah, S., and Dutta J. 2010. Superhydrophobic surfaces using selected zinc oxide microrod growth on ink-jetted patterns. *Journal of Colloids and Interface Science* 156(1–3): 810–15.

Sapkota, A., Anceno, A.J., Baruah, S., Shipin, O.V., and Dutta, J. 2011. Zinc oxide nanorod mediated visible light photoinactivation of model microbes in water. *Nanotechnology* 22(21): 5703.

Singapore Agency for Science Technology and Research (SASTR). 2010. *A\*STAR SIMTech Nanotechnology in Manufacturing Initiative (NiMI) to Overcome Challenges to Tap Market Potential*. Singapore Agency for Science, Technology and Research. http://www.a-star.edu.sg/Media/News/PressReleases/tabid/828/articleType/ArticleView/articleId/1363/Default.aspx (accessed May 11, 2011).

Sugunan, A., Thanachayanont, C., Dutta, J., and Hilborn, J.G. 2005. Heavy-metal ion sensors using chitosan-capped gold nanoparticles. *Science and Technology of Advanced Materials* 6(3–4): 335–40.

Ul Islam, N., Ahmed, K.F., Sugunan, A., and Dutta, J. 2007. Forensic fingerprint enhancement using bioadhesive chitosan and gold nanoparticles. *Proceedings of the Second IEEE International Conference on Nano/Micro Engineered and Molecular Systems (IEEE NEMS 2007)*. Bangkok, Thailand, January 16–19, pp. 411–15.

Ullah, R. and Dutta, J. 2008. Photocatalytic degradation of organic dyes with manganese-doped ZnO nanoparticles. *Journal of Hazardous Materials* 156(1–3): 194–200.

## Endnote

1. A well-known example of unsafe storage of hydrogen was provided by the space shuttle *Challenger* in 1986.

# Section III

# Appropriateness

# 7

## Nanotechnology and Global Health

**Deb Bennett-Woods**

### CONTENTS

When it comes to global health, there is no "them" ... only "us."

**—Nils Daulaire (2001)**

### 7.1 Introduction

As human society has evolved and adapted to a wide range of environments, the ability to move beyond mere survival to thrive in a specific environment has often been mediated by the discovery and application of new knowledge and increasingly advanced technologies. At the same time, not all technological advances result in automatic benefits for all of humankind. Emerging technologies pose both significant opportunities and threats to the collective health of the human community. Nanoscience and its related technologies have the potential be among the most sophisticated advances in human understanding and mastery of the environment. Nanotechnology allows the manipulation of matter at the atomic scale, yielding an unprecedented understanding of nature itself as well as enabling its application to a plethora of human challenges. In terms of global health, the breadth and scope of that application are limited only by how we ultimately define health, the health-related goals we set to achieve, and the strategic insight and collective will to realize those goals.

The nature of the human condition is characterized by a collective, perpetual struggle to achieve and retain a functional level of health amid famine, lack of shelter, disease, violent conflict, overcrowding, aging, and other natural and self-made barriers. This functional level of health has, for most of human history, allowed the species to survive and even to thrive when circumstances allow. Health is perhaps the most valuable commodity we can possess as human beings. At a fundamental level, health enables all other human activity—economic, political, social, and cultural—and is a key component of both individual and communal well-being. In this most basic sense, health is not a North–South or East–West issue. All human populations are challenged to maintain a minimum standard of health and suffer the loss of potential when that minimum is not met. If anything, global issues such as climate change and emerging infectious disease have revealed conditions of risk and uncertainty that are increasingly shared by all human communities, regardless of economic wealth, level of development, political stability, cultural factors, or access to natural resources.

In addition, the determinants of human health are complex and all-encompassing, reaching far into every domain of human existence. Poverty, war, environmental degradation, violent crime, illiteracy, homelessness, exploitation of women and children, the drug trade, economic and political instability, and every other identifiable social challenge ultimately end up in a country's respective health care system or in the hands of health-related NGOs in some form or another. Physical injury, emotional trauma, infectious disease, chronic illness, mental illness, stress-related disorders, and early death are too often the outcomes of shortcomings and failures elsewhere in society. For all our dramatic advances and successes in medicine specifically and in public health generally, the category of human health remains the repository and a key indicator of all that is not quite right in the larger society. Traditional approaches to global health, which tend to focus almost solely on a biomedical paradigm, suffer from this lack of systemic consideration and the appropriate resources to mitigate the social foundations of ill health (Maclean and Maclean 2009; Commission on Social Determinants of Health 2008).

My purpose in this chapter is to explore the relationship between the emerging potential of nanotechnology and a working conception of human health. I give consideration to ways in which nano-enabled technologies might operate to improve human health on a global scale and am particularly interested in the impact on the Global South. My consideration includes the direct effects of technology as well as the more indirect, and the unintended or unanticipated effects technology can have on both individual and collective measures of health and health care systems across the globe. I pose a set of strategically focused questions based on both moral and practical considerations, illustrating the need to better prioritize scarce resources, assume a more collaborative posture, and promote a more informed, anticipatory, and systems-oriented approach to global health.

## 7.2 Definition of Health

Health is not a particularly straightforward concept. Perhaps the most widely quoted definition of health is that found in the Constitution of the World Health Organization (WHO). Adopted in 1946, the WHO defines health as "a state of complete physical, mental and social well-being and not merely the absence of disease or infirmity" (World Health Organization 1946/2006). Although clear and concise, how one might operationalize this definition is quite another matter. It has been variously criticized as too broad, too idealistic, oversimplified, and generally ill defined (Larson 1999). Nonetheless, the WHO definition has widespread acceptance and for good reason. The concept of health has broad implications, its value is subject to individual and cultural interpretations, and the systemic relationship of its determinants is poorly understood. Larson characterizes the WHO definition as one model of health and contrasts it with three competing models. A closer look at each of these four models gives some insight into the complexity of addressing global health with targeted technologies. First, the *medical model* defines health narrowly in terms of disease or disability based on objective measures. Its approach is focused on cause, prevention, and cure, and it represents the predominant approach to medical research. Because the medical model tends to structure itself around specific diseases or disabilities, the priorities tend to be technical and strongly directed at developing new drugs and technologies, improving existing drugs and technologies, and encouraging use of these interventions (Ranson and Bennett 2009). For example, much of the work in medicine involving nanotechnology has been in the realm of cancer research where there have been notable advances in imaging technologies for earlier and better diagnosis, targeted drug development, and more effective drug delivery systems (Hamdy, Alshamsan, and Samuel 2009).

Such technological interventions are well suited to an established and effective health care system capable of the full spectrum of high-tech diagnosis, treatment, and follow-up by well-trained health professionals. For many nations in the Global South, cost and lack of a fully developed infrastructure for health care delivery impedes the diffusion of these highly targeted, nano-enabled advances (Liao 2009). For example, while some specific disease eradication programs have had a measure of success, such as with the elimination of smallpox, they have largely failed to address the overall disease burden for southern populations, rendering the traditional medical model insufficient (Magnussen, Ehiri, and Jolly 2004). As Liao (2009) notes, disease-specific interventions are frequently funded on the basis of short-term goals measured by numerical targets such as number of vaccinations administered; lack sound assessment methods for efficacy or sustainability; end abruptly with no forward strategy; and rarely include local citizens in decision making. It could be noted that this approach holds relatively true for socioeconomically disadvantaged populations in both the North and South.

A second model can be termed the *wellness model*, which focuses on a more general and subjective state of well-being that enables a person to overcome disease via a positive outlook, healthy lifestyle, and supportive social and spiritual networks (Larson 1999). While technology may have a limited direct effect in this less technologically oriented model, mental and physical challenges that can occur secondary to poverty, economic displacement, and the breakdown of community and other social determinants of health are highly relevant. To the extent that emerging technologies either increase or decrease opportunities for physical, emotional, economic, and social well-being, they affect individual and communal measures of health. For example, stable employment and a living wage clearly contribute to one's well-being. Nanotechnology is often characterized as a global opportunity from which southern countries with basic technical capacities may directly benefit with respect to economic development and accompanying increases in the standard of living. However, technological advances that undermine existing industries in southern countries could have the opposite effect by a net reduction in employment opportunities (Invernizzi, Foladori, and Maclurcan 2008).

Third, the *environmental model* relates health to the ability of the individual to "maintain a balance with ... [his or her] environment, with relative freedom from pain, disability, or limitations, including social abilities" (Larson 1999). This model more overtly recognizes the dynamic interplay between biological and sociological elements of health as they exist in a specific environment. With respect to health in the Global South, one strength of defining health in this way is that it allows consideration of the context-specific challenges found in each country and specific locale. For example, in a location where waterborne disease is a primary source of illness and disability, a nano-enabled technology that supports health promotion via clean water may directly contribute to an individual's capacity to maintain a functional balance of health in that particular environment.

Returning to the WHO definition of health as "a state of complete physical, mental and social well-being and not merely the absence of disease or infirmity" (World Health Organization 1946/2006, 1), it seems clear that each of the three prior models contributes something to an understanding of health in its broadest definition. However, it should also be clear that health promotion strategies based on each individual approach to defining health will be quite different and might even work at cross purposes.

The social causes of poor health and health-related inequalities are referred to as the social determinants of health and are increasingly recognized as fundamental barriers to health in both the North and South (Venkatapuram 2010). The WHO Commission on Social Determinants of Health identifies a framework that includes social position, education, occupation, income, gender, ethnicity, and race. It is mediated on one side by a country's socio-economic and political context, as well as governance, social and economic policy, and cultural and societal norms and values on the other. Taken together, these factors have a mutually significant impact on an individual's material circumstances, social cohesion, psychological factors, behaviors, and biology (Commission on Social Determinants of Health 2008). The inter-action between social determinants is complex and can leave researchers and health workers feeling like they are playing the Whack a Mole carnival game in which addressing one issue allows another two to pop up in its place. The common wisdom is that limited resources preclude addressing all issues simultaneously; however, fragmentation of efforts and resources creates its own problems. For example, immunization programs have been successful in reducing infant and child mortality from infectious disease, yet leave children in need of basic food, shelter, education, and economic security, all of which are also important determinants of health over a lifetime. As aptly quoted by one Haitian health worker: "giving people medicine for TB and not giving them food is like washing your hands and drying them in the dirt" (Kidder 2004, 34). New technologies are often, if not always, introduced with little regard for the many contextual factors that determine the ultimate utility of the technology. At the very least, the strategic introduction of any nano-enabled technology should take a systems-oriented approach that appreciates the full spectrum of potential health effects and the specific social context.

## 7.3 Historical Perspective on Technology and Health

Technology has increasingly come to be seen as a panacea for human health, particularly in the North where robust economies support high-tech health care infrastructure and public health systems. On the one hand, such faith in technology is understandable given the medical advances of the past century. However, the results of medical advances have not diffused at anywhere near the same rate across the world. One can look with some amazement at the speed with which cell phone technology reached the far edges of the globe in less than a generation, enabling some of the poorest, most isolated, and least tech-nologically advanced societies a measure of global communication. Yet those same populations still suffer extraordinarily high rates of morbidity and mortality from what have long been preventable conditions in the Global North.

Prior to the past century, infectious disease was the primary health risk for most human populations. In recent decades, technologies that enabled public sanitation, pasteurization and preservation of food, vaccinations, and antibiotics have effectively shifted the burden of disease in the North from infectious disease to noninfectious chronic disease (Cartwright 1972). In response, modern industrialized societies have evolved sophisticated health care systems that are highly reliant on advanced technologies for diagnostics and treatment along with an extended health care infrastructure that includes settings for primary care, urgent care, acute care, transitional and long-term care, home care, hospice care, and independent allied health services such as physical therapy.

Nanotechnology has the potential to take health care delivery to an extraordinary new level. In a 2010 report, the Freedonia Group estimated that demand for health care products using nanotechnology in the United States alone would reach $75 billion in 2014 and then double to $149 billion by 2019 (Freedonia Group 2010). The bulk of products are specifically nano-medicines, although nanodiagnostics and other classes of therapeutics are also growing. Significant new advancements are already in use with many more in clinical trials. The category of nanotechnology in health care represents a range of medical applications that commonly include pharmaceuticals, medical diagnostics, and medical devices and implants.

Nano-agents are being investigated directly in the treatment of disease. For example, silver nanoparticles have been found to impede viral replication in HIV without inducing resistance and with the potential to act as a powerful preventive viricide (Lara et al. 2010). It is clear that pharmaceuticals will drive a major contribution toward nano-enabled applications that simplify, speed up, and reduce the costs of drug development and testing as well as increased drug safety and efficacy (Ferrari and Downing 2005). Biocompatible nanoparticles are providing new platforms for drug delivery, including alternative routes of administration for existing drugs that will minimize drug degradation, allow site targeting, and reduce side effects (Kubik, Bogunia-Kubik, and Sugisaka 2005). In particular, biodegradable polymer nanoparticles and other nanomaterials appear to be ideal candidates for cancer therapy, vaccine delivery, contraceptives, and targeted antibiotic delivery (Hamdy, Alshamsan, and Samuel 2009; Peek, Middaugh, and Berkland 2008). Similarly, nano-applications in pharmacogenomics and pharmacogenetics are opening the door to a truly personalized medicine with chemotherapeutics designed to the specifications of the individual cancer cell, potential replacement or repair of defective or nonfunctional genes, and the possibility of genetic immunization with DNA vaccines (Tibbals 2011).

Another area in which nanotechnology is likely to create new opportunities is in medical diagnostics. These include advances in medical imaging (Mazzola 2003) and screening microarrays for everything from rapid genotyping and genetic analysis to early diagnosis and monitoring of cancer, genetic epidemiology, tissue typing, microbial identification in infectious disease, and drug validation (Campo and Bruce 2005). For example, the new nano-enabled field of metagenomics, the genomic study of microorganisms, is beginning to produce more effective vaccines, diagnostics, and antibiotics. Other important technological advances in early diagnosis and monitoring of disease with particular application in the Global South include molecular recognition nanosensors, disposable microfluidic diagnostics, and isothermal gene amplification (Tibbals 2011).

Finally, nanofabrication tools and techniques are expected to revolutionize the construction of medical devices and implants that have improved biocompatibility and longevity (Van den Bueken, Walboomers, and Jansen 2005). For example, catheters with antimicrobial silver nanoparticle coatings to prevent common catheter-related infections have been introduced, and nano-engineered scaffolding and encapsulation techniques are being used in clinical trials for central nervous system regeneration or restoration in diseases including Huntington's disease and amyotrophic lateral sclerosis (Tibbals 2011).

Clearly, the potential for nanotechnology to have beneficial impacts on human health is great. On the other hand, medical technologies have not always provided an unqualified benefit. Recent problems with so-called blockbuster drugs in the U.S. market have raised questions about the adequacy of clinical trials and safety assessment. Similarly (and as will be discussed in Chapter 10), there is much concern raised about the potential safety of engineered nanoparticles.

In addition, most advanced medical technologies have been shown to increase the cost of health care. For example, nanotechnologies have enabled rapid advances in robotic surgery. In a review of cost studies of robot-assisted procedures, the additional variable cost of the procedure rose $1600 and, when the amortized cost of the robot is included, it rose $3200 (Barbash and Glied 2010). Interestingly, there is currently no evidence that long-term outcomes of robot-assisted surgery are better than conventional surgical procedures (Barbash and Glied 2010). Even more startling is the recent approval of sipuleucel-T, trade name Provenge, a nano-enabled treatment for prostate cancer. The drug increased median survival by 4.1 months but at $31,000 per treatment with a normal course of three treatments. While some nanotechnologies have the potential to reduce the cost of health care in targeted areas such as vaccine production and administration, they are also likely to increase costs in many areas, at least initially. For example, pharmaceutical companies routinely cite the high costs of research and development (R&D), including the costly clinical trial and Food and Drug Administration approval process, to justify the high pricing of new drugs.

Finally, while nano-enabled advances in medicine hold great promise for global health concerns, there is a strong likelihood that investment and R&D will be largely driven by northern markets (as per discussion in Chapter 4). In some cases, such as with vaccine development, technologies developed for targeted application in the North should easily transfer to the South. In other cases, high costs and infrastructural needs will impede transfer or simply fail to target health issues endemic to the South.

## 7.4 Promise of Emerging Technologies

The promise of emerging nanotechnologies on human health is broad, including mitigation of current issues in energy, agriculture and food, water, air pollution, and construction as well as a wide range of direct outcomes from medicinal treatment through health monitoring (Singer, Salamanca-Buentello, and Darr 2005). Although the medical model of health is limited in its ability to address important social determinants of health, it is crucial to the direct management of disease burden. Medicine is perhaps the arena in which nanotechnology currently has the most to offer, with a number of nano-based advances that could potentially reduce the disease burden in the South.

### 7.4.1 Nanotechnology and Medicine

Much attention has been directed toward immunization as a major intervention for health in the global South. For example, the Bill and Melinda Gates Foundation and its partners launched the Grand Challenges in Global Health to fund innovations that address global health issues (Grand Challenges in Global Health 2010). Two of the seven long-term goals and six of the fourteen grand challenges are related to improvement of existing vaccines and development of new ones. More than one million children die annually from vaccine-preventable diseases. Among the technical challenges to effective immunization are the requirement for multiple doses, the need to refrigerate the vaccine, cost and infections associated with needles and injections, the need for more reliable tests for evaluating live attenuated vaccines, more effective antigen design for immunity, and a better understanding of immunological responses (Grand Challenges in Global Health 2010).

Nanotechnology is currently being applied to these issues with early success in several areas. For example, mucosal vaccines can be delivered orally or nasally, do not require trained personnel, eliminate the use of needles, and improve compliance. Nanotechnology is being used to design nanoparticles into which vaccine components can be encapsulated along with mucosal adjuvants to maximize the immune response (Chadwick, Kriegel, and Amiji 2010). In some cases, nanoparticles are being designed to serve as adjuvants (Reddy et al. 2007). In 2009, BioSante Pharmaceuticals announced the successful use of calcium phosphate nanoparticles as adjuvants in its H1N1 and H5N1 nasal flu vaccines. It was also reported that the strength of the vaccine would allow for lower doses, stretching the quantity of available vaccine (BioSante Pharmaceuticals 2009). Another approach to mucosal immunity involves nano-emulsion vaccines delivered nasally; researchers at the University of Michigan demonstrated immunity in animal studies to smallpox, HIV, anthrax, hepatitis B, and influenza. In late 2009, these researchers and their partner, NanoBio Corporation, received a $9 million grant to expand their work to other pathogens, and they are currently in Phase 1 of a clinical trial testing a nano-emulsion-based intranasal vaccine for influenza (University of Michigan Health System 2009). In another approach to vaccine delivery by Harvard researchers, a novel aerosol version of the common tuberculosis vaccine sprayed directly to the lungs appeared to result in significantly better protection than the traditional injection. In addition, the rapid drying process used to create micro- and nanoscale particles allows the vaccine to be stored without refrigeration. It is predicted that these and other nano-enabled approaches to vaccine production and delivery, several of which are currently in clinical trials, will result in novel vaccines for viral and parasitic infections including hepatitis, HIV, malaria, and even cancer (Peek, Middaugh, and Berkland 2008).

Another of the Grand Challenge long-term goals is to measure health status; the fourteenth and final Grand Challenge is the development of technologies that allow assessment of multiple conditions and pathogens at point of care (Grand Challenges in Global Health 2010). Infectious disease diagnostic techniques have changed little in the past half century. They remain expensive, limited by slow analysis, require skilled workers, have low detection thresholds, and are unable to detect more than one strain of an infectious agent (Hauck et al. 2009). However, nanotechnology is enabling the development of new and better assays that potentially offer inexpensive, rapid, multiplexed disease detection at the point of care. Lab-on-a-chip technologies using quantum dots, metallic nanostructures, and other nanoparticles are under development. For example, with tuberculosis there is a need to be able to detect both latent TB infection and concurrent HIV infection. Rapid HIV testing with early detection and viral load data could minimize the spread of the disease and facilitate treatment. Finally, the ability to differentiate specific malarial strains and similar diseases will improve targeted treatment and slow the emergence of treatment-resistant parasites. There remain a number of technical challenges and far too little work is being dedicated specifically toward infectious diseases of the South; the potential for point-of-care diagnosis and enhanced pathogen identification, however, does exist (Grand Challenges in Global Health 2010). In addition to point-of-care diagnostics, optical biosensors using similar nano-enabled assays could be used in monitoring the quality of food, water, and air given adequate investment in an effective infrastructure (Ligler 2009).

### 7.4.2 Nanotechnology, the Environment, and Climate Change

As noted, climate change poses a number of risks to human health. The promise of nanotechnology to create more sustainable options for clean energy provides long-term hope that climate change can be slowed over time. In a similar vein, nanotechnology holds great promise for increasing health outcomes associated with access to clean water. There are approximately one billion people in the world with no access to potable water, and more than twice that many lack access to adequate sanitation. Nearly 50 percent of the population in southern countries suffers from one or more of the primary waterborne diseases including diarrhea, intestinal helminth infections, dracunculiasis, schistosomiasis, and trachoma (Bartram et al. 2005). A tremendous amount of work has been done in the area of developing nanomembranes, particles, and other materials for water treatment and remediation, sensing and detection, and pollution prevention (Theron, Walker, and Cloete 2008).

## 7.5 Shadow Side of Technology and Moral Consideration

Nanotechnology raises at least two distinct issues with respect to negative impacts on health. The first relates to its potential environmental impact on patients, the workforce, and the larger ecosystem. The second issue involves accessibility, cost, and the potential for further inequity in the distribution of health care resources and will be explored here.

If they can be proven cost effective when compared with traditional applications, nano-enabled medical applications will reduce health care costs and most likely improve user access. On the other hand, if high-cost interventions developed in response to chronic illnesses in the North are adopted by health care systems in the South that are faced with an increase in chronic disease throughout the population, we may see an overall shift in focus and reduction in resources available for prevention as well as a reduction of treatment for infectious disease.

Following the WHO Constitution's definition of health, there is a statement that "the enjoyment of the highest attainable standard of health is one of the fundamental rights of every human being without the distinction of race, religion, political belief, economic or social condition" (World Health Organization 1946/2006, 1).

In 2000, the UN Committee on Economic, Social and Cultural Rights further adopted a General Comment on the Right to Health that specifies, "[The] right to health extends not only to timely and appropriate health care but also to the underlying determinants of health, such as access to safe and potable water and adequate sanitation, an adequate supply of safe food, nutrition and housing, healthy occupational and environmental conditions, and access to health-related education and information, including on sexual and reproductive health" (Committee on Economic, Social and Cultural Rights 2000).

The question of rights is an overtly moral one. A right to the basic conditions necessary for health is generally argued on the basis of justice and is limited to what is reasonable in a particular context. There exists a realistic concern that nanomedicine may prove, at least initially, very expensive. While many would argue that equal access to medical advances can be justified solely on the basis of a moral principle of justice (Ebbesen 2009), such arguments utterly fail to account for the pragmatic realities of scarce resources and the current economic, technological, and political divides between the Global North and South. If nano-enabled technologies provide a potential path to the achievement of health as a matter of social justice, then strategic questions regarding how to pursue and implement those technologies must be considered in light of technology as a scarce resource to be justly distributed. Unfortunately, while relatively easy to defend in principle, the concept of health or health care as a human right has proven difficult to operationalize or enforce.

In another take, one might also argue that a right to health has a strong utilitarian value that serves a greater good including the self-interest of all nations. In the same way that the burden of disease, climate change, environmental degradation, and political, economic, and social instability fuel health inequities in the Global South, those same issues pose emerging threats to the long-term health of the Global North. Newly emerging diseases such as SARS, hemorrhagic fevers, mad cow, and Hanta virus, along with reemerging diseases including malaria, tuberculosis, and drug-resistant microbes, have been identified as serious threats on a global scale. Coupled with bioterrorism, the burden of emerging disease becomes a matter of national and international security with direct threat of morbidity and mortality, as well as threats to economic and political stability (Morens, Folkers, and Fauci 2004).

Another moral issue to consider is that of consent. Much attention has been paid in countries like the United States and the United Kingdom to public engagement in nanotechnology. Public responses tend to focus around four basic positions (Doubleday 2007). The first is optimism that nanotechnology will yield positive benefits with respondents particularly supportive of its use in medicine. The second is a general skepticism regarding the ability of governments and business to regulate new technologies and, in particular, the inability to effectively thwart unintended consequences and risks. Similarly, the third and fourth concerns are that not all technologies will be socially beneficial in terms of health, the environment, and the economy, and that public involvement won't be considered in decision making and policy. For example, Musee, Brent, and Ashton (2010) point out that South Africa's strategic pursuit of nanotechnology as a means of addressing critical development challenges, such as clean water and health care, has not included a research strategy to investigate the environmental, health, and safety risks. If public engagement is challenging in the North, the populations of southern countries are likely to have even less voice in how nanotechnology is introduced into their respective environments. Recent controversies over pharmaceutical testing and translational research in the South illustrate the relative vulnerability of populations who are poor, functionally illiterate, and have few options when it comes to protecting or improving their health (Bhutta 2002; Sofaer and Eyal 2010).

## 7.6 Framework of Practical Issues and Strategic Priorities

Given the scope, complexity, and moral dimensions of health issues, strategic prioritization is complex at best. There is a "chicken-or-the-egg" quality to considerations of where best to invest energy and resources to get the maximum return with respect to improved health outcomes. Three pragmatic questions are central to the strategic development and deployment of new technologies: what to fund, when to transfer knowledge, and how to regulate (Bennett-Woods 2008; the last of these will be more adequately addressed in Chapter 12). The tendency is to answer the questions in relative isolation at separate points along the technology development continuum; however, commentators in nanotechnology have long called for more "upstream" consideration of the potential impacts and necessary safeguards, given its relatively rapid evolution and broad scope of impact on so many aspects of human knowledge and control (Bennett-Woods 2007; Ihde 1993; Roco and Bainbridge 2001; Royal Society 2004).

### 7.6.1 What to Fund?

The worldwide economic recession increases the need to urgently respond to global health issues while also decreasing the available resources and political will to act. Funding for innovation in global health comes from a variety of governmental, philanthropic, and corporate sources. This breadth is likely both an asset and a detriment in that it allows for a broad range of viewpoints, goals, and differently targeted approaches while also lacking a cohesive and comprehensive strategy for maximizing the outcomes of those investments. More commonly, there is a lack of coordination between funding initiatives, research priorities, and demonstrated societal needs (Malsch 2008). Although there is likely something to be gained in all efforts, funding goals are often narrowly targeted, sacrificing deep change for short-term measures of success. Social determinants of health—in particular, socioeconomic status and class—require integrated approaches with a focus on sustainable improvements in the foundations of health. Funding priorities should effectively guide industry priorities. Priorities must also actively seek a combination of short- and long-term goals, consistent with the goals of the population itself, while overtly addressing systemic issues.

## 7.6.2 When to Transfer Knowledge?

The bottom line with any new technology is that it is only as useful as it is accessible and affordable. In the case of health-related applications, the impact of obstacles to access can be literally measured in lives lost. Nine million children die every year in the Global South from largely preventable disease. The questions around when and how to transfer intellectual property such as patents are legally complex. There has been a flurry of activity within the past couple of years around technology transfer as it relates to medical patents. There exists a significant gap in access to lifesaving drugs in the Global South. Although there are a number of barriers to access, one of the most striking is the role of patents in preventing lower-cost, generic production of drugs in southern countries. A vast amount of basic nanoresearch in the United States is publicly funded via research universities that, prior to the Bayh–Dole Act of 1980, did not typically apply for patents on biomedical discoveries. However, a recent survey reports that U.S. universities own patents on almost 20 percent of the most innovative new drugs, with most of the remainder held by pharmaceutical companies (Sampat 2009). While ideas for overcoming these barriers is explored more fully in Chapter 9, effective strategies for increasing access to drugs in southern markets might include some combination of compulsory and open licensing that grants access to lower-cost producers while protecting incentives for R&D. In particular, there is a risk of reducing incentives for innovation in the area of those "neglected" diseases that are prevalent in lower- and medium-income countries but with limited markets in higher-income countries (Flynn, Hollis, and Palmedo 2009).

---

## 7.7 Conclusion

He who has health, has hope. And he who has hope, has everything.

**—Arabian proverb**

A strategic approach to nanotechnology must combine instrumental technological potential with an appreciation for the systemic context within which the technology may be applied. Nowhere should this be more evident than in the multifactorial context of health and its social determinants. If we consider only nanomedicine, primary barriers contributing to inequality in access and development include the current patenting system for drugs that benefits the North, the less-homogeneous nature of southern countries that requires a wider range of approaches, and the lack of targeted R&D (Tyshenko 2009). However, one could argue that some version of each of these barriers can easily be extended to the challenges of all health-related opportunities presented by nanotechnologies. For example, access to nano-enhanced water filtration systems will likely be mediated by patent restrictions; country-specific geographic, political, cultural, and economic realities; and prioritization of resources for development, implementation, and maintenance.

In other words, the critical element driving the ability of nanotechnology to have a substantial impact on global health in the near future is addressing the contextual issues of health ahead of the instrumental potential of technology. The technology is advancing by leaps and bounds while our appreciation of the complex context of global health lags far behind, resulting in fragmentation of efforts and wasted resources despite even the best intentions. The first step in understanding the context is public engagement supported by a research agenda that clarifies the critical systemic links between social determinants of health and the traditional biomedical paradigm that characterizes most current efforts to mitigate global health challenges. Such a strategy requires a strong model of collaboration and accountability (Liao 2009; Missoni and Foffani 2009). Public engagement in the process builds trust, awareness, and practical knowledge, allowing new technologies to be introduced most effectively based on a realistic assessment of existing infrastructure and social or political barriers. A collaborative, multidisciplinary research agenda can reduce fragmentation and better target scarce resources by informing and directing government, academia, various industrial sectors, and other international stakeholders in more closely coordinated efforts that more fully address upstream causes (Benatar, Gill, and Bakker 2011). The Adelaide Statement on Health in All Policies (WHO, 2010) proposes a similar strategic awareness of the interrelationships between the larger concept of health and issues of economics, security, education, agriculture, basic infrastructure, environmental considerations, housing, community services, land, and culture. Although difficult to accomplish, an effective North–South collaboration includes a full range of stakeholders who appreciate our common interests in global health issues and at least holds the promise of innovative action that can generate a more organic and systemic level of change by addressing multiple determinants of health simultaneously.

Ronald Labonte (2008) identifies five distinct approaches to global health policy in which health is defined variously as a matter of security, development, global public good, commodity, or rights. Each approach has its own embedded set of priorities and limitations. For example, health as security can rapidly mobilize resources for some immediate threat such as a newly emerging virus with pandemic potential but may naturally lose focus on longer-term development priorities such as reductions in maternal and infant morbidity and mortality. Health as a global public good places ongoing pressure for shared financing and regulation, while rights-based approaches naturally attend to the broad spectrum of underlying social inequities. Health as a commodity accurately represents the bottom line, cost/benefit calculation that accompanies any investment in basic research or newly applied technologies. Although Labonte argues that health as a commodity is unjustified, a more pragmatic approach is to accept that each of the five views holds some strategic insight into how nano-enabled technologies will unfold and be applied to the challenges of global health. Nanotechnology is already revolutionizing the process and outcomes of medicine and health. The overriding goal of advocates for global health should be to ensure that existing health gaps between North and South are not further skewed by rapid and focused development in the North, so as to be accompanied by a fragmented and marginally effective diffusion across the South. Consideration of a broad range of the social determinants of health is critical to an effective strategy for generating comprehensive, systems-oriented approaches that employ resources effectively while attending to the moral considerations posed by current global inequities in health.

## References

Barbash, G.I. and S.A. Glied. 2010. New technology and health care costs—The case of robot-assisted surgery. *New England Journal of Medicine* 363, no. 8: 701–4.

Bartram, J., Lewis, K., Lenton, R., and A. Wright. 2005. Focusing on improved water and sanitation for health. *Lancet* 365 (February): 810–12.

Benatar, S.R., Gill, S. and Bakker, I. 2011. Global health and the global economic crisis. *American Journal of Public Health* 101, no. 4: 646–653.

Bennett-Woods, D. 2007. Integrating ethical considerations into funding decisions for emerging technologies. *Journal of Nanotechnology Law and Business* 4, no. 1.

Bennett-Woods, D. 2008. *Nanotechnology: Ethics and Society*. Boca Raton, Fla.: CRC Press, 2008.

Bhutta, Z.A. 2002. Ethics in international health research: A perspective from the developing world. *Bulletin of the World Health Organization* 80, no. 2: 114–19.

BioSante Pharmaceuticals. 2009. BioSante Pharmaceuticals News Releases. *BioSante Pharmaceuticals*. August 17. http://www.biosantepharma.com/News-Releases.php?ID=081709 (accessed January 2, 2010).

Campo, A. and I.J. Bruce. 2005. Diagnostics and high throughput screening. In *Biomedical Nanotechnology*, edited by N.H. Malsch. Boca Raton, Fla.: Taylor & Francis.

Cartwright, F.F. 1972. *Disease and History*. New York: Dorset Press.

Chadwick, S., Kriegel, C. and M. Amiji. 2010. Nanotechnology solutions for mucosal immunization. *Advanced Drug Delivery Reviews* 62, no. 4/5: 394–407.

Chambers, J.D. and P.J. Neumann. 2011. Listening to Provenge—What a costly cancer treatment says about future Medicare policy. *New England Journal of Medicine* 10, no. 1056.

Clift, R. 2006. Risk management and regulation in an emerging technology. In *Nanotechnology: Risk, Ethics and Law*, edited by G. Hunt and M. Mehta, 140–53. London: Earthscan.

Colvin, V.L. 2003. The potential environmental impact of engineered nanomaterials. *Nature Biotechnology* 21, no. 10: 1166–70.

Committee on Economic, Social and Cultural Rights. 2000. *Substantive Issues Arising in the Implementation of the International Covenant on Economic, Social and Cultural Rights: General Comment No. 14 (2000)*. Committee, United Nations, Geneva: United Nations Economic and Social Council, 18.

Commission on Social Determinants of Health. 2008. *Closing the Gap in a Generation: Health Equity through Action on the Social Determinants of Health (Final Report).* World Health Organization.

Daulaire, N. 2001. Fighting terror with hope: Global health in the new reality. Speech sponsored by the Yale School of Medicine and Department of Epidemiology and Public Health, October, 2001. *Yale Medicine* (Winter).

Doubleday, R. 2007. Risk, public engagement and reflexivity: Alternative framings of the public dimensions of nanotechnology. *Health, Risk and Society* 9, no. 2 (June): 211–27.

Ebbesen, M. 2009. The principle of justice and access to nanomedicine in national healthcare systems. *Studies in Ethics, Law, and Technology* 3, no. 3: 15.

Faunce, T.A. 2007. Nanotechnology in global medicine and human biosecurity: Private interests, policy dilemmas, and the calibration of public health law. *Journal of Law, Medicine and Ethics* (Winter): 629–42.

Ferrari, M. and G. Downing. 2005. Medical nanotechnology: Shortening clinical trials and regulatory pathways? *Biodrugs* 19, no. 4: 203–10.

Flynn, S., Hollis, A., and M. Palmedo. 2009. An economic justification for open access to essential medicine patents in developing countries. *Journal of Law, Medicine and Ethics* (Summer): 184–208.

Freedonia Group. 2010. *Nanotechnology in Healthcare.* Market Research, Freedonia Group, Abstract Only.

Glenn, J.C., Gordon, T.J., and E. Florescu. 2009. *2009 State of the Future.* Washington, D.C.: Millennium Project.

Grand Challenges in Global Health. 2010. *Grand Challenges in Global Health..* http://www.grandchallenges.org/Pages/BrowseByGoal.aspx (accessed January 12, 2010).

Hamdy, S., Alshamsan, A., and J. Samuel. 2009. *Nanotechnology for Cancer Vaccine Delivery.* Vol. X, chap. 17 in *Nanotechnology in Drug Delivery*, edited by M. de Villers, P. Aramwit, and G.S. Kwon, 519–43. New York: Springer.

Hauck, T.S., Giri, S., Gao, Y., and C.W. Chan. 2009. Nanotechnology diagnostics for infectious diseases prevalent in developing countries. *Advanced Drug Delivery Reviews*, online version.

Ihde, D. 1993. *Philosophy of Technology: An Introduction.* New York: Paragon House.

Invernizzi, N.I., Foladori, G., and D. Maclurcan. 2008. Nanotechnology's controversial role for the South. *Science Technology and Society* 13, no. 1: 123–48.

Kidder, T. 2004. *Mountains beyond Mountains: The Quest of Dr. Paul Farmer, a Man Who Would Cure the World.* New York: Random House.

Kimbrell, G.A. 2007. The potential environmental hazards of nanotechnology and the applicability of existing law. In *Nanoscale: Issues and Perspectives for the Nano Century*, edited by N.M. de S. Cameron and M.E. Mitchell, 211–38. Hoboken, N.J.: John Wiley & Sons.

Kohler, A.R. and C. Som. 2008. Environmental and health implications of nanotechnology: Have innovators learned the lessons from past experiences? *Human and Ecological Risk Assessment* 14: 512–31.

Kubik, T., Bogunia-Kubik, K., and M Sugisaka. 2005. Nanotechnology on duty in medical applications. *Current Pharmaceutical Biotechnology* 6: 17–33.

Labonte, R. 2008. Global health in public policy: Finding the right frame? *Critical Public Health* 18, no. 4 (December): 467–82.

Lara, H.H., Ayala-Nunez, N.V., Ixtepan-Turrent, L., and C. Rodroguez-Padilla. 2010. Mode of antiviral action of silver nanoparticles against HIV-1. *Journal of Nanobiotechnology* 8, no. 1.

Larson, J.S. 1999. The conceptualization of health. *Medical Care Research and Review* 56:123–36.

Liao, N. 2009. Combining instrumental and contextual approaches: Nanotechnology and sustainable development. *Journal of Law, Medicine and Ethics* (Winter): 781–89.

Ligler, F.S. 2009. Perspective on optical biosensors and integrated sensor systems. *Analytical Chemistry* 81, no. 2: 519–26.

Maclean, S.J. and D.R. Maclean. 2009. A "new scramble for Africa": The struggle in Sub-Saharan Africa to set the terms of global health. *Round Table* 98, no.402: 361–71.

Magnussen, L., Ehiri, J., and P. Jolly. 2004. Comprehensive versus selective primary health care: Lessons for global health policy. *Health Affairs* 23, no. 3 (May/June): 167–76.

Malsch, I. 2008. Which research in converging technologies should taxpayers fund? Exploring societal aspects. *Technology Analysis and Strategic Management* 20, no. 1 (January): 137–48.

Mazzola, L. 2003. Commercializing nanotechnology. *Nature Biotechnology* 21, no. 10: 1127–43.

McMichael, A.J., and G. Ranmuthugala. 2007. Global climate change and human health. In *Globalization and Health*, edited by I. Wamala and S. Kawachi, 81–97. New York: Oxford Press.

Millennium Project. 2009. Global challenges facing humanity. *The Millennium Project.* http://www.millennium-project.org/millennium/Global_Challenges/chall-08.html (accessed September 15, 2009).

Missoni, E. and G. Foffani. 2009. Nanotechnologies and challenges for global health. *Studies in Ethics, Law, and Technology* 3, no. 3.

Morens, D.M., Folkers, G.K., and A.S. Fauci. 2004. The challenge of emerging and re-emerging infectious diseases. *Nature* 430 (July): 242–49.

Musee, N., Brent, A.C., and P.J. Ashton. 2010. A South African research agenda to investigate the potential environmental, health and safety risks of nanotechnology. *South African Journal of Science*, 106, no. 3/4.

Norheim, O.F. and Y. Asada. 2009. The ideal of equal health revisited: Definitions and measures of inequity in health should be better integrated with theories of distributive justice. *International Journal for Equity in Health* 8, no. 40.

Peek, L.J., Middaugh, C.R., and C. Berkland. 2008. Nanotechnology in vaccine delivery. *Advanced Drug Delivery Reviews* 60: 915–28.

Ranson, M.K. and Bennett, S.C.. 2009. Priority setting and health policy and systems research. *Health Research Policy and Systems* 7, no. 27.

Reddy, S.T., van der Vlies, A.J., Simeoni, E., Angeli, V., Randolph, G.J., O'Neil, C.P., Lee, L.K., Swartz, M.A. and J.A. Hubbell. 2007. Exploiting lymphatic transport and complement activation in nanoparticle vaccines. *Nature Biotechnology* 25, no. 10 (October): 1159–64.

Roco, M. and Bainbridge, W.S. (Eds). *Nanotechnology: NSET Workshop Report.* National Science Foundation. (2001). Available: http://www.wtec.org/loyola/nano/NSET.Societal.Implications/nanosi.pdf (accessed September 20, 2011).

Royal Society and the Royal Academy of Engineering. 2004. *Nanoscience and Nanotechnologies: Opportunities and Uncertainties.* London: The Royal Society.

Sampat, B.N. 2009. Ensuring policy and laws are both effective and just: Academic patents and access to medicines in developing countries. *American Journal of Public Health* 99, no. 1 (January): 9–16.

Singer, P.A., Salamanca-Buentello, F. and A.S. Darr. 2005. Harnessing nanotechnology
    to improve global equity. *Issues in Science and Technology* (Summer): 57–64.
Sofaer, N. and Eyal, N. 2010. The diverse ethics of translational research. *American
    Journal of Bioethics* 10, no. 8: 19–30.
Stuckler, D. 2008. Population causes and consequences of leading chronic diseases:
    A comparative analysis of prevailing explanations. *Milback Quarterly* 86, no. 2:
    273–326.
Sweet, L., and B. Strohm. 2006. Nanotechnology: Life-cycle risk management. *Human
    and Ecological Risk Assessment* 12:528–51.
Theron, J., Walker, J.A. and T.E. Cloete. 2008. Nanotechnology and water treatment:
    Applications and emerging opportunities. *Critical Reviews in Microbiology*
    34:43–69.
Tibbals, H. 2011. *Medical Nanotechnology and Nanomedicine*. Boca Raton, Fla.: CRC Press.
Tyshenko, M.G. 2009. The impact of nanomedicine development on North-South
    equity and equal opportunities in healthcare. Press release. *Studies in Ethics,
    Law, and Technology* 3, no. 3: 19.
University of Michigan Health System. 2009. $9.3 million award to boost nanotech
    vaccine research. October 13. http://www2.med.umich.edu/prmc/media/
    newsroom/details.cfm?ID=1334 (accessed December 2, 2009).
Van den Bueken X., Walboomers, F., and J.A. Jansen. 2005. Implants and prostheses. In
    *Biomedical Nanotechnology*, edited by N.H. Masch. Boca Raton, Fla.: Taylor & Francis.
Venkatapuram, S. 2010. Global justice and the social determinants of health. *Ethics and
    International Affairs* 24, no.2: 119–30.
World Health Organization. Adelaide Statement on Health in All Policies. 2010.
    http://www.who.int/social_determinants/hiap_statement_who_sa_final.pdf
    (accessed September 20, 2011)
World Health Organization. 1946/2006. Constitution of the World Health
    Organization. *World Health Organization*. April/June. http://www.who.int/
    governance/eb/who_constitution_en.pdf (accessed August 3, 2009).
World Health Organization. 2009. *World Health Statistics 2009*. World Health Organization.

# 8

## Toward Pro-Poor Nano-Innovation (Zimbabwe, Peru, Nepal)

David J. Grimshaw

### CONTENTS

## 8.1 Introduction

Technology has long been recognized as playing an important role in human development. Its destructive nature has also been well documented. British economist E. F. Schumacher founded an international nongovernmental organization (NGO) now called Practical Action to promote the use and adoption of "intermediate technologies"[1] to reduce poverty in the world. It is now thirty-eight years since the publication of Schumacher's seminal work, *Small Is Beautiful: Economics as if People Mattered*. Despite the best efforts of many people, poverty is still widespread and many poor people do not have access to the appropriate technologies that could help them.

New science-led technologies present some specific challenges, including the perceptions of high cost, risk, complexity, and the lack of knowledge about what technologies are available. Yet new technologies also present new opportunities. Although older technologies are inevitably entrenched in existing systems of patents, production, and markets, there is an opportunity with new technologies to do things differently and see different outcomes.

How can science-led new technologies be made available to poor people? A report by the World Bank (2008, 1) concludes that "growth is a necessary, if not sufficient, condition for broader development, enlarging the scope for individuals to be productive and creative." Yet there is often an assumption that growth will simply trickle down toward what Prahalad and Hammond (2003) call "the bottom of the pyramid." There have been suggestions, however, that the provision of information and an accompanying workflow model may facilitate the adoption of new technologies in the Global South; Wilson (2007), for example, draws attention to the importance of knowledge networks and the need for respecting local knowledge while harnessing new knowledge.

From this starting point, in this chapter I articulate the case for doing things differently. I suggest that identifying the attributes of new technology that need changing is a first step toward the kinds of technology-related actions and policies that will ultimately contribute to the fulfillment of human needs. The example of nanotechnologies suggests that the key attributes to work on are power, price, promise, and poverty. I present a way of gaining insight into the kinds of actions that need to be embedded in international development efforts that aim to challenge poverty by the use of new technology. It is hoped that the approach I advocate will be useful for organizations engaging in dialogues with key stakeholders about the diffusion of new technologies.

As one small step toward a new way of doing things, I present a case study of "nanodialogues" I helped facilitate in Zimbabwe as a team leader at Practical Action. The dialogues connected the needs of poor people with scientists who are in the process of developing new nanotechnology applications to produce clean drinking water. The experience was then used to reflect on how to take steps beyond dialogue toward more engagement between scientists and relevant stakeholders. Dialogues held in Peru led to continued engagement as demonstrated at the Nanotechnologies in Peru website.[2] I then explore what is required for the ultimate step of moving beyond involvement and innovation into delivery of appropriate new technologies to poor communities through review of a public consultation surrounding the use of nanosensors for detecting arsenic in water throughout Nepal. I also suggest actions and policies that seek to ensure human needs are met with the assistance of technology. I propose a framework for enabling new technologies to address human needs, based on my experiences in Zimbabwe, Peru, and Nepal. My framework elaborates on the concepts of dialogue, engagement, and delivery that are further illustrated by two case studies.

## 8.2  Toward Pro-Poor Innovation

As discussed throughout this book, the predominant, traditional view of technology has been based on technological determinism, a position where, as Langdon Winner suggests, "the adoption of a particular technical system requires the creation of a particular set of social conditions as the operating environment of that system" (1986, 32). Such thinking leads to a "technological push" philosophy as embodied in the motto of the 1933 Chicago World Fair: "Science explores, technology executes, man conforms" (Fox 2002, 1). The worldview on which this philosophy is based is predominantly northern, where the power is vested in global enterprises with large research and development (R&D) budgets and where markets have consolidated to one or two main suppliers. An example of this is the domination of Microsoft in the market for personal computer software, where Windows has about a 90 percent market share. Practical Action views technology as not only meaning the hardware or technical infrastructure but also the information, knowledge, and skills that surround it and the capacity to organize and use them.

An alternative view of technology is clearly required. Robin Grove-White et al. (2000) have suggested that the development of technologies needs to be seen as a social process. This alternative view must recognize the role of the user (the southern poor) and the context provided by the cultural and political environment in which the user is based. The distinction being made here has been labeled by Edgerton (2006) as "technology in use," with the subsequent argument that the historical emphasis on technological innovation is misleading. Much technology that is in use in the world is adapted or imitated rather than innovated.

New science-led technologies, by definition, have not been employed or used before, so there is an opportunity to do things differently and see different outcomes. For example, the Project on Emerging Nanotechnologies[3] articulated a high ideal when recently stating that nanotechnology should be green from the outset (Maynard 2008).

New technology is often driven by a "push" model. Practical Action seeks to ensure that the adoption of new science-led technology is enabled by local demand and made in response to human needs. The fundamental underpinning of the new technological program is to support a shift from new technologies being market led toward enabling the fulfillment of human needs. For example, rather than nanotechnologies providing products for high-price markets (such as sunscreen), Practical Action seeks to enable nanotechnologies to provide for human needs such as clean water or low-cost sustainable energy. The case studies I present in this chapter are examples of this approach in practice. Although each case study is related to issues around the application of nanotechnologies, it is first important to understand some generic characteristics of new technologies.

## 8.3 Generic Characteristics of New Technology

Each new wave of technology increases the gap between rich and poor or, as Knell (2010, 139) puts it, "within each cycle, income distribution tends to

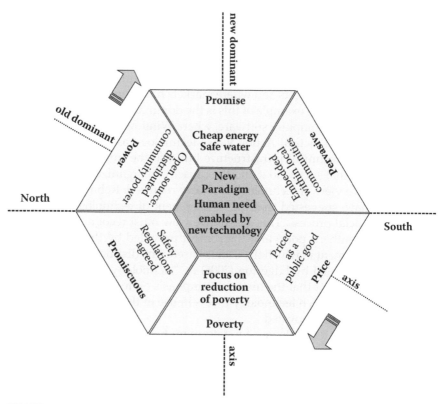

**FIGURE 8.1**
Seven-P framework of new technologies.

widen as new technologies create new financial opportunities." What do we know about the generic characteristics of new technologies that might help us to understand how to promote the application of nanotechnologies to benefit the poor? In this section I discuss the Seven P's of new technology that serve as a framework for our thinking: power, price, promise, poverty, pervasive, promiscuous utility, and paradigm (see Figure 8.1).

The *power* of a new technology stems from two sources—from the power held by the scientist or expert who "knows" more than the user, and the multinational corporations that have power in the market. This may be a power that is protected by intellectual property rights. An example of the power of scientists and multinationals to set the agenda for the exploitation of new technologies is the neglect of tropical diseases. Trouiller et al. (2002, 2188) found that, "of 1,393 new chemical entities marketed between 1975 and 1999, only 16 were for tropical diseases." However, the Global Health initiative provides a good example of coordinated action, which has led to many drugs being marketed to the Global South at a price close to cost.

The *price* at which new technologies enter the marketplace is often high. The price may be driven by a need to recoup the investment in the R&D processes that led to the launch of the new product. The market may also support premium prices, given that new technologies, when introduced, are often in short supply. In the Global South, new technologies such as information and communication technologies (ICTs) are often too expensive to adopt. Those that are introduced by donors as part of development projects are sometimes unsustainable when donor funding is withdrawn. High R&D costs can also act as a barrier for the Global South to start research activities.

The hype and *promise* surrounding a new technology create a level of expectations that is rarely fulfilled—such as the paperless office that was promised with the introduction of ICTs into office environments in the 1970s. In the context of development, ICTs were often introduced to share information with poor people. Yet the expectations of improving livelihoods via increasing knowledge have rarely been fulfilled. At the other extreme, one of the most successful technologies—the radio—was initially not thought to have much promise at all.

As previously mentioned, the *poverty* gap between rich and poor widens with each wave of new technology. For example, although the speed of diffusion of technology is increasing—it took seventy-four years for the telephone to reach 50 million people, but only four years for the Internet to do the same—there is dramatically unequal distribution in its diffusion (EIU 2004). Developing countries take a long time to build the capacity to innovate with new technologies (Fong 2009) and often lack appropriate social and organizational structures needed to exploit their potential (Avgerou 1999).

The word *pervasive* is often applied to a universal network of computers (Amor 2002). The nature of new technologies—for example, biotechnology, nanotechnology, and ICTs—tends to be much more pervasive than that of earlier generations of technologies, such as rail travel, which had impacts on defined geographical areas and economic systems. Many ICT devices are connected by wireless technologies and can connect across different platforms. Cairncross (1997) suggests that a major impact of ICTs is the "death of distance," as communications can be utilized seemingly anytime, anywhere. But this is a contested view as is evident by Disdier and Head (2008) and Carrère and Schiff (2005) who argue that distance remains an important determinant of trade.

New technologies typically have many uses. For example, a computer can be used to provide in-car navigation or to run the complex financial systems of a large business. So *promiscuous utility*, a term used by Amor (2002), denotes the varied uses of new technology. Conceptually, pervasive and promiscuous utilities are linked and that is why they are portrayed at opposite ends of the new technology web (see Figure 8.1). Unlike the other attributes in the framework, both pervasive and promiscuous are characteristics of the technology rather than social constructs.

The *paradigm* term was introduced into scientific literature by Kuhn (1961). The definition used here is an adaptation from Ritzer (1975, 4) who states that "a paradigm serves to define what should be studied, what questions should be asked, and what rules should be followed in interpreting the answers obtained. The paradigm is the broadest unit of consensus within a science and serves to differentiate one scientific community from another."

The paradigm is the way in which we frame technology. There are many different ways in which we can do this. One example is advanced by Leach and Scoones (2006) who argue that citizen engagement is vital to ensure that technology responds to the challenges of international development.

In the following case studies I illustrate a way forward based on stakeholder dialogues as the first part of the three-stage process of dialogue, engagement, and delivery. The first case study achieves dialogue and engagement and the second moves toward delivery.

## 8.4 Case Study 1: Nanodialogues in Zimbabwe and Peru

In 2006, researchers from the U.K. think tank Demos, the University of Lancaster, some of my colleagues from Practical Action, and I collaborated on a process designed to engage Zimbabwean community groups and scientists from both the North and South in debates about new nanotechnologies (Grimshaw et al. 2006).

Collectively referred to as the nanodialogues, the dialogue was one of four[4] experiments in public engagement with nanotechnologies, and was funded by the U.K. Office of Science and Technology's *Sciencewise* program. *Sciencewise* was created to foster interaction between scientists, government, and the public on impacts of science and technology, and is primarily focused on delivering benefits to the U.K. economy.

As has been outlined by others throughout this book, the potential benefits of the applications of nanotechnologies in the Global South are exciting. But the conversation linking the needs of people in the Global South to the resources and scientific knowledge of researchers around the world needs to be nurtured.

In the United Kingdom, the present awareness about nanotechnology is seen as an opportunity to have an earlier and more open debate about emerging technologies, so as to avoid the antagonism and distrust generated with genetically modified (GM) foods. The government is supporting the Royal Society and Royal Academy of Engineering's call for "a constructive and proactive debate about the future of nanotechnologies ... at a stage when it can inform key decisions about their development and before deeply entrenched or polarized positions appear" (Royal Society 2004, 82).

The nanodialogues were a set of opportunities for early public debate in parts of the Global South. The first of these dialogues aimed to engage communities in Zimbabwe in discussions about nanotechnology.

A study about the relevance of nanotechnology's application areas for poor people found the top three applications that could help the South were energy storage, production, and conversion; agricultural productivity enhancement; and water treatment (Salamanca-Buentello et al. 2005).

The Practical Action team conducted the nanodialogue over a two-week period in July 2006, involving local individuals, scientists from the North and South, and policymakers. We chose water treatment as a focus for our dialogue because, first, in development terms, it is a well-established priority, and second, because technology is at a stage where it may be able to make a significant contribution to filtration and decontamination (Hille et al. 2006). The Millennium Development Goal is to halve the proportion of people without sustainable access to safe drinking water and basic sanitation by 2015. Our dialogue sought to introduce the views and values of people for whom clean water is an everyday problem into debates about responses that might involve nanotechnology in some shape or form. By involving scientists who are engaging in leading research, the debates moved "upstream." One of Practical Action's hopes was for a sustained dialogue between scientists and end users that would enable new technology to deliver on human needs rather than be driven by market wants.

Approaches to providing water for poor communities have often been driven by either economics or technology. The economics route might typically center on the importance of regulations, institutions, and open markets, while the technology approach might focus on designing a water pump, filter system, or novel application of nanotechnology. Failure to address the issue of water provision might also be seen as a cultural, political, or managerial problem.

In recognition of these characteristics of the problem domain, Practical Action took a systemic approach, building upon our experience of engaging people in the Global South in debates about new technologies (see Rusike 2005). Figure 8.2 depicts the problem situation in the form of a more holistic picture. During the first day of the workshop, this rich picture was drawn by the organizers as a reflection of the problem presentation so as to convey relationships and connections much more clearly than prose.

The picture shows that there is a need to bridge the knowledge gaps between local and global scientists, a need to listen to local people and understand the context and dimensions of need, and a need for new business models to produce products that will provide for human needs such as clean water, rather than those driven by consumer wants. Areas of potential conflict are also illustrated; the main areas are the affordability of solutions and

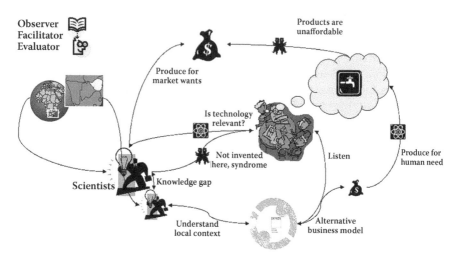

**FIGURE 8.2**
Holistic picture of the problem situation.

the "not invented here" syndrome, which can easily lead to a lack of owner-ship and adverse consequences for the sustainability of the technology.

One outcome of the meeting was a call for poor communities to be involved in debates about whether nanotechnologies can contribute to social and eco-nomic development. The way forward will need to take into account the risks and costs in addition to the opportunities for real benefits to poor people. As this dialogue has taken place before many products using nanotechnologies have become established in the market, the hope is that such early discus-sions with scientists will enable them to consider the needs of the poor. This might go some way to delivering public value from science (Grimshaw 2007).

On its own, dialogue and involvement will not deliver meaningful results without a process of innovation by the key stakeholders. Practical Action's work aims to bring those key stakeholders together and facilitate processes that allow community input to design, and later to test prototypes, to ensure technologies are appropriate. Building on our experience in Zimbabwe, in November 2007 we held a seminar in Lima, Peru. This was followed by a much smaller and more interactive workshop in April 2008, also in Lima. The main outcome of the workshop was a series of community-driven goals: to identify priority water problems that might be assisted by an application of nanotechnology; to establish a network of scientists involved in nanotech-nology in the Andean region of Latin America; and to develop a website to foster a community of interest about nanotechnology and water in Peru.

There is an ongoing engagement of key stakeholders focused on the prob-lems of water pollution around small mines in the Andean region of Peru. Typical among the problems is the pollution of water courses caused by mer-cury as a by-product of gold-mining activities.

## 8.5 Case Study 2: Arsenic Sensors in Nepal

The challenge in this case was to move beyond dialogue (Zimbabwe) and engagement (Peru) to delivery. The UN has estimated that around 1.4 million people are at risk from arsenic contamination in Nepal (UNICEF 2006). Testing of the 400,000 tube wells in the Terai—the lowest outer foothills of the Himalayas—is the first essential step. The current cost and accuracy of existing technologies presents a challenge to the important ongoing cycle of testing. Grimshaw and Beaumont (2007) recognized that, although much scientific effort had gone into the problem of filtering arsenic out of drinking water, much less effort had gone into developing cheap, reliable, and accurate arsenic sensors. Discussions with key stakeholders in Nepal confirmed that this was a key area where improved technology was needed.

An arsenic sensor workshop was held in May 2009 in Kathmandu, Nepal. It was attended by twenty-three key stakeholders, including local scientists, community members and policymakers, a scientist from the United Kingdom, NGO representatives, the Nepalese Department for Water Supply and Sanitation, and UNICEF. In Nepal water testing is organized centrally by the Department for Water Supply and Sanitation in collaboration with UNICEF. The approach developed was to undertake blanket testing of wells every year. Many issues arise from this approach. First, given new wells every year, the number of wells keeps changing. Currently the cost of a new well is around Nr2000 and this is within reach of people who want a water source closer to home. Ideally there needs to be testing of the well at the time of drilling. Second, due to seasonal and other factors, the levels of arsenic may vary with times of year and at different depths. To ensure the health of the population therefore requires more frequent testing, perhaps at least every six months. Third, there are wide variations in operating temperatures—up to 45°C—which presents a design challenge for the technology developer. Fourth, there is also an issue as to who is responsible for conducting the tests.

The workshop group responded to this challenge by exploring whether a sensing function could be embedded into filtering technology. Participants thought that this might make it more likely that people would change the filter unit when the effectiveness of the unit was low. UNICEF noted that test technology had to go through a testing protocol before it can be used in practice. It was therefore suggested that more detailed knowledge of that protocol might assist scientists in developing a suitable sensor.

A question arose as to how much detailed operational knowledge users would require before using the devices. For example, if it was a black box[5] approach, there might be no need to understand anything about nanotechnology in order to operate it; as one participant observed, "testing should be as easy as turning on TV and selecting a channel." It seemed a commonly held belief in discussions that the user should be a responsible person in the local community. Ideally the sensor should be able to detect not only arsenic but also other unwanted elements in the water, such as nitrates or biological contaminants.

In the final session of the workshop we had an open discussion about future directions. There was willingness, even eagerness, to make progress by working together to improve arsenic sensor technologies utilizing nanotechnology.

There was general support for an innovation process that involved all key stakeholders to develop technologies that met local people's needs. All stakeholders expressed a willingness to be involved and stressed the importance of building local capabilities. Recognition was also given to the need to test any prototype technology both in the laboratory and in the field in a responsible manner. The result was emphasis on the need for having public awareness and communications with local communities built into any project; understanding more about nanotechnologies and their potential; integration into other water awareness approaches; regular interaction with stakeholders; and questioning how technology can be made available and be reliable.

## 8.6 Technology as an Enabler

These case studies illustrate the role that stakeholders can play in the dialogue, engagement, and delivery processes. I now discuss the broader framework of technology as an enabler (as illustrated in Figure 8.3), one that suggests a feasible alternative to the usual technologically determinist, profit-driven motivations.

Practical Action's vision, as outlined in the first section of this chapter, is to enable new science-led technologies to deliver products that fulfill human needs rather than consumer wants. Rather than nanotechnologies providing products for high-price markets, we seek to enable nanotechnologies to provide for human needs such as clean water or low-cost sustainable energy.

I start by discussing the top part of Figure 8.3. Emerging and potentially disruptive technologies need to be monitored (Bower and Christensen 1995). The technologies that have the potential to disrupt existing patterns of technology adoption and forecasts are those that need to be identified as early as possible. Unfortunately, it is a characteristic of a disruptive technology that it is not foreseen and therefore not planned for. For example, mobile phones now have a larger market share in many African countries than do older, more-established landline technologies. This also illustrates the way in which new technologies can offer the potential for countries to leapfrog their expected development situation (Davison et al. 2000). Some of these new technologies may also be intermediate technologies, as described by Schumacher—for example, open source software (as discussed in Chapter 9; see also Grimshaw 2004, 2006). For the majority of new technologies that are science led, there needs to be a new way of enabling them to meet the needs of poor people.

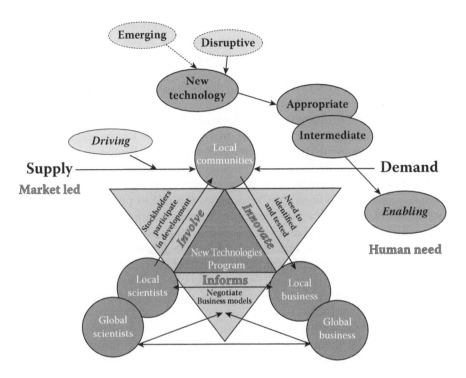

**FIGURE 8.3**
Technology as an enabler.

The lower part of Figure 8.3 shows the nature of interventions that need to be made to enable technology to fulfill Practical Action's vision. We have always started with local communities to identify their needs. This is shown in the right hand part of the triangle. Additionally there needs to be a connection made between local communities, local scientists, and global scientists. This is the "involve" part of the triangle in the model. The final part of the triangle illustrates the need to negotiate new business models that will enable the technologies to diffuse to people who have a need. This will require involvement of local and global business as well as scientists.

Practical Action's pro-poor innovation model raises a number of issues and questions. The first relates to how new technologies can deliver social and economic progress. Technology has consistently failed to meet the needs of the poor, with 2.7 billion people living on less than US$2 per day (UN 2006). According to Jeffrey Sachs, "the single most important reason why prosperity spread, and why it continues to spread is the transmission of technologies and the ideas underlying them" (2005, 41). Yet, as we discussed earlier, the national diffusion of technologies does not always lead to local economic growth and certainly does not lead to equitable distribution of the benefits of the technology. Based on his work at Practical Action, Sharad Rai (2007) has illustrated participatory innovation development through a case study in Nepal where he focused on the human, social, and economic factors that inhibit and promote pro-poor innovation. But is the scale of economic growth implicit in approaches such as this appropriate to a world challenged by climate change?

A further related concern arises from the problem of developing new business models or processes that support the development of science-led new technologies that fulfill real needs. Andrew Adwerah (2007) discusses the social entrepreneurship model based on experience in Kenya and finds that income can be generated and poverty reduced by combining the local provision of new, socially focused technologies, such as brick and oil presses and manual water pumps, with basic business practices to support individual livelihoods. Wilson (2007) discusses the role of public–private partnerships and stresses the importance of professional challenge with respect to creating a better world as a key motivator for private-sector participants. Also in support of new business models, the former head of the U.K's Sustainable Development Commission, Jonathon Porritt (2006), has argued that, to enable sustainable development, people need to understand market mechanisms and innovation processes and then work with key stakeholders to enable business models that will deliver on human needs rather than on consumer wants. With existing technologies, this becomes a challenge because the business models, including the supply chain logistics, are already well established. In the case of new technologies, there is a window of opportunity before products are released into the market to negotiate new business models. However, how do we move on from pilot projects? Many science-led new technologies, especially those using information communication technologies (ICTs), are tested using donor funding for limited time periods. For example, Young and Matthews (2007) discussed recent research from the U.K.-based Overseas Development Institute, suggesting there is a need for more systematic analysis of the impact of ICTs on poverty reduction in rural areas. The challenge is to capture the learning, adapt the business model, and implement sustainable change.

## 8.7  Policy Implications

The key message is that to achieve the vision of enabling new science-led technologies to deliver products that fulfill human needs rather than consumer wants we need to reengineer the market for new technologies. Put like this, it seems like an insurmountable obstacle. However, policy can play a role in making these innovations achievable. Key needs include collaboration rather than competition, because in this way the designers and innovators are connected to the users throughout the technology development process, thereby ensuring the continued involvement of scientists and local communities; ensuring that the assessment of the appropriateness of new technologies will take account of the risks and costs in addition to the opportunities for real benefits to poor people; that alternative business models, such as social entrepreneurship, are encouraged; that local communities should choose the technologies they wish to adopt, participating in the process of innovation; and that upstream dialogues between scientists and local communities are encouraged as a way of engaging scientists in the provision of needs-based development.

## 8.8 Conclusions

New technologies present many challenges to those concerned with how to reduce poverty in the world. The promise of many new technologies has been high, yet the ability to deliver sustainable change in the lives of poor people has been limited. At the same time, the very models and assumptions underpinning much of international development have been economic growth. There is a need to move away from the old paradigm, which is supply driven and delivers products to a market at a price so as to maximize profits for the owners of the intellectual capital. The arguments I have presented in this chapter amount to a case for using a new paradigm based on enabling choices to be made that fulfill the needs of people.

Implementing this new paradigm for technology development will not be easy, just as changing the culture of an organization is not easy. Identifying the attributes of new technology that need changing is a first step toward the kinds of actions and policies that will ultimately lead to the fulfillment of human needs being enabled by new technology. The example of nanotechnologies suggests that the key attributes to work on are power, price, promise, and poverty. In this chapter I have presented a way of gaining insight into the kinds of actions that need to be embedded in international development efforts that aim to challenge poverty through the use of new technologies. I hope the approach I have advocated will be useful for organizations that are engaging in dialogues with key stakeholders about the diffusion of new technologies.

## References

Adwerah, A. 2007. Social entrepreneurship in Kenya, *id21 insights* 68 (September): 5.
Amor, D. 2002. *Internet Future Strategies: How Pervasive Computing Services Will Change the World*. Saddle River, N.J.: Prentice-Hall.

Avgerou, C. 1999. How can IT enable economic growth in developing countries? *Information Technology for Development* 8(1): 15–28.

Bower, J.L. and Christensen, C.M. 1995. Disruptive technologies: Catching the wave. *Harvard Business Review* (January–February): 43–53

Cairncross, F. 1997. *The Death of Distance: How the Communications Revolution Will Change Our Lives*. Boston, Mass.: Harvard Business School Press.

Carrère, C. and Schiff, M. 2005. On the geography of trade: Distance is alive and well. *Revue Economique* 56(6): 1249–74.

Checkland, P.B. and Scholes, J. 1990. *Soft Systems Methodology in Action*. Chichester: John Wiley.

Danowitz, A.K., Nassef, Y., and Goodman, S.E. 1995. Cyberspace across the Sahara: Computing in North Africa. *Communications of the ACM* 38(12): 23–28.

Davison, R., Vogel, D., Harris, R., and Jones, N. 2000. Technology leapfrogging in developing countries—An inevitable luxury? *Electronic Journal on Information Systems in Developing Countries* 1(5): 1–10.

Disdier, A. and Head, K. 2008. The puzzling persistence of the distance effect on bilateral trade. *Review of Economics and Statistics* 90(1): 37–48.

Economist Intelligence Unit (EIU). 2004. *Reaping the benefits of ICT: Europe's productivity challenge*, http://graphics.eiu.com/files/ad_pdfs/MICROSOFT_FINAL.pdf (accessed 12 November 2008).

Edgerton, D. 2006. *The Shock of the Old: Technology and Global History since 1900*. London: Profile Books.

ETC Group. 2003. *The Big Down: From Genomes to Atom*. ETC Group. http://www.etcgroup.org (accessed 7 September 2003).

Fong, W.M.L. 2009. Digital divide: The case of developing countries. *Issues in Informing Science and Information Technology* 6:471–78.

Fox, N. 2002. *Against the Machine: The Hidden Luddite Tradition in Literature Art and Individual Lives*. Washington, D.C.: Shearwater Books.

Grimshaw, David J. 2004. *The Intermediate Technology of the Information Age*. New Technology Briefing Paper 1, 2004, Practical Action: Rugby.

Grimshaw, D.J. 2006. FOSS as an intermediate technology for the information age. *Proceedings of the FOSSFA Conference*. Nairobi, Kenya (January).

Grimshaw, D.J. 2007. Nano-dialogues, helping scientists to meet poor people's needs. *id21 insights* 68 (September): 3.

Grimshaw, D.J. and Beaumont, A.E. 2007. *How Can New Science-Led Technologies Contribute to Mitigation of and Adaptation to Arsenic in the Water of Bangladesh*. Science for Humanity, unpublished paper.

Grimshaw, D.J., Stilgoe, J., and Gudza, L.D. 2009. How can new technologies fulfill the needs of developing countries? In *Nanotechnology Applications—Solutions for Improving Water*, edited by M. Diallo, J. Duncan, N. Savage, A. Street, and R. Sustich, 535–50. Norwich, N.Y.: William Andrew.

Grove-White, R., Macnaghten, P., and Wynne, B. 2000. *Wising Up: The Public and New Technologies*. Research Report, Centre for the Study of Environmental Change, Lancaster University.

Hazeltine, B. and Bull, C. 1999. *Appropriate Technology: Tools, Choices, and Implications*. New York: Academic Press, 3.

Hille, T., Munasinghe, M., Hlope, M., and Deraniyagala, Y. 2006. *Nanotechnology, Water and Development*. Washington, D.C.: Meridian Institute. Available from http://www.merid.org/nano/waterpaper (accessed 10 January 2007).

Knell, M. 2010. Nanotechnology and the sixth technological revolution. In *Nanotechnology and the Challenges of Equity, Equality and Development*, edited by S.E. Cozzens and J.M. Wetmore, 127–43. Dordrecht: Springer.

Kuhn, T.S. 1961. *The Structure of Scientific Revolution*. Chicago: University of Chicago Press.

Leach, M. and Scoones, I. 2006. *The Slow Race: Making Technology Work for the Poor*. London: Demos.

Maynard, A. 2008. Quoted in Bloom, J. 2008. Tiny particles hit the big time. *Guardian*, 4 September, p. 6.

McConnell, M. 1988. *Challenger: A Major Malfunction*. London: Unwin Hyman.

Porritt, J. 2006. *Capitalism as if the World Matters*. London: Earthscan.

Prahalad, C.K. and Hammond, A. 2003. Serve the world's poor profitably. *Harvard Business Review* (September).

Rai, S. 2007. Supporting local innovation in Nepal. *id21 insights* 68 (September): 3.

Ritzer, G. 1975. *Sociology: A Multiple Paradigm Science*. Boston: Allyn and Bacon.

Royal Society and Royal Academy of Engineering. 2004. *Nanoscience and Nanotechnologies: Opportunities and Uncertainties*. London: Royal Society. www.royalsoc.ac.uk (accessed 20 July 2004).

Rusike, E. 2005. Exploring food and farming futures in Zimbabwe: A citizens' jury and scenario workshop experiment. In *Science and Citizens*, edited by M. Leach, I. Scoones, and B. Wynne, 249–55. London: Zed Books.

Sachs, J.D. 2005. *The End of Poverty*. London: Penguin Books

Salamanca-Buentello, F., Persad, D.L., Court, E.B., Martin, D.K., Daar, A.S., and Singer, P.A. 2005. Nanotechnology and the developing world. *PLoS Medicine* 2(4): 300–303.

Schumacher, E.F. 1973. *Small Is Beautiful: A Study of Economics as if People Mattered*. London: Abacus.

Schumacher, E.F. 1979. *Good Work*. London: Jonathan Cape.

Scoones, I. 2006. *Science, Agriculture and the Politics of Policy: The Case of Biotechnology in India*. Delhi: Orient Longman.

Stilgoe, J. 2007. *Nanodialogues: Experiments in Public Engagement with Science*. London: Demos.

Trouiller, P., Olliaro, P., Torreele, E., Orbinski, J., Laing, R., and Ford, N. 2002. Drug development for neglected diseases: A deficient market and a public-health policy failure. *Lancet* 359:2188–94.

UN. 2006. *Fast Facts: The Faces of Poverty*. UN Millennium Project. Available at http://www.unmillenniumproject.org/resources/fastfacts_e.htm (accessed 5 May 2011).

UNICEF. 2006. *Diluting the Pain of Arsenic Poisoning in Nepal*. http://www.unicef.org/infobycountry/nepal_35975.html (accessed 29 July 2011).

Wilson, G.A. 2007. Threats, opportunities and incentives for pro-poor innovation. *id21 insights* 68 (September): 6.

Winner, L. 1986. *The Whale and the Reactor*. Chicago: University of Chicago Press.

World Bank. 2008. *The Growth Report: Strategies for Sustained Growth and Inclusive Development*. Commission on Growth and Development, http://cgd.s3.amazonaws.com/GrowthReportComplete.pdf (accessed 22 September 2008).

Young, J. and Matthews, P. 2007. Enhancing rural livelihoods: The role of ICTs. *id21 insights* 68 (September): 5.

## Endnotes

1. The notion of intermediate technology has now largely been subsumed under the appropriate technology movement that is generally recognized as "encompassing technological choice and application that is small scale, labor intensive, energy efficient, environmentally sound and locally controlled" (Hazeltine and Bull 1999, 3).

2. Site developed by Practical Action, Latin America—http://www.nanotecnologia.com.pe.

3. The Project on Emerging Nanotechnologies was established in April 2005 as a partnership between the Woodrow Wilson International Center for Scholars and the Pew Charitable Trusts. The project is dedicated to helping ensure that as nanotechnologies advance, possible risks are minimized, public and consumer engagement remains strong, and the potential benefits of these new technologies are realized (http://www.nanotechproject.org/about/mission/).

4. The other three were all conducted in the United Kingdom. Further details can be found in Stilgoe (2007).

5. In science and engineering, a black box is a device, system, or object that can be viewed solely in terms of its input, output, and transfer characteristics without any knowledge of its internal workings, that is, its implementation is opaque (black) (http://en.wikipedia.org/wiki/Black_box).

# 9

## Open Source Appropriate Nanotechnology

Usman Mushtaq and Joshua M. Pearce

### CONTENTS

## 9.1 Introduction

The question of who benefits from nanotechnology is worth asking, especially if nanotechnology is to serve as a means for addressing aspects of global poverty and its associated challenges. After all, the lack of clean water and other resources in small villages around the world does not occur because the technology is unavailable but because the technology is out of the financial reach of the marginalized community. Even if the technology is financially accessible, it may remain operationally inaccessible. For example, the residents of a small village may collect taxes through a community Water Board to buy a nanofiltration membrane sterilization system to have clean water, but they may still not have the operational and technical know-how to set up and maintain the system because that knowledge is proprietary. Similarly, while nanotechnology may hold options for addressing challenges of the marginalized everywhere, the benefits of nanotechnology may not be shared equally, thus propagating the same inequities in access to resources that currently exist or even further disfranchising the poor (Bruns 2004). Nanotechnology knowledge is currently locked away in property systems, making it inaccessible for those who need it the most. How-to knowledge, as well as actual nanotechnologies, are also locked away in the proprietary system. This means that even if an alternate process to create a nanotechnology is developed, the inventors would not be legally able to use or produce that technology. This form of locking away knowledge is clearly against the best interests of the poor, as it tends to concentrate knowledge into the hands of wealthy corporations—as knowledge that once belonged to the intellectual commons is being privatized for the purposes of profit making even though that knowledge is often produced using public funding (ETC Group 2005). This restricts public research and service organizations from spreading the fruits of nanotechnology research to the general public (Sampat 2003). In this way, the gap between the ability of experts and the general public to access nanotechnology knowledge increases physically, intellectually, and financially. Nanotechnology cannot simply provide ideas for addressing poverty without taking into account the social context in which it is introduced and the political process through which it is developed (Foladori and Invernizzi 2005).

If issues of who has access to knowledge and who has the rights to develop that knowledge are addressed, then it may be possible for nanotechnology to benefit everyone—including the poor (Bruns 2003). In this chapter we propose the application of the open source concept from software development to nanotechnology as a method to open up nanotechnology for everyone. Already, the field of appropriate technology has benefited from the application of open source principles. Open source appropriate technology (OSAT) has been proposed as a method to increase access to appropriate technologies for the marginalized and also to increase the rate of innovation and localization for those technologies (Pearce and Mushtaq 2009; Buitenhuis, Zelenika, and Pearce 2010). As seen in the open source software movement in products like Linux,[1] participation can be extended to previously marginalized groups if knowledge is opened up. This open source concept can be applied not only to the software of nanotechnology (modeling software) but also to the hardware of nanotechnology (novel nanostructures). Opening up the development of nanotechnology will not only help innovation by introducing more perspectives to the development (Raymond 1999) but will also bring in new communities to provide feedback and testing (Butler et al. 2007). In this chapter we explore why we believe nanotechnology should be open sourced and how the open sourcing of nanotechnology could help drive equitable and sustainable futures.

## 9.2 What Does Nanotechnology Have to Do with Patents?

### 9.2.1 Life in the Anti-Commons

The currently framed intellectual property (IP) system, under which nanotechnology is largely produced, must be understood to appreciate the barriers that are preventing nanotechnology from assisting in equitable sustainable development. The concept of intellectual property arose during the twelfth and thirteenth centuries among the medieval urban guilds of artisans and craftsmen (Long 1991). Seeking to keep the commercial value of their products, medieval guilds worked within their cities to create patent laws. By the fifteenth century, guild members in European cities like Venice were enjoying the full protection of patents on their intellectual products. These early craftsmen realized that the value of their original and unique work had to be protected; otherwise, they would not gain social or economic credit for their work.

This early IP system gradually evolved into a complex set of nationally and internationally administrated laws and concepts. The World Intellectual Property Organization (WIPO) defines intellectual property as the "legal rights which result from intellectual activity in the industrial, scientific, literary and artistic fields" (2004, 3). The goal of the modern IP system is to safeguard the products of industrial, scientific, literary, and artistic labor by granting the creators a time-limited right to control the use of their creations. This means that any processes or products that are unique and original can be removed from the public domain by the creator. If that information is no longer in the public domain, one can only access the information through the creator. This ensures that creators have a motivation to innovate as they can take advantage of their creations by supplying the market with a product not available elsewhere. However, this can too easily lead to a situation where the creator obtains a permanent monopoly and restricts innovation and creation in others. For this reason, a time limit on modern IP protection allows for the eventual release of protected information into the public domain, recognizing the need to disseminate innovation.

The intellectual property system has been criticized from a wide variety of viewpoints. First, some have accused pharmaceutical companies of biopiracy, locking knowledge about indigenous fauna and flora away from people in the Global South (Shiva 1997; Mgbeoji 2006; Robinson 2010). In cases like this, patents are seen as a new form of colonialism—a tool for the removal of indigenous knowledge from those who have "owned" it for centuries. Meanwhile, others have argued that, in an area like software, patents limit innovation by locking away basic and obvious algorithms (Garfinkel, Stallman, and Kapor 1999). Much like musical notes for symphonies, algorithms are the building blocks of any complicated software program. However, the patenting of these algorithms has led to locking the basic means of coding away from common use. This has led software companies to give up on features in their programs that infringe on patent rights or even scrub development of a software program because it cannot proceed without using a basic algorithm that has already been patented. In biomedical research, patents have limited innovation by restricting, through exorbitant access costs, the use of building-block technologies that may be used in downstream research and development (R&D) (Heller and Eisenberg 1998). An innovator may even need to secure multiple patents from various patent holders before proceeding, which clearly delays innovation. Academia, which has always had a vibrant culture of information exchange, is also being affected by the privatization of knowledge. Since the 1980s, American universities have been moving toward patenting not only their applied research but also their research tools, especially in fields such as biotechnology and molecular biology. This has caused the rise of transaction costs for information exchange in academia, as universities must now deal with licensing schemes, royalty payments, and administrative complexity (Mowery et al. 2001). Obviously, higher transaction costs limit innovation as academics share less knowledge.

In addition to limiting knowledge sharing and innovation, IP systems lead to knowledge commodification and a self-perpetuating narrowing of knowledge access. Due to the high transaction costs associated with obtaining patent rights, only corporate entities can access and make use of the knowledge within IP systems. To recoup their losses in obtaining property rights, these corporate entities must commodify knowledge, thereby making it inaccessible to almost everyone—but especially the already marginalized (Baber 2001). In South Africa before 1998, patent protections made drugs for HIV/AIDS too expensive for the average person. This forced the South African government to circumvent international patent law just to provide life-saving drugs for their people—to which pharmaceutical companies responded by trying to sue the South African government for patent infringement. After an avalanche of negative publicity, however, the companies dropped their lawsuit in 2001, giving victory to the government (Barnard 2002). While there is enough of a market for AIDS drugs in the Global North such that pharmaceutical companies will continue to develop those drugs, at the moment, the same cannot be said of drugs that have limited potential in terms of economic returns, irrespective of their social value in the Global South (Shantharam 2005).

There are also many inefficiencies within the current IP protection system, often resulting from "overpatenting." For example, despite the increasing rate of patent applications, many patents developed by companies are not even used. This might happen if companies are focusing on a different industry from the patent at the time, if there is a lack of resources to take advantage of the patent, or when patents are gained during business acquisitions (Chesbrough 2006). All of these IP problems are unfortunately being repeated in the field of nanotechnology.

As of 2008, the U.S. Patent and Trademark Office (USPTO) has granted over 7,406 nanotechnology patents, with the number of nanotechnology patents it has granted having increased exponentially over the past thirty years (Chen et al. 2008). Despite this phenomenon, the current system of patents is ill suited for exploring the potential of nanotechnology because it limits innovation for many of the same reasons it limits innovation in other fields (Vaidhyanathan 2006). The patenting of basic nanotechnologies is leading to higher transaction costs, locking away obvious or basic knowledge and removing knowledge from the public domain of the Global South (Schummer 2007).

In addition to these disadvantages, the uniqueness of nanotechnology (Burgi and Pradeep 2006) makes patenting ill suited for several more reasons. Since most firms do not have viable commercial applications for their nanotechnologies, they must prove their business models by showing the number of patents they hold, even if those patents may never have any viable commercial applications. The result is that, in the field of nanotechnology, technology and business interests are leading science as opposed to the other way around. A large number of these nanotechnology patents present new knowledge about basic science in the quantum field, which therefore raises questions about ownership of science (Einsiedel and Goldenberg 2004). By holding these patents, firms are essentially locking away information about the building blocks of nanotechnology that could spur innovation in other sectors (Vaidhyanathan 2006). What would have happened to our understanding of the universe if Einstein, who once worked at a patent office, or a private firm had patented the theory of relativity or the photoelectric effect? In contrast to other emerging fields at similar stages in their emergence, nanotechnology is faced with a unique situation in which the basic science of the technology is patented. While technologies in other fields had time to develop due to a lack of patenting or legal restrictions on the patenting of basic science, nanotechnologies are locked away in the private sphere, limiting the sharing of knowledge in the field (Lemley 2005). This will limit downstream innovation as innovators will not be able to gain access to basic knowledge to develop more complex technologies or the transaction costs of such access will be too high. However, we envision a more equitable alternative for nanotechnology—open source.

### 9.2.2 Open Source As an Alternative

The open source movement has its roots in the software world. The term *open source* emerged during a strategy session between several hackers of the early open software movement (Bretthauer 2002). The term was used to describe software that allowed users to access and change the source code of a software program. Eric Raymond, in *The Cathedral and the Bazaar* (2001), argued that open source was a fundamentally new way to create and design technology by relying on the eyeballs of the many instead of the minds of the few. Using the cathedral and bazaar analogy, Raymond claimed that open source—drawing from the rich, nonhierarchical, gift-based culture of hackers—was similar to the bazaar where everyone could access and contribute equally in a participatory manner. Open source treats users as developers by encouraging contribution, recognizing good work through peer approval, and propagating improved code (Weber 2001). In open source development, there is no distinct line between users and developers, and they are often one and the same. This is opposed to the closed, industrial, and proprietary approach of "the cathedral," where a few programmers in some form of hierarchy worked in seclusion from their user community.

Linux is perhaps the best example of an open source project created by the masses. Linus Torvalds originally developed Linux in 1994 as a little hacking project to replace a program called Minix, which was a teaching tool in computer science courses. He released the source code to everyone. The hacker and the software developer community at large immediately took a liking to Linux. The program went from a little side project to a full PC-based operating system to which more than 3,000 developers across ninety countries and five continents contributed. In the first few years of its development, more than 15,000 people submitted code or feedback to the Linux community. Linux went from a few hundred to several million lines of code. Despite its rapid growth and large developer community, the reliability and quality of this operating system rank very highly (Moon and Sproull 2000). This coincides with Raymond's claim that "given enough eyeballs, all bugs are shallow" (Raymond 2001, 30).

Today, various versions of Linux exist, driven by user-developers who were not content with just one version of Linux. Although the basics of open source have remained the same since the start of the movement, the current definition of open source has been expanded to include such criteria as free redistribution rights and no discrimination against people or groups in accessing source code (Open Source Initiative 2011). The open source movement has continued to grow and thrive, as is evident by the myriad open source software applications available, begging the question: What motivates people to contribute to open source projects? The open source community runs on a gift economy, which rewards contributors through a process of peer review. If the contributions of an individual are of merit, the contributor will have a greater say in the decision-making process of design. This type of reward is key to the continued success of the open source movement as it provides motivation for users to become excellent developers and frequent contributors (Raymond 1998). This gift culture actually has similarities to the research culture found in academia, where peer review and peer recognition drive excellence and contribution (Bergquist and Ljungberg 2001). In both cultures, recognition of contributions determines social status, so contributors are motivated to add to the knowledge of their community in meaningful ways. In contrast to the IP system, in an open source framework it is actually self-serving for an individual or group to share.

Due to the success of the open source software movement, the concept behind open source has spread beyond the software community. Open source has been theorized to have application in areas such as appropriate technology (Pearce and Mushtaq 2009), design (Vallance, Kiani, and Nayfeh 2001), grassroots media (Gillmor 2006), education (Long 2002), culture (Rosenzweig 2006), and even medicine (Maurer, Rai, and Sali 2004). This has resulted in initiatives like Science Commons, Biobricks, OpenWetWare, and Appropedia.[2] The latter, for example, is a wiki (collaborative website) that allows designs of appropriate technology to be open and accessible to all. Developers can post directions on creating an appropriate technology and users can access those directions and even suggest improvements.

In light of this history, we propose the open source design of nanotechnology as an alternative to the IP system. There is a raft of associated advantages that we shall now explore. From the outset, open-sourcing nano design would help to overcome the limitations of the IP system discussed above by reducing the potential for monopolies, increasing the speed of innovation through collaborative production, and by making knowledge open and accessible to a larger community (Thakur 2008). If knowledge is freely and openly available to all, monopolies are more difficult to create and there would be fewer roadblocks and lower transaction costs to innovation, as knowledge would not be privatized. Potentially, new communities of research may emerge where everyone has the same access and rights to knowledge and contributes back to the community with new knowledge. Eventually, the open source model of nanotechnology would allow for users of nanotechnology to also become developers of nanotechnology, thereby enlarging the community of contributors. Of course, as with any open technology, lay developers will not contribute to it until it can be shown to be relevant to their concerns. This is especially important if nanotechnology is to be opened up to marginalized communities. Therefore, designers of nanotechnology should address the real needs—such as those that deal with water, health, poverty, or food security—of people.

The enlargement of the nanotechnology developer community through the addition of user-developers would create a greater drive toward application because of the introduction of more user "eyeballs" (Bruns 2001). Of greatest relevance to our chapter, no longer just driven by profit or corporate interests, the nanotechnology development community could actually develop nanotechnologies that are geared to human needs, instead of aiming to just patent the newest slice of knowledge that offers the greatest financial return. However, developing relevant nanotechnologies would entail being able to localize technologies to a particular context. Fortunately, the user-centric focus of open source nanotechnology would allow for customization at the user level, which would ultimately lead to reduced costs, less dependence on a single supplier or information source, and a reduced risk of obsolescence (Bruns 2001). Such local customization could also add to the innovation of knowledge in a community.

Open-sourcing nanotechnology would also help lower its potential risks. Since the open source community would be responsible for the development of any open source nanotechnology, there would be greater oversight of the project as the community would be responsible for making sure unsafe technologies were not released to the public. The open project would also have greater transparency than the development process for a closed technology, since it would allow for external monitoring. Anyone could have access to the documentation and source files of an open nanotechnology project.

Open source nanotechnology would be especially helpful for marginalized communities throughout the world. In the current IP model, companies must invest in R&D or pay the transaction costs associated with obtaining IP rights. In an open source model, companies would have open access to knowledge in the nanotechnology community and could rely on user-developers for R&D. Since companies would no longer have to recoup the losses associated with transaction costs or pay into R&D, research into nanotechnologies that do not yield large profit margins could be undertaken. This would help create nanotechnology responses to areas that are not commercially feasible but socially just, whether that is the provision of cheap medicine in the Global South or water filtration devices for poor rural communities. In addition, the cost of entry for businesses into the nanotechnology market would also be lowered, opening up opportunities for commercial innovation to be more globally widespread (Thakur 2008). Not only could companies then enter the nanotechnology market but so could lay user-developers. This broadening of stakeholders in nanotechnology communities would do much to fight against the privatization of knowledge in the field of nanotechnology, which threatens to marginalize communities by denying them self-determination.

## 9.3  Open Nanotechnology: To Boldly Go Where No One Has Gone Before

Due to its interdisciplinary nature, the field of nanotechnology is a combination of information (for example, chemical formulas), software (for example, modeling tools), and hardware (for example, the atomic force microscope). In addition, there are actual nanotechnologies and then there are the tools to create or manipulate those technologies. For example, carbon nanotubes are specific devices consisting of information and hardware, but a scanning electron microscope is a tool consisting of software and hardware used in the field of nanotechnology. The open source concept can be applied to the field of nanotechnology in all of these areas.

### 9.3.1 Open Information

Both technical and nontechnical information on nanotechnology should be openly available for all to share and use so it can be reproduced by another party. For this to occur, it is crucial that the directions to re-create a technology be detailed enough for reproduction, otherwise that technology cannot be truly open source. However, information sharing should not just be limited to nanotechnology design. Information about potential materials, the implications of nanotechnology, and novel applications should also be made open to the public. This will allow information to be localized for a specific context—just as Wikipedia has been localized for different languages (Ortega, Gonzales-Barahona, and Robles 2008).

Software platforms such as NanoHub.org already exist for members of the nanotechnology community to share information (Klimeck et al. 2008). As of March 2011, NanoHub.org had 172,225 users accessing and sharing open source content on nanotechnology. The content ranges from simulation tools to education material and uses a user-rating system to review submitted content for appropriateness, completeness, and quality.

### 9.3.2 Software

Software in the field of nanotechnology ranges from image rendering and microscope control software to molecular modeling tools. Just like other software applications, the source code for these applications can be made open and available to all. Not only would this allow people to access basic tools to become user-developers, it would but also increase the speed of innovation. Some molecular modeling tools have already been made open by releasing the source code, such as OpenRasMol—a molecular graphics program that allows for the visualization of proteins, nucleic acids, and other small molecules. Giving users the opportunity to display, show, and share models of molecules upon which nanotechnology could be modeled, the software has been particularly well received by the science community. Several derivatives of OpenRasMol have now been released and have consequently seen heavy public use in the science community (Bernstein 2000).

### 9.3.3 Hardware

Faced by inaccessible proprietary hardware, nanotechnology hobbyists and innovators have created home-manufactured open source hardware for working with nanotechnology, which is useful as the materials and equipment are low cost and publicly available (Bruns 2001). This is not a novel approach by any means, since the open hardware initiative has already been working toward making items such as circuit boards more openly accessible. However, the open hardware concept should be further expanded, following recent developments in the nanotechnology space. One example is the scanning tunneling microscope (STM) produced by the SXM Team, which can be used to examine and build molecular structures (SXM Team 2011). Using low-cost materials and equipment (when compared to commercial STMs) and using publicly available designs, this group's STM can be produced for under €1,000—in contrast to commercial STMs that can cost up to €100,000.

### 9.3.4 Open Standards and Regulations

To enforce copyright restrictions and limit their liability for nonfunctional software, early software developers created custom licenses that were contracts between the user and developer. These expensive contracts were typically created by the legal representatives of both parties for each piece of software. However, as software products became more ubiquitous and the cost of these contracts became prohibitive, developers adopted the End User License Agreement (EULA) as the primary form of licensing. Such agreements treat the software as a copyrighted product to which the user has only certain rights. Unlike custom licenses, EULAs can be applied to every piece of software without much modification, allowing for high-volume distribution of software.

Licenses are just as important in the world of open source software. However, contrary to proprietary software, early open source software did not include any licenses because there was no need to enforce copyright restrictions. Released into the public domain, software was free to be redistributed and modified as users wished. This allowed some users to repackage modified open source software as proprietary software. Software that had once been in the commons became private property. At the same time, the lack of open source contracts encouraged companies like IBM that worked on common software projects like UNIX to assert their intellectual property rights on common software (Lerner and Tirole 2002). To prevent this movement of software from the commons, Richard Stallman, a Massachusetts Institute of Technology programmer and supporter of open source, called for developers to release their software under the GNU license rather than into the public domain (Lerner and Tirole 2005).

The GNU license required that a program's source code be freely available to and modifiable by all. This ensured that programs released under the GNU license could not become proprietary. In addition, the GNU license prevented the commercialization of software by requiring that derivative works be subject to the same license conditions as the original program and that open and closed software could not be mixed under the license (Free Software Foundation 2007). While the GNU license was widely used in the 1980s, alternatives to the GNU license were created in the 1990s to overcome some of its restrictions, like the mixing of open and closed code (Lerner and Tirole 2002). Today, a wide variety of open source software licenses exist that fall into three major categories: unrestrictive, restrictive, and highly restrictive (Lerner and Tirole 2005). While highly restrictive licenses do not allow

**TABLE 9.1**

Comparison of Open Source License Classes

| | Unrestrictive (Examples: BSD, X11, MIT, Python) | Restrictive (Examples: MPL, LGPL, IBM, Apple) | Highly Restrictive (Example: GPL) |
|---|---|---|---|
| Can be relicensed? | No | No | No |
| Can distribute derived works without disclosing modifications? | Yes | No | No |
| Can incorporate in a combined work with closed source files? | Yes | Yes | No |

*Source:* Adapted from Bruns, B., *Nanotechnology* 12, 3, 2001.

for any mixing of open and closed code or any transfer of open code to closed code, unrestrictive licenses allow for the commercialization of code. Table 9.1 shows the major differences in the three classes of open source software licenses. In the case of unrestrictive licenses, the line between open and closed software licenses is close, but distinct.

However, open source licenses are not limited to software; they have spread to documentation, art, hardware, and academic journals. These licenses have the same major characteristics as open source software licenses—namely, public availability of the source code and the right to free redistribution. Some of these licenses could easily be adapted for open source nanotechnologies (Bruns 2001). An open nanotechnology license would allow for the modification and distribution of open source nanotechnologies in a responsible manner. The open hardware license would be a great basis as a license for open source nanotechnology due to its focus on physical artifacts (TAPR 2007). However, it would still need to be adapted to address the software or information component of open source nanotechnology. In addition, the new license would have to allow for the mixing of open and closed licenses, since a great deal of nanotechnology research has already been patented.

A novel strategy to deal with these patents would be to create patent pools through agreements between multiple patent holders to share each other's patents freely (Shapiro 2000). These patent pools would greatly reduce the transaction costs of gaining the rights to all the patents for a single product. For example, the Manufacturers Aircraft Association was formed in 1917 by various aerospace patent holders to simplify the process of building airplanes. Similarly, all the patents for a particular piece of nanotechnology could be placed in a patent pool to simplify the process of building that technology. Patent pools could work well with less restrictive or unrestrictive open nanotechnology licenses since those licenses would allow for the mixing of open and closed technologies. Another approach to patent pools is articulated by Vandana Shiva and Radha Holla-Bhar, who propose the use of collective patents as a way to recognize local common rights over community knowledge (Shiva and Holla-Bhar 1997). Through collective patents, a group of traditional or indigenous knowledge-holders can choose to hold a patent or set of patents collectively for their community as opposed to private corporations. This allows those communities to exercise control over their own knowledge. Similarly, lay developers of open nanotechnology could hold collective patents.

Another licensing strategy for nanotechnology could be hybrid licensing models—a combination of closed and open licenses. A program could be released under a closed commercial license at first but then later released as open source. For example, this strategy has been employed by a PC game, *FreeSpace 2*, which was initially released as a commercial product but the source code was later released to the wider community. The project has been so successful that several versions of the game based on the original source code have been created.[3] Using the hybrid model, nanotechnology designs could be developed under a closed license and then released to the public.

In short, an open source nanotechnology license would have to deal with both hardware and software. It would have to work well with proprietary software or patented technology so that the license was not too restrictive. To be worthwhile, this license would also have to encourage the creation of patent pools to decrease overall transaction costs if innovators were working with both open and patented nanotechnology. Of course, open licenses are not the only way to create open nanotechnology. Like developers earlier in the open source software movement, open nanotechnology designers could just publish all material into the public domain (Bruns 2001). In theory, this would ensure that nanotechnology knowledge released into the public domain could not be patented. However, history has shown that the release of knowledge into the public domain does not prevent that knowledge from being patented if enough modifications are made to it for it to then be considered novel. If no licensing scheme is set up, open nanotechnology could move to the anti-commons over time and limit future innovation.

One problem with open source licenses remains the question of whether the licenses can be legally enforced (McGowan 2001). Since open source products can become the products of multiple authors under multiple licenses— and each with its own level of restrictiveness—there would be difficulties in determining authors and their legally recognized contributions. There would also be questions from the contract law and property rights domain, since open licenses do not strictly adhere to either type of legal domain. Currently, open licenses would have to give way to the Copyright Act and its intellectual property rights in a court of law (McGowan 2001).

### 9.3.5 Appropriate Business Models

American software activist Richard Stallman is fond of stating that open software is "free as in free speech, not free beer" (Lessig 2004, xiv). Open source software can be profitable (although "profitable" does not necessarily have to equate to "for profit"; for more, see Chapter 13). To prove Stallman's point, a wide variety of commercial open source software products like Red Hat and VA Linux are available to users (Lerner and Tirole 2002). Even companies that have typically produced proprietary software are moving toward making some of their products open source. They are adopting hybrid models of production, where open and closed parts exist together or where closed parts eventually become open. Companies like Microsoft and IBM have launched projects to develop and use open source software (Lerner and Tirole 2002). Clearly, the for-profit business model is (financially) compatible with open source software. Can it be compatible with open source nanotechnology?

Arduino, an Italian technology firm, provides a business model appropriate for open nanotechnology. Arduino designs open hardware circuit boards. All data relating to this company's hardware—from software to circuit schematics—is posted on the Internet. Thus, users can access all the knowledge needed to build their own circuit boards. It is financeable, as users have been willing to pay Arduino for the cost of materials and labor, even if they have access to all the information needed to create a product. Arduino staff are also hired by companies and individuals for their expertise in building open hardware circuit boards. In this case, users are paying for the expertise and attention of the Arduino staff on custom work. While Arduino may not be creating capital through knowledge, it is still a sustainable enterprise by creating capital from staff expertise and manufacturing capacity (Thompson 2008). Nanotechnology firms that open up their knowledge resources could use a similar business model based upon provision of manufacturing capacity and staff expertise. These companies could offer customization work as a service while their product is openly available in a standard form. Other services they could provide include technical support for their products, consulting with clients and third parties to create integrated approaches to various challenges, or manuals and classes for clients to learn the ins and outs of a technology (Behlendorf 1999).

Companies using open nanotechnology could even create profit from providing access to the specialized tools needed for nanotechnology R&D. This approach would provide users with the necessary tools to create and modify nanotechnology. Using these specialized tools, user-developers could design, simulate, prototype, and evaluate custom products (von Hippel and Katz 2002). The toolkit model works by dividing up the design of a technology into needs-based and solutions-based parts. The solutions-based parts of a technology tend to be standardized and therefore fall within the domain of the manufacturer. However, the needs-based parts depend on the requirements of the user, meaning users could certainly have greater input into and control into this aspect of design. The semiconductor industry has used the toolkit model successfully by allowing users to customize semiconductor chips after the semiconductor firm has built the standard devices on the chips, such as transistors (von Hippel and Katz 2002). Nanotechnology vendors could provide low-cost fabrication labs similar to the MIT FAB LAB (Mikhak et al. 2002), where users would pay for the nanofab lab, but have open access to a library of nanotechnology designs they can create and modify. Not only would this accelerate innovation by connecting users directly with development, but it would also allow companies to avoid costly customization work.

Meanwhile, other entrepreneurs have brought the ideas of peer-to-peer (P2P) software to banking through the Open Source Hardware Bank (Ganapati 2009). The bank connects small investors willing to make micro-investments to specific open hardware projects. In this financing model, investors fund specific projects hoping for a share of the return, estimated at 5 to 15 percent. The cost of project failure is minimized for each investor as each needs only to invest a small amount of money. However, multiple investors may need to fund larger hardware projects as, unlike open source software projects, there is a need for start-up capital to buy materials and physical artifacts. Open nanotechnology projects could be funded in the same way through micro-investments. The use of this P2P funding model would reduce the barrier of capital for open nanotechnology innovators, thus freeing them up to concentrate on their projects. Nanotechnology vendors could open-source the software for their physical artifacts but keep the artifacts closed to make a profit from them. They could even give away design knowledge but, at the same time, provide stable and high-quality versions of their products. Alternatively, vendors could provide support, education, and material resources while open-sourcing their core technologies. Therefore the use of unrestrictive licenses (licenses that allow for mixing of closed and open licenses) would mean that vendors could open-source certain parts of their technology but keep the rest proprietary.

## 9.4 Open Nanotechnology Applications

Open source nanotechnology is not just a dream but an actuality. It has very real implications for the world; in the coming decades, open nanotechnology will have an impact on how humanity uses water, constructs materials, and obtains energy.

### 9.4.1 Clean Water for All

Humanity is facing the threat of dwindling safe water supplies, denial of individual livelihoods due to contaminated water, and increasing privatization of remaining water sources (Barlow and Clarke 2002). Water purification, like methods of filtration, and its capacity to deliver safe drinkable water for all is perhaps the greatest promise of open nanotechnology (Savage and Diallo 2005). This makes open nanotechnology water purification projects like the one at Open Source Nanotechnology (OS Nano) very important in ensuring everyone has access to clean water, which is a major problem in the poorest areas of the world (Hashmi and Pearce 2009). The project at OS Nano uses magnetite nanocrystals to remove arsenic from water. The website has directions on how to create these nanocrystals from everyday household products. In a 2010 paper, the OS Nano group further proved their concept by using edible oils, vinegar, drain openers, water, rust, and heating to form nanoparticles that were able to remove more than 99 percent of arsenic from drinking water (Yavuz et al. 2010). Not only is the process accessible at an informational level, but it is also accessible from a materials/equipment level as it requires no specialized items to be replicated.

### 9.4.2 Efficient Solar Power

As discussed in Chapter 2, nanotechnologies have enabled the creation of novel materials for use in the production and storage of energy. It is now well established, for example, that second-generation thin-film solar cells operate at higher stabilized efficiencies when incorporating a small density of nanocrystals (Myong et al. 2006; Pearce et al. 2007). These nanocrystal-containing, thin-film solar cells were developed as part of the Thin Film Partnership Program (TFPP), funded by the National Renewable Energy Lab (NREL). The TFPP was an open source collaboration started in 1992 that aimed to improve "the efficiency and reliability of emerging thin-film PV technologies through collaboration among industry, national laboratories, and universities" (Margolis, Mitchell, and Zweibel 2006, 1). The program focused on photovoltaic materials that are all now rapidly expanding production: amorphous silicon (a-Si), copper indium diselenide (CIS) and related materials, and cadmium telluride (CdTe). Throughout the partnership, team meetings were held where research institutions presented findings and shared challenges. Companies and university centers also submitted quarterly reports to NREL staff that were to be shared among the other team members. The TFPP clearly had a large impact on the U.S. photovoltaic industry, and a similar program is poised to assist newer, third-generation cells, which use nanoparticles directly and have shown high levels of power output, even when using production processes similar to ink-jet printing (Fairley 2004). In such a process, nanoparticles are "printed" onto a layer of film, at which point the particles assemble themselves to create solar cells. Technology startups are already attempting to commercialize this idea. With these self-assembling solar cells, anyone could create a solar cell anywhere using basic materials, which has not been possible using conventional technologies; however, this will only hold true if knowledge regarding these self-assembling nanoparticles remains in the public domain. The U.S. Department of Energy is aware of this benefit and is already trying to leverage the power of open source to drive development in the energy sector and photovoltaic devices in particular. Its new site, OpenEI.org, is a free, open source, knowledge-sharing platform providing access to data, models, tools, and information that accelerate the transition to clean energy systems.

### 9.4.3 New Construction Materials

Aerogels are solids made up of nanoparticles arranged in highly porous three-dimensional networks (Malanowski, Heimer, and Luther 2006). This allows aerogels to be some of the lightest solids known. These low-density solids have interesting properties such as extremely low thermal/acoustic conductivity, optical translucency, high porosity, and a low dielectric constant. They can be used as insulation for buildings due to their low thermal conductivity, as sensors due to their translucency, as drug delivery/regulation systems due to their porosity, and in various electronic applications due to their low dielectric constants. They could also potentially be used as implant materials for humans, as shock absorbers, and as filters for purifying water. Aerogels have even been used to build lithium batteries that can store more energy than batteries made of other materials (Prasad 2008). In short, aerogels may revolutionize everything from how we build to how we store energy.

Although aerogels can be created using several different methods,[4] these techniques have never been openly accessible outside the academic community. Aerogel.org, created by Stephen Steiner and Will Walker, aims to change that by opening up the literature and work around making and using aerogels (Thakur 2011). Through podcasts and blog posts on their website, Steiner and Walker show how various types of aerogels can be created. They also reach into the open hardware space with do-it-yourself instructions on how to create equipment needed for aerogel creation, such as a supercritical dryer. These detailed directions allow for anyone with minimal funding, an understanding of basic chemistry, and an Internet connection to make custom aerogels. The blog format of the site allows for users to collaborate with each other in the process. The site is also licensed under the Creative Commons Attribution 3.0 License, which ensures everything on the site is open to all.

---

## 9.5 Conclusion: Where Could Open Nanotechnology Go from Here?

We agree with the views of many skeptics who argue that nanotechnology could help address the technical challenges facing marginalized populations, but only if existing social inequities are also addressed. Closed proprietary systems, like the current IP system, favor the dominant over the marginalized, since only those with resources can access closed systems. Developing nanotechnology in an open source model has the potential to redress many of our global challenges, since developments would be a common good for all, shared by all.

However, to be successful, open nanotechnology will have to work closely with existing intellectual property protection systems. Therefore, an unrestrictive or moderately restrictive open license that allows for the mixing of open and closed technologies and the commercialization of those technologies will be most appropriate in the short term. Many of the recent developments in the field of nanotechnology have already been patented, so future innovators cannot avoid working with patents. An open nanotechnology license would have to treat all the different components of a nanotechnology artifact separately. The Berkeley software distribution (BSD) license, a restrictive license with few conditions, or the slightly more restrictive GNU lesser general public license would be ideal licenses to adapt for open nanotechnology. They could be used to license the documentation, source code, or firmware associated with the physical device. The physical device would be better off being licensed under an open hardware license like the TAPR open hardware license. Meanwhile, nanotechnology firms could start including sunset provisions in the licenses for their products. This would mean that those products would become open after a certain amount of time, as determined by the firm.

Firms could employ various business models to commercialize open nanotechnology like widget frosting[5] and consulting—the case of Red Hat Linux demonstrates that providing services for stable, open products can work as a business model. Red Hat has become one of the most successful software companies in the world by providing support, customization, and extensions for Linux, an open source operating system. While its base software is freely available, Red Hat has made its money through helping other businesses and individuals that use Linux. Using open business models, nanotechnology firms would release the source code for their technologies under various licenses, thereby decreasing the costs of R&D. Websites like Opensourcenano.net or NanoHub.org could be used by firms, academia, and individuals to release their nanotechnology designs into the public domain. Since established research firms like IBM and Samsung are keeping their nanotechnology research private (Spurgeon 2001), new startup firms have an opportunity to enter an untouched market. By releasing their research into the public domain, these firms would have an advantage over the older companies. Furthermore, there are strong social benefits to government and civil society in encouraging open nanotechnology.

Despite large government investments into nanotechnology R&D, worldwide the public has been left out of having any say in the design of the technology, giving rise to concerns of nanotechnology haves and have-nots (Roco and Bainbridge 2001). Open source nanotechnology may be the people's way back in, striking at the heart of design. In 2007, the Canadian Institutes of Health Research (CIHR) established a policy requiring that the public should have open access to research results.[6] The CIHR policy states that, as its research is publicly funded, the results of such research should be disseminated as widely as possible. For this same reason and because nanotechnology has the potential to affect every aspect of our lives (Moore 2002), we need to hold nanotechnology research to the same standard. We contend that governments and civil society organizations should encourage open nanotechnology so that public funding does not automatically translate into support for private profits. Governments would also benefit from public oversight of government-funded nanotechnology projects, so that civil society organizations and the public can externally monitor and govern those projects. Not only could this reduce public fears about the consequences of technologies, but it would also help the government in monitoring whether public funds are used legally and ethically.

Throughout this chapter, we have argued why open nanotechnology is needed and showed how it would work. This is an urgent matter, as action must be taken before too much basic knowledge gets locked into proprietary systems. Richard Stallman, one of the founders of the open source movement, sums up the choice we have in developing open or closed technology: "The easy choice was to join the proprietary software world, signing nondisclosure agreements and promising not to help my fellow hacker. Most likely I would also be developing software that was released under nondisclosure agreements, thus adding to the pressure on other people to betray their fellows too. I could have made money this way, and perhaps amused myself writing code. But I knew that at the end of my career, I would look back on years of building walls to divide people, and feel I had spent my life making the world a worse place" (Stallman 1999).

---

# References

Baber, Z. 2001. Globalization and scientific research: The emerging triple helix of state-industry-university relations in Japan and Singapore. *Bulletin of Science Technology Society* 21(5): 401–8.

Barlow, M., and T. Clarke. 2002. *Blue Gold: The Fight to Stop the Corporate Theft of the World's Water.* New York: New Press.

Barnard, D. 2002. In the High Court of South Africa, Case No. 4138/98: The global politics of access to low-cost AIDS drugs in poor countries. *Kennedy Institute of Ethics Journal* 12(2): 159–74.

Behlendorf, B. 1999. Open source as a business strategy. In *Open Sources: Voices from the Open Source Revolution*, edited by C. DiBona, S. Ockman, and M. Stone, 149–70. Sebastopol, Calif.: O'Reilly & Associates.

Bergquist, M., and J. Ljungberg. 2001. The power of gifts: Organizing social relationships in open source communities. *Information Systems Journal* 11: 305–20.

Bernstein, H.J. 2000. Recent changes to RasMol, recombining the variants. *Trends in Biochemical Sciences* 25(9): 453–55. doi: 10.1016/S0968–0004(00)01606–6.

Bretthauer, D.W. 2002. Open source software: A history. *Information Technology and Libraries* 21(1): 3–10.

Bruns, B. 2001. Open sourcing nanotechnology research and development: Issues and opportunities. *Nanotechnology* 12(3): 198–210.

Bruns, B. 2003. *Participation in Nanotechnology: Methods and Challenges.* Presented at the annual conference of the International Association for Public Participation, Ottawa, Canada, May 19–22, 2003. Retrieved from http://www.bryanbruns. com/bruns-p2nano.pdf (accessed August 3, 2011).

Bruns, B. 2004. Applying nanotechnology to the challenges of global poverty: strategies for accessible abundance. Presented at the first conference on Advanced Nanotechnology: Research, Applications, and Policy, Washington, D.C. Retrieved from http://www.foresight.org/Conferences/AdvNano2004/ Abstracts/Bruns/BrunsPaper.pdf (accessed August 3, 2011).

Buitenhuis, A.J., I. Zelenika, and J.M. Pearce. 2010. Open Design-based strategies to enhance appropriate technology development. *Proceedings of the 14th Annual National Collegiate Inventors and Innovators Alliance Conference*, pp. 1–12. Retrieved from http://nciia.org/sites/default/files/pearce.pdf (accessed August 3, 2011).

Burgi, B.R., and T. Pradeep. 2006. Societal implications of nanoscience and nanotechnology in developing countries. *Current Science* 90(5): 645–58.

Butler, B., L. Sproull, S. Kiesler, and R. Kraut. 2007. Community effort in online groups: Who does the work and why? In *Leadership at a Distance: Research in Technologically Supported Work*, edited by S.P. Weisband, 171–94. New York: Lawrence Erlbaum Associates.

Chen, H., M.C. Roco, L. Xin, and Y. Lin. 2008. Trends in nanotechnology patents. *Nature Nanotechnology* 3:123–25.

Chesbrough, H. 2006. *Open Business Models: How to Thrive in the New Innovation Landscape.* Boston: Harvard Business School Press.

Einsiedel, E.F., and L. Goldenberg. 2004. Dwarfing the social? Nanotechnology lessons from the biotechnology front. *Bulletin of Science Technology Society* 24(1): 28–33.

ETC Group. 2005. *Nanotech's Second Nature Patents: Implications for the Global South.* http://etcgroup.org/upload/publication/54/02/com8788specialpnanomar-jun05eng.pdf (accessed April 18, 2011).

Fairley, P. 2004. Solar-cell rollout. *Technology Review* 107(6): 35–40.

Foladori, G., and N. Invernizzi. 2005. Nanotechnology for the poor? *PLoS Medicine* 2(8): e280.

Free Software Foundation. 2007. The GNU general public license. *GNU Project.* http://www.gnu.org/licenses/gpl.html (accessed November 18, 2009).

Ganapati, P. 2009. Open source hardware hackers start P2P bank. *Wired Magazine*, March 18. http://www.wired.com/gadgetlab/2009/03/open-source-har/ (accessed November 19, 2009).

Garfinkel, S.L., R.M. Stallman, and M. Kapor. 1999. Why patents are bad for software. In *High Noon on the Electronic Frontier: Conceptual Issues in Cyberspace*, edited by P. Ludlow, 35–46. Cambridge, Mass.: Massachusetts Institute of Technology.

Gillmor, D. 2006. *We the Media: Grassroots Journalism by the People, for the People.* Sebastopol, Calif.: O'Reilly Media, Inc.

Hashmi, F. and J.M. Pearce. 2009. Viability of small-scale arsenic-contaminated water purification technologies for sustainable development in Pakistan. *Sustainable Development.* doi: 10.1002/sd.414.

Heller, M.A., and R.S. Eisenberg. 1998. Can patents deter innovation? The anticommons in biomedical research. *Science* 280(5364): 698–701.

Klimeck, G., M. McLennan, S. Brophy, G. Adams, and M. Lundstrom. 2008. nanoHUB.org: Advancing education and research in nanotechnology. *Computing in Science and Engineering* 10(5): 17–23.

Lemley, M.A. 2005. Patenting nanotechnology. *Stanford Law Review* 58:601–30.

Lerner, J., and J. Tirole. 2002. Some simple economics of open source. *Journal of Industrial Economics* 50(2): 197–234.

Lerner, J., and J. Tirole. 2005. The scope of open source licensing. *Journal of Law, Economics, and Organization* 21(1): 20–56.

Lessig, L. 2004. *Free Culture: How Big Media Uses Technology and the Law to Lock Down Culture and Control Creativity.* New York: Penguin Press.

Long, P.D. 2002. OpenCourseWare: Simple idea, profound implications. *Syllabus* 15(6): 12–14.

Long, P.O. 1991. Invention, authorship, "intellectual property," and the origin of patents: Notes toward a conceptual history. *Technology and Culture* 32(4): 846–84.

Malanowski, N., T. Heimer, and W. Luther. 2006. *Growth Market Nanotechnology an Analysis of Technology and Innovation.* Weinheim: Wiley-VCH.

Margolis, R., R. Mitchell, and K. Zweibel. 2006. *Lessons Learned from the Photovoltaic Manufacturing Technology/PV Manufacturing R&D and Thin-Film PV Partnership Projects.* National Renewable Energy Lab Technical Report NREL/TP-520–39780.

Maurer, S.M., A. Rai, and A. Sali. 2004. Finding cures for tropical diseases: Is open source an answer? *PLoS Medicine* 1(3): e56.

McGowan, D. 2001. Legal implications of open-source software. *University of Illinois Law Review* 241–304.

Mgbeoji, I. 2006. *Global Biopiracy: Patents, Plants and Indigenous Knowledge.* Vancouver: University of British Columbia Press.

Mikhak, B., C. Lyon, T. Gorton, N. Gershenfeld, C. McEnnis, and J. Taylor. 2002. Fab lab: An alternate model of ICT for development. *Development by Design.* http://phm.cba.mit.edu/papers/02.00.mikhak.pdf (accessed April 18, 2011).

Moon, J.Y., and L. Sproull. 2000. Essence of distributed work: The case of the Linux kernel. *First Monday* 5(11).

Moore, F. 2002. Implications of nanotechnology applications: Using genetics as a lesson. *Health Law Review* 10(3): 9.

Mowery, D.C., R.R. Nelson, B.N. Sampat, and A.A. Ziedonis. 2001. The growth of patenting and licensing by U.S. universities: An assessment of the effects of the Bayh-Dole act of 1980. *Research Policy* 30(1): 99–119. doi: 10.1016/S0048–7333(99)00100–6.

Myong, S.Y., S.W. Kwon, J.H. Kwak, K.S. Lim, J.M. Pearce, and M. Konagai. 2006. Good stability of protocrystalline silicon multilayer solar cells against light irradiation originating from vertically regular distribution of isolated nano-sized silicon grains. *Fourth World Conference on Photovoltaic Energy Conversion Proceedings*, p. 492.

Open Source Initiative. 2011. The open source definition. http://opensource.org/docs/definition.html (accessed April 18, 2011).

Ortega, F., J.M. Gonzales-Barahona, and G. Robles. 2008. *On the Inequality of Contributions to Wikipedia*. Presented at the Hawaii International Conference on System Sciences 2008, Waikoloa, January 7–10.

Ostrom, E. (1990). *Governing the Commons: The Evolution of Institutions for Collective Action*. New York: Cambridge University Press.

Pearce, J.M., and U. Mushtaq. 2009. *Overcoming Technical Constraints for Obtaining Sustainable Development with Open Source Appropriate Technology*. Presented at the 2009 IEEE Toronto International Conference, Toronto, Canada, September 26–27.

Pearce, J.M., N. Podraza, R.W. Collins, M.M. Al-Jassim, K.M. Jones, J. Deng, and C.R. Wronski. 2007. Optimization of open-circuit voltage in amorphous silicon solar cells with mixed phase (amorphous + nanocrystalline) *p*-type contacts of low nanocrystalline content. *Journal of Applied Physics* 101(11): 114301.

Prasad, S.K. 2008. *Modern Concepts in Nanotechnology*. New Delhi: Discovery.

Raymond, E.S. 1998. Homesteading the noosphere. *First Monday* 3(10).

Raymond, E.S. 1999. The cathedral and the bazaar. *Knowledge, Technology, and Policy* 12(3): 23–49.

Raymond, E.S. 2001. *The Cathedral and the Bazaar: Musings on Linux and Open Source by an Accidental Revolutionary*. Sebastopol, Calif.: O'Reilly & Associates.

Robinson, D.F. 2010. *Confronting Biopiracy: Challenges, Cases and International Debates*. Washington, D.C.: Earthscan.

Roco, M.C. and W.S. Bainbridge. 2001. *Societal Implications of Nanoscience and Nanotechnology*. Dordrecht: Kluwer Academic.

Rosenzweig, R. 2006. Can history be open source? Wikipedia and the future of the past. *Journal of American History* 93(1): 117–46.

Sampat, B.N. 2003. Recent changes in patent policy and the "privatization" of knowledge: Causes, consequences, and implications for developing countries. In *Knowledge Flows and Knowledge Collectives: Understanding the Role of Science and Technology Policies in Development*, edited by B. Bozeman, D. Sarewitz, S. Feinson, G. Foladori, M. Gaughan, A. Gupta, B. Sampat, and G. Zachary, 39–81. Global Inclusion Program of the Rockefeller Foundation.

Savage, N. and M.S. Diallo. 2005. Nanomaterials and water purification: Opportunities and challenges. *Journal of Nanoparticle Research* 7(4): 331–42.

Schummer, J. 2007. The impact of nanotechnologies on developing countries. In *Nanoethics: The Ethical and Social Implications of Nanotechnology*, edited by F. Allhoff, P. Lin, J. Moor, and J. Weckert, 291–307. Hoboken: Wiley.

Shantharam, Y. 2005. The cost of life: Patent laws, the WTO and the HIV/AIDS pandemic. *Undercurrent* 2(2): 48–56.

Shapiro, C. 2000. Navigating the patent thicket: Cross licenses, patent pools, and standard setting. *Innovation Policy and the Economy* 1:119–50.

Shiva, V. 1997. *Biopiracy: The Plunder of Nature and Knowledge*. Cambridge, Mass.: South End Press.

Shiva, V. and R. Holla-Bhar. 1997. Piracy by patent: The case of the neem tree. In *The Case against the Global Economy and for a Turn toward the Local*, edited by J. Mander and E. Goldsmith, 146–59. San Francisco: Sierra Club Books.

Spurgeon, B. 2001. Nanotechnology firms start small in building big future. *New York Times*, January 29. http://www.nytimes.com/2001/01/29/business/worldbusiness/29iht-btnano.2.t.html?pagewanted=1 (accessed November 20, 2009).

Stallman, R. 1999. The GNU operating system and the free software movement. In *Open Sources: Voices from the Open Source Revolution*, edited by C. DiBona, S. Ockman, and M. Stone, 53–70. Sebastopol, Calif.: O'Reilly & Associates.

SXM Team. *Scanning Tunneling Microscope Construction Kit.* http://sxm4.uni-muenster.de/stm-en/ (accessed April 13, 2011).

TAPR. 2007. *TAPR Open Hardware License.* http://www.tapr.org/TAPR_Open_Hardware_License_v1.0.txt (accessed November 19, 2009).

Thakur, D. 2008. *The Implications of Nano-Technologies for Developing Countries—Lessons from Open Source Software.* Presented at the Workshop on Nanotechnology, Equity, and Equality, Center for Nanotechnology in Society, Arizona State University, November 20.

Thakur, D. 2011. Open access nanotechnology for developing countries: Lessons from open source software. In *Nanotechnology and the Challenges of Equity, Equality and Development*, edited by S.E. Cozzens and J.M. Wetmore, 331–48. Dordrecht: Springer.

Thompson, C. 2008. Build it. Share it. Profit. Can open source hardware work? *Wired Magazine* 16(11). http://www.wired.com/techbiz/startups/magazine/16-11/ff_openmanufacturing (accessed November 19, 2009).

Vaidhyanathan, S. 2006. Nanotechnologies and the law of patents: A collision course. In *Nanotechnology—Risk, Ethics and Law*, edited by M. Mehta and G. Hunt. London: Earthscan.

Vallance, R., S. Kiani, and S. Nayfeh. 2001. *Open Design of Manufacturing Equipment.* Presented at the first International Conference on Agile, Reconfigurable Manufacturing, Ann Arbor.

von Hippel, E., and R. Katz. 2002. Shifting innovation to users via toolkits. *Management Science* 48(7): 821–33.

Weber, S. 2001. The political economy of open source software. In *Tracking a Transformation: E-commerce and the Terms of Competition in Industries*, edited by BRIE-IGCC E-conomy Project, 406–34. Washington, D.C.: Brookings Institution.

World Intellectual Property Organization (WIPO). 2004. *The Concept of Intellectual Property.* http://www.wipo.int/about-ip/en/iprm/pdf/ch1.pdf (accessed April 18, 2011).

Yavuz, C., J. Mayo, C. Suchecki, J. Wang, A. Ellsworth, H. D'Couto, E. Quevedo, A. Prakash, L. Gonzalez, C. Nguyen, and V.L. Colvin. 2010. Pollution magnet: Nano-magnetite for arsenic removal from drinking water. *Environmental Geochemistry and Health* 32(4): 327–34.

## Endnotes

1. Linux is an open source operating system. For its history and context, see Glyn Moody, *Rebel Code: The Inside Story of Linux and the Open Source Revolution* (Cambridge, Mass.: Perseus, 2002).

2. Science Commons (http://creativecommons.org/science), Biobricks (http://biobricks.org), OpenWetWare (http://openwetware.org), Appropedia (http://appropedia.org).

3. See: http://scp.indiegames.us/.
4. See "Classic Aerogel Papers" at Aerogel.org (http://www.aerogel.org/?p=1196).
5. A hardware supplier may choose to provide open source software (the widget) for use with their patented hardware. In this case, users are free to modify, use, and share the software, so the company makes no profit from that. However, it indirectly profits through the free distribution and maintenance of software designed to run on its hardware. Widget frosting may also refer to services, add-ons, and extensions provided by a software supplier for its open software. While its base software may be open source, the add-ons (the frosting) would be the profit makers for the company. For more on this, see the "Open Source Case for Business" at the Open Source Initiative (http://www.opensource.org/advocacy/case_for_business.php).
6. See CIHR "Policy on Access to Research Outputs" (http://www.cihr-irsc.gc.ca/e/34846.html).

# Section IV

# Governance

# 10

## Nanotechnology and Risk

Fern Wickson

### CONTENTS

## 10.1 Introduction

The discourse of risk dominates modern industrialized societies and is particularly prominent in discussions relating to the acceptability of new technologies such as nanotechnology. For responsible political decision making on technological advances, it is typically claimed that what is needed is "comprehensive scientific risk assessment," with the most important questions being "What are the potential risks?" and "How can we manage these risks to avoid negative impacts?" In this chapter I introduce readers to important concepts and critical literature around the notion of risk—specifically around risk analysis as a decision-aiding tool. I then present emerging scientific research on risks relating to nanotechnology, highlighting how the discourse of risk narrows the frame of discussion about the desirability of nanotechnology development. Finally, I indicate alternative decision-aiding tools that might begin to push discussions about nanotechnology beyond a narrow discourse of risk.

## 10.2 Discourse of Risk

### 10.2.1 Risk Society

In a widely cited social science thesis, Ulrich Beck has argued that risk has become the central organizing concept of modern industrialized societies (Beck 1986). According to Beck, people in industrialized societies have become increasingly aware that the application of science and technology is often accompanied by unintended adverse effects. Or in Beck's words, that "the sources of wealth are 'polluted' by growing 'hazardous side effects'" (1986, 20). This, he argues, has led to an increasing focus on how to handle the risks of modern industrial development. Beck describes this "Risk Society" as representing a new phase of modernity, a phase in which the primary concern is no longer with the production and distribution of goods but with the production and distribution of "bads." Broadly, it is argued that as basic needs (and excessive desires) have largely been catered for in modern industrialized societies, scarcity has subsided as the issue of primary concern. The focus is no longer solely on controlling nature for the production of useful goods but on how to handle the problems resulting from technological and economic development—that is, on the production and distribution of risks.

### 10.2.2 Risk Analysis As a Decision-Making Tool

Risk can be defined as the likelihood that an undesirable event will occur, multiplied by the extent of its consequences. For new technological developments, the process of risk analysis has become the dominant tool for informing and aiding decision making (Winner 1986). The process of risk analysis has typically been described as consisting of three stages—risk assessment, risk management, and risk communication. *Risk assessment* involves scientists identifying potential adverse effects and calculating the probability of their occurrence. This risk assessment from scientific experts is then passed to policymakers who perform the second stage, *risk management*, where decisions are made about the relative importance of the risks in question and how they will be managed so as to minimize potential harm. Finally, once decisions have been made, *risk communication* occurs so that the general public is informed about the risks involved and the chosen management initiatives. A crucial characteristic of this conventional approach is that it assumes a clear distinction between the stages of technical/fact-based risk assessment and normative/values-based risk management.

When employed as a decision-aiding tool, risk analysis generally adopts a realist concept of risk (Adams 1995; Robins 2002). This is the idea that risks exist "out there" and can be accurately and objectively quantified by experts. Using the technical definition that risk equates to the probability of a hazard occurring multiplied by the magnitude of its impact, a realist concept suggests that scientific experts are able to calculate the real degree of risk associated with any technology. Other positions or opinions about the risks involved are then deemed to be false and/or irrational and stemming from a lack of knowledge. As a tool to aid decision making, realist risk analysis can therefore be viewed as a technocratic approach because it suggests that, by relying on scientific experts, rational, objective, and politically neutral decisions can consequently be made. As governments employing a regulatory discourse of risk overwhelmingly use this realist understanding and its technical definition, they also tend to adopt an approach to decision making that privileges scientific knowledge and the advice of experts. This realist approach to risk and the resulting unquestioned authority of scientific expertise in political decision making on new technologies has, however, come under heavy criticism from the social sciences.

### 10.2.3 Challenges to the Technical Understanding of Risk

In general terms, there are three key fields of social science research that have challenged the adequacy of the technical or realist approach to risk. These are: psychometric research (from psychology), cultural theory (from anthropology), and typologies of uncertainty (from science and technology studies). These fields suggest that realist risk discourses fail to account for important factors involved, such as the characteristics of the risk in question, the influence of divergent worldviews, and various forms of uncertainty.

Psychometric research has challenged the realist notion of risk by suggesting that there are a number of characteristics (beyond likelihood and magnitude) that influence how risks are evaluated by people. These include whether the risks are voluntarily taken, how familiar they are, how controllable they are, and whether they have catastrophic potential or the potential to have an impact on future generations (Slovic, Fischhoff, and Lichtenstein 1982; Slovic 1987, 1991). This is said to explain apparent contradictions, such as why people might happily drive a car but protest the development of nuclear power, even though the level of risk associated with driving has been calculated as much higher. That is, people find the risks associated with nuclear power unacceptable because of characteristics such as a catastrophic potential, potential to affect future generations, lack of familiarity, and lack of voluntary undertaking. By highlighting the importance of such characteristics, psychometric research has suggested that while experts tend to assess risks solely on a statistical basis in relation to probabilities and mortality rates, the public is often sensitive to nonstatistical considerations in their assessments and tend to perform a more contextual assessment of the risks posed by a particular technology (Otway 1987). This suggests that there are important factors of technological risk not captured during formal processes of risk analysis and that these should be incorporated into decision making on new technologies (Otway 1980; Slovic 1998).

Cultural theory argues that risk debates are actually not primarily about physical risks at all but rather relate to different underlying views on social organization and the nature of nature. Cultural theory presents a typology characterizing beliefs in preferred forms of social organization as being individualist (preference for freedom from constraints), hierarchist (supporting hierarchical social organization), egalitarian (strong group loyalties but not supporting externally imposed rules), or fatalist (no support for organized groups or belief in individual control) (Douglas and Wildavsky 1982; Schwarz and Thompson 1990; Thompson et al. 1990). For environmental risk debates, another fourfold typology is presented to characterize different views on nature. In general terms, these are characterized thus: nature as robust, nature as fragile, nature as tolerant, and nature as unpredictable (Adams and Thompson 2002). By providing typologies that can be used to characterize different worldviews, the aim of cultural theory is to allow the different premises underlying and framing debates about physical risks to be made explicit.

While psychometric approaches emphasize the importance of individual psychology and cultural theory emphasizes the importance of social commitments, both represent constructivist rather than realist understandings of risk. According to these approaches, there is not a "real" degree of risk that can be captured through risk assessment as performed by experts. The process is considered inadequate either because it fails to take into account the nature of the risks involved—for example, whether they are familiar, controllable, or reversible—or because it fails to consider the way in which judgments about risks can be differentially framed by varying beliefs about society and nature.

A final challenge to realist notions of risk that I wish to describe is the emergence of typologies of uncertainty from science and technology studies (Wynne 1992; Stirling 1999; Stirling and Gee 2002; Felt and Wynne 2007; Funtowicz and Ravetz 1993). While the emerging typologies differ in how they draw boundaries of distinction, some general patterns can be synthesized as follows:

- *Risk* always implies uncertainty to some extent. According to the emerging typologies, however, the term *risk* is defined as specifically relevant to those situations where both the potential outcomes and the probabilities associated with those outcomes can be well characterized.

- *Incertitude* is a term that can be applied to situations where there is some agreement about the potential outcomes or impacts that may occur, but the basis for assigning probabilities is not strong. This is due to a lack of relevant information that can be reduced through further research.

- *Indeterminacy* refers to a type of uncertainty that exists because of the complexity associated with predicting outcomes (and probabilities) within various open and interacting social and natural systems. This complexity means that our knowledge will always be incomplete because science is simply unable to take every factor of a dynamic system into account.

- *Ambiguity* refers to a type of uncertainty resulting from contradictory information or the existence of divergent framing assumptions and values. This type of uncertainty arises because there are, for example, different approaches to generating knowledge, different interpretations of the significance of generated knowledge, different ways of evaluating the quality and strength of knowledge, and different understandings of how to act in light of knowledge.

- *Ignorance* refers to our inability to conceptualize, articulate, and consider outcomes and causal relationships that lie beyond our current frameworks of understanding. This type of uncertainty can be described as the things "we don't know that we don't know" and represents an inability to ask the right questions rather than a failure to provide the right answers.

By focusing only on a quantification of risk, realist discourses generally fail to take into account ambiguity, indeterminacy, ignorance, and even in some cases, incertitude (Wynne 1992; Stirling and Gee 2002). These differing forms of uncertainty create a space through which diverse views, values, and assumptions shape divergent perceptions and assessments of risk. Failure to explicitly and transparently handle uncertainties, as well as the values and assumptions that operate through them, means that the psychological and social factors influencing risk assessments (as described by psychometric and cultural theory) remain implicit and hidden during decision making. This also means that decision-making processes based on realist risk analysis are likely to remain the subject of ongoing debate as people continue to emphasize different characteristics of the risks in question, and argue from competing premises in relation to social and biological organization. The ability to transparently handle uncertainties in decision making and incorporate broader social and cultural criteria into the assessment process is seen as dependent on a reconceptualization of the role for expertise (Otway 1987), the establishment of a two-way path of communication between the public and experts (Funtowicz and Ravetz 1994), and the encouragement of increased public participation in decision-making processes (Wynne 2001).

## 10.3 Risk and Nanotechnology

### 10.3.1 Potential Risks from Nanosciences and Nanotechnologies

In the past five years, there has been a dramatic growth of interest in the potential health and environmental impacts of nanosciences and nano-technologies[1] (nanoST). This is partly because it cannot be assumed that the "novel properties" that are characteristic of the nanoscale will only be novel for the good but also because (as suggested above) considering new technologies in terms of potential risks has become a characteristic feature of industrialized societies. However, the answer to the question "What are the risks of nanoST?" remains hotly contested, with the level of knowledge described in terms such as rudimentary (Balbus et al. 2007) and the level of uncertainty as extreme (Kandlikar 2007). The so-called gaps in knowledge resemble something more akin to gaping chasms. Research is, however, starting to emerge on this topic, and some of the key findings and future challenges will be summarized here.

### 10.3.2 Could Carbon Nanotubes Be the New Asbestos?

Carbon nanotubes share a general structural similarity to asbestos in that both are small, stiff, needle-like fibers. A pressing line of enquiry has been concerned with the issue of whether carbon nanotubes pose risks similar to asbestos. The answer emerging from early scientific research appears to be that this is indeed possible. In the research currently available, carbon nanotubes have been shown to cause inflammation and granulomas (scarlike lesions) (Poland et al. 2008; Muller et al. 2005; Shedova et al. 2005; Warheit et al. 2004; Ma-Hock et al. 2009; Lam et al. 2004), which is the same bodily response that results from exposure to asbestos and precedes the development of cancers such as mesothelioma. It has been demonstrated that carbon nanotubes also have the potential for skin-cell toxicity through dermal exposure and genotoxicity (toxicity at a molecular level), including the ability to damage DNA (Shedova et al. 2003; Zhu et al. 2007; Balbus et al. 2007).

Carbon nanotubes are carbon atoms arranged into a cylindrical structure and can be either single-walled (a single layer of atoms composing the cylindrical shape) or multiwalled (cylinders existing within cylinders). They come in a range of different sizes, tend to agglomerate, and can contain different amounts of residues from metal catalysts used during their production. All of these factors have the potential to have an impact on their observed toxicity. The studies cited above have been conducted with carbon nanotubes of different types (single-walled, multiwalled), lengths (short, long), preparations (single tubes, tangled agglomerates, ground pieces), test systems (for example, cell cultures, mice, rats), and exposure methods (inhalation, injection, dermal deposition). Although this diversity in the studies makes them difficult to compare, it is certainly possible to interpret the totality of the findings as indicative of an emerging pattern of potential for carbon nanotubes to cause harm.

In order to understand the extent that any potential for harm will actually manifest, we need to understand the extent to which workers, consumers, and the environment will be exposed. If we are to believe the optimists who champion the notion that nanoST will have a revolutionary impact touching upon every aspect of our lives, we might reasonably expect exposure to nanoST to be high. Carbon nanotubes also represent one of the boom areas of nanoST, with a predicted global market of products worth US$2.6 trillion by 2014 (Holman and Lackner 2006). Currently, however, there is extremely limited information available on levels of exposure to carbon nanotubes. The information that is available suggests that workers in nanotechnology industries and R&D facilities are exposed, to some degree, through both inhalation and dermal deposition (Maynard et al. 2004; Bergamaschi 2009). However, there are few references to public and environmental exposures, nor does research take into account the expected increase in future levels of exposure as use becomes more widespread. In studies of toxicity and exposure, it is also crucial to take into account the incredible persistence of carbon nanotubes, which represent one of the most biologically nondegradable man-made materials currently available (Lam et al. 2004). It is therefore particularly important to consider full life cycles and the likelihood of extended time delays between exposure and effect.

Challenges to understanding the risks posed by carbon nanotubes relate not only to the limited information available but also to a deep-level debate about which paradigms, methods, and approaches are appropriate for testing. For example, in studies designed and conducted according to a fiber-toxicology paradigm, long nanotubes appear most pathogenic. However, if tested according to the methods and approaches most relevant for particles, short nanotubes may also demonstrate significant toxicity (Poland et al. 2008). While relevant for nanotubes, the latter paradigm is largely being applied to develop an understanding of the toxicity of free nanoparticles.

### 10.3.3 Engineered Nanoparticles in Biological Systems

There is general consensus within the scientific community that the toxicity of engineered nanoparticles[2] cannot be derived from our understanding of their bulk counterparts (Donaldson et al. 2006; Oberdörster, Oberdörster, and Oberdörster 2005; SCENIHR 2006). Nanoparticles of materials (such as titanium dioxide, zinc oxide, and silver) are likely to be *more* toxic than their bulk counterparts. This is because as smaller particles have a larger surface area and surfaces are generally more reactive, nanoscale particles can interact with biological systems in ways that differ from their bulk counterparts. Additionally, because engineered nanoparticles are so small, they have the ability to penetrate cell membranes, enter the bloodstream and lymphatic system, and move throughout the body, including into the heart, nervous system, bone marrow, brain, and fetus (Oberdörster, Oberdörster, and Oberdörster 2005; SCENIHR 2006; Oberdörster 2004; Lockman et al. 2004; Takeda et al. 2009). Engineered nanoparticles also have the ability to act as vectors, able to not only bind and carry other chemicals and pollutants as they move throughout biological systems but also to enhance their toxicity and biological availability (Handy and Owen 2008; Baun et al. 2008; Royal Commission on Environmental Pollution 2008)—the so-called Trojan horse effect. All of these features have resulted in calls for nanoscale materials, and particularly engineered nanoparticles, to be treated and evaluated as new substances in toxicity testing and regulation.

### 10.3.4 Comprehensive Scientific Risk Assessment Provided As the Answer

Although the potential for harm to human health and the environment is increasingly recognized for nanoST, the most common response to the potential dangers is to emphasize the need for comprehensive scientific risk assessment. It is assumed that as long as comprehensive scientific risk assessment is performed, these potential harms can be identified, managed, and kept within tolerable limits. There are, however, problems facing this proposed strategy for controlling the development of nanoST.

### 10.3.5 Substantial Challenges of Nanotoxicology

The complexity of scientifically assessing the risks of nanomaterials for humans and the environment should not be underestimated. The novel properties that characteristically emerge on the nanoscale mean that we cannot extrapolate an understanding of nanomaterial toxicity from our experience with the material in bulk form. The most important factors for understanding toxicity are also not necessarily the traditional dose metrics of mass or number but rather characteristics such as surface area, surface charge, length, shape, agglomeration state, and solubility. These features differ for different nanomaterials as well as for different forms or species of the same nanomaterial. Additionally, all of the above characteristics important for understanding toxicity can be altered through interaction with environmental factors such as pH, salinity, water hardness, and the presence of organic matter (Handy and Owen 2008). This means that the relevant properties can change throughout the life cycle of a product (Balbus et al. 2007). Additionally, not only does the variety of nanomaterial products suggest the possibility of multiple exposure pathways (including ingestion, inhalation, injection, and dermal exposure) (Oberdörster, Oberdörster, and Oberdörster 2005), it may also be relevant to consider different routes at different stages throughout a product's life cycle (Bergamaschi 2009). On top of this, the mobility of nanoparticles means they may translocate to parts of an organism's body that may not be indicated as relevant by the initial route of exposure. Different species also have very different susceptibilities to nanomaterials, meaning that tests done with a single species are insufficient for understanding environmental risks (Baun et al. 2008). Furthermore, understanding ecological risk requires an awareness of the fate and behavior of nanomaterials in the environment (for example, their movements through soil, air, water, and the various organisms of an ecosystem). Adding to this already incredibly complex picture is the point that testing should ideally be done not just on acute effects but also on the potential for chronic effects, multitrophic effects (through the food chain), bioaccumulation, and sublethal impacts, such as behavioral change and reduced immunity or reproductive fitness (Owen and Handy 2007). This means that, in order to scientifically assess the risks posed by nanomaterials to human health and the environment, we need research that:

- Documents the various physicochemical characteristics of each nanomaterial throughout the different stages of a product's life cycle;
- Considers multiple routes of organism exposure;
- Tests various species of organisms (including micro-organisms);
- Examines various body parts of exposed organisms (including cell components);
- Reflects on potential movements through complex ecosystems;
- Uses an extended timeframe; and
- Is sensitive to a range of different possible impacts beyond acute toxicity and death.

While this already poses a difficult challenge, making matters significantly worse is the fact that, for nanomaterials, we have not yet developed the appropriate methods and instrumentation necessary to perform the testing required (Grierger, Hansen, and Baun 2009). This extends to the very fundamental level of lacking ways to detect, measure, characterize, and therefore monitor nanoparticles across a range of different media (Balbus et al. 2007; SCENIHR 2006; EFSA 2009). This means that not only is there acutely limited research on the toxicity of nanomaterials, there is also an inability to achieve adequate testing in the short term. It has been suggested that toxicity testing on just currently commercially available nanomaterials would take decades to complete and require the investment of over US$1 billion (Choi, Ramachandran, and Kandlikar 2009). Despite regular statements on the clear importance of (eco)toxicology research, the funding currently available is extremely limited (Editor 2008). Funding for health and environmental nanotechnology research is often in combination with that available for ethical, legal, and social aspects and, in total, both are typically only awarded at around 3 to 5 percent of the budgets available for nanoST development.

This situation creates a number of critical paradoxes. Firstly, toxicological specificities arguably require that nanomaterials be assessed on a case-by-case basis, but this is practically impossible (Walker and Bucher 2009). We also need physicochemical characterization for nanomaterials throughout their life cycles, but good methods for this are not currently agreed upon. There is a critical need for information on exposure levels, but new methods and equipment are required to adequately detect and measure nanoparticles (SCENIHR 2006). Furthermore, mass and number alone are insufficient as dose metrics, but alternative factors, such as surface area and surface chemistry, are not a part of safety assessment regulations. Finally, while further research is urgently needed, there is a lack of standardized testing procedures and reference materials that would enable coordinated development and comparison across studies.

As a result, some environmental organizations (Friends of the Earth 2006; ETC Group 2009) and politicians (Lucas 2003; Anonymous 2007) have called for a moratorium on commercialization, and significant scientific organizations have recommended that environmental release of engineered nanoparticles should be avoided as much as possible (Royal Society and Royal Academy of Engineering 2004). Although the suggestion of a moratorium on commercialization until more information can be gathered has been controversial, what does seem clear is that the traditional decision-making tool of scientific risk assessment is crippled by the profound lack of information on nanotoxicology. The vast seas of uncertainty and the decades required to conduct the necessary research mean that the idea of decision making through a scientific assessment of risks should be recognized as not currently possible. There is, therefore, a need to move away from a sole focus on scientific risk assessment as a decision-making tool and toward the exploration of approaches that enable a deliberative negotiation of uncertainty. This is *not* to suggest that risk assessment based on the best available nanotoxicology research should not be carried out to help inform decision making—just that this is not sufficient on its own.

The extreme uncertainties involved in nanoST make it particularly vital to recognize the importance of social and ethical aspects, including the crucial role played by visions, values, and beliefs in pushing this particular technological trajectory forward. Negotiating situations of uncertainty

**TABLE 10.1**

Some of the Necessary Choices in Nano(eco)toxicology

- Which nanotechnology to study, e.g., which nanoparticle is considered most relevant, interesting or important
- Which test subject to use, e.g., what organism or what part of an organism to test effects on
- Which test system, methods, tools or paradigm to use, e.g., which dose metric to focus on—mass of particles, number of particles or surface area of particles
- Which route of exposure to examine, e.g., choosing how test subjects will be exposed
- Which endpoints to observe, e.g., whether to observe deaths, lesions, white blood cell counts, protein activity levels and so on
- How to interpret the results, e.g., to what extent they are related to the nanomaterial under study, impurities/contaminants in the test sample, the specificities of the organism involved, limitations of the method, and so on

involves making choices without complete knowledge, which means that these choices cannot help but be influenced by values, beliefs, assumptions, and worldviews. In the model of risk analysis that usually informs decision making on science and technology, the role of values is generally recognized only in the stage of risk management. It is, however, important to realize that values permeate all stages, including that of conducting scientific research. For example, in (eco)toxicology research on nanoST, scientists have to make a range of choices that cannot help but be based on values, beliefs, and assumptions as outlined in Table 10.1.

The enormous uncertainties involved mean that scientists must make choices in generating their knowledge and that all of these choices are inevitably informed by values and remain open to legitimate debate through alternative framing and interpretation. When science is used to inform political decision making or public understanding, recognizing uncertainty and the importance of different choices and assumptions becomes paramount.[3] It is, however, important to note here that while deliberative interrogation of "end-of-pipe" science for policy is particularly important for nanoST, so are deliberative negotiations around the allocation of funding and the sociotechnical trajectories being pursued. This is particularly important for nanoST because public-sector investment has been primarily responsible for stimulating the development and institutionalization of the field (Schummer 2007; Bürgi 2006; and as shown in Chapter 4).

### 10.3.6 Risk As the Only Legitimate Social Concern

The predominance of risk as a way of structuring discussion on nanotechnology and its role in our futures means that physical harm to human and environmental health appears as the only legitimate social concern. Other concerns relating to social and ethical issues are relegated to the margins of the debate and are given no real substantive value when it comes to decision making. For decision makers to take action—such as to steer funding away

from nanotechnology or to erect regulatory barriers to slow its rapid development—concrete evidence of serious risk to human health or the environment
is required. This is despite the debate around genetically modified organisms (GMOs) clearly demonstrating the possibility for concerns extending
beyond those of physical risk (de Melo-Martin and Meghani 2008). For example, in the debate about the use of GMOs in agriculture, public concerns can
relate to social issues such as corporate control over the food chain or ethical
questions about organismal integrity and the crossing of species or kingdom boundaries (for example, incorporating animal or microbial genes into
a plant). Physical risks to human health and the environment are certainly
important for nanotechnology, but they are not the *only* legitimate issues of
social concern. As highlighted in previous chapters, there are other relevant
social questions to be explored, such as potential impacts on national and
international labor markets through product replacement, as well as ethical
questions, such as how the development of nanoST supports particular concepts of the human relationship to the biological world.

If nanoST truly represents a revolutionary new field with the potential to
affect every aspect of our daily lives, we would be wise to interrogate some of
the broader social and ethical questions at stake, especially if we are to think
in terms of enduring sustainability. This includes asking questions such as:

- What are the underlying assumptions and visions driving nanoST
  forward?
- Do we support these assumptions and visions?
- How might applications change our societies and communities
  in practice?
- How might applications affect fundamental concepts such as
  human/nature relations?
- Who will be the winners and who will be the losers?
- How might ethical values of justice and fairness be applied?
- Can we choose to pursue certain aspects and not others?
- Who or what is controlling where the field is going?
- How should we control it?
- Can we steer nanoST in sustainable or socially beneficial directions?
- How can nanoST contribute to our concept of the good life?
- In what exactly is it that we want to prioritize and invest our time
  and money?

Some of the fundamental beliefs supporting the currently permissive position on nanotechnology commercialization (despite the lack of comprehensive toxicological research) include assumptions that economic growth is the
highest good; all innovation contributes to economic growth and is thereby

good; progress is equivalent to technological advance; and technological fixes to any future problems are possible. Although these beliefs may hold true for some people, legitimate alternatives clearly exist and these should rightfully be subject to open negotiation and debate.

Underlying all questions of science policy is the extent to which we believe citizens and social institutions have control over the trajectories of science and technology. If we take the technologically determinist view that technology determines its own path of development and, subsequently, also our social structures and cultural beliefs, there is little we can do except try and minimize the impact of any negative consequences that may arise. If, however, we believe that social, political, and economic factors (such as funding bodies, legal constraints, regulatory institutions, patterns of consumption, and cultural values) play a role in determining what science and technology are pursued, arguably we have the power to take a more active role and guide science and technology in directions that seem most beneficial and desirable according to our social goals and ethical frameworks. This entails a shift from risk governance to innovation governance (Felt and Wynne 2007). If we wanted to take this a step further and think about not only our own goals and needs but those of the larger biological community of which humans are a part, we may begin to discuss a shift from innovation governance to ecological governance. In what follows, I outline a number of alternative decision-aiding tools that exist for new technologies that aim to integrate the consideration of social and ethical issues more directly. These alternative approaches could help policymakers move beyond questions of technical risk and open up for the consideration of multiple futures.

## 10.4 Beyond Risk

### 10.4.1 Benefits Assessment

Perhaps most pressing when talking about the need to move beyond a sole focus on assessing risks, is the paradox for decision making that currently exists. When employing a discourse of risk, decision makers argue that they require concrete evidence of physical harm before taking action against the commercialization of nanotechnologies. At the same time, however, governments are supporting nanotechnology development based on largely hypothetical and unproven claims to benefits. If we continue to require comprehensive risk analysis to stop nanotechnology development, it could be argued that perhaps we at least need to balance this with comprehensive benefit analysis to support it. For example, do the benefits of nanosilver socks really outweigh the risks that the silver nanoparticles and ions pose to human health and the environment? While this may be seen as just advocating cost–benefit analysis over risk analysis, a danger exists that without subjecting the benefits of particular nanotechnology applications to equally rigorous scrutiny and standards of proof for decision making, a paradox emerges that will always favor the commercialization of (potentially dangerous) new technologies.

The current (and usually implicit) default approach is that what is considered socially beneficial and desirable is increased economic growth and, therefore, as long as nanoST perpetuates this, the field is supported and encouraged (with the market ultimately taking care of what will be developed or not). Other approaches that place more emphasis on specific social goals and ethical priorities are, however, starting to emerge for nanoST. For example, in recent years the European Commission has released a code of conduct for nanoST research (European Commission 2008), which is intended to be used by researchers to help develop projects and by funders to decide which projects to finance. In the code, responsible nano research is presented as that which is precautionary, makes a contribution to the achievement of a sustainable society and the millennium development goals, and is transparent and comprehensible to all. In consultation with stakeholders, the commission also identified areas of restriction for nanoST research, including a prohibition on human enhancement research and a selective moratorium on developing products involving intrusion into the human body (for example, food and cosmetics) (European Commission 2008). This means that when thinking about which research fields should be prioritized and given public funding, the European Commission is advocating a more active steering of developments in socially desirable and publicly negotiated directions.

### 10.4.2 Knowledge Assessment

It has been suggested that when stakes are high, values in dispute, and decisions urgent (as is indeed the case for nanoST), a new type of science for policy is needed—a "postnormal science" (Funtowicz and Ravetz 1993). This is a science that acknowledges the uncertainties involved and allows for a process of extended peer review. Such a process would evaluate the quality of the knowledge employed through deliberations involving multiple scientific disciplines, stakeholders, and members of the public. While different disciplines within the scientific community could debate issues such as the methodological soundness and alternative interpretations for particular scientific studies, the broader community could deliberate over issues such as relevant protection goals, endpoints for the assessment and the weight that should be given to different types of studies. A broad-based deliberative process allows the strength and quality of any evidence for decision making to be tested by exploring how it could be differentially framed and interpreted and what degree of support different choices and assumptions attain within different communities. The approach of pedigree assessment offers one established tool for conducting such processes.

### 10.4.2.1 Pedigree Assessment

The concept of developing "pedigrees" of science for policy was first proposed by Funtowicz and Ravetz (1990). The tool of pedigree assessment aims to make explicit the different value-laden choices and assumptions involved in developing knowledge and open these up for broadly deliberative evaluation. Pedigrees aim to provide "an evaluative account of the production process of information" (van Der Sluijs et al. 2005, 482), with the assumption that a transparent identification and negotiation of the various choices and assumptions involved will "enhance the quality and robustness of the knowledge input in policy-making" (Craye, Funtowicz, and van Der Sluijs 2005, 216). In the first instance, pedigrees of knowledge are concerned with identifying the crucial aspects in the production of scientific knowledge—that is, those places where choices and assumptions are made. These crucial aspects include issues such as how the scientific problem is defined, which method is chosen for the research, which endpoints and indicators are used, which statistical tools are applied, how the results are interpreted, to what kind of review the study has been subjected, and how the findings have been communicated. Ideally, sets of critical questions and a qualitative evaluative scale are then developed for each crucial aspect. For example, if examining the negative health effects of a particular nanomaterial on an organism, one crucial aspect may be the choice of indicators (the measurable elements that determine an effect) where there is a choice between measuring observed fatalities, number of tumors, white blood cell counts, or strength of fingernails. A critical question may ask: "How well does the selected indicator cover the effect one wants to have knowledge about?" A qualitative scale might progress from exact measure, good fit, well correlated to weakly correlated, where strength of fingernails might be seen as weakly correlated and fatalities an exact measure. The pedigree is the resulting matrix of the assessment of the knowledge based on the qualitative ranking of the various crucial aspects (presentable in a range of forms; see Craye, Funtowicz, and van Der Sluijs 2005; van Der Sluijs et al. 2005; Wickson 2009). By selecting crucial aspects, critical questions, and qualitative scales, and performing the evaluation, emphasis is typically placed on negotiation in workshops involving a range of stakeholders—particularly those with different perspectives, values, and interests. Obviously, the development of a pedigree of knowledge will usually involve various levels of dissent and debate, and this underlies its usefulness—the process enables the importance of value-laden choices and assumptions in science for policy to be made apparent through contestation, as well as the meaningfulness of these to be assessed and discussed from various perspectives. In this way, it opens up by making clear the impact of multiple perspectives rather than closing down by trying to provide an objective basis for decision making.

### 10.4.3 Technology Assessment

Technology assessment is an approach that is specifically concerned with identifying the potential social impacts and implications of new technologies. Although the concept and practice has been around for decades, recent developments in the field have sought to alter the structure and aims of former approaches. For example, in its original form, technology assessment took the technology as a given, was primarily aimed at predicting social impacts, and was a process directed at providing information to policymakers. In new approaches, however, much more weight is given to the idea of a co-production between science and society. The specific aim then becomes to integrate social science research and public values into research and development (R&D) processes more directly. The two most widely known examples of this approach are constructive technology assessment and real-time technology assessment.

#### 10.4.3.1 Constructive Technology Assessment

Constructive technology assessment (CTA) has largely been developed in the Netherlands, with Arie Rip the most high-profile proponent of the approach. CTA begins from a belief in the co-production of science and society and is particularly focused on the construction of technologies, with an aim to broaden the design of new technologies so that societal aspects are explicitly included as important design criteria (Scot and Rip 1997). The focus is also specifically on modulation through iterative interaction between social and technological actors. In this sense, the idea is to have technology developers modifying designs according to social criteria, trialing new approaches, engaging in broad-based dialogues, modifying designs further if required, trialing again, and so on. While the actual process and tools used may be diverse, there are three basic elements considered key to CTA approaches: firstly, a process of sociotechnical mapping in which the dynamics of technology development are mapped out in combination with social views and preferences; secondly, early and controlled experimentation with technologies to identify unintended impacts and allow for modulation in design; and finally, dialogue between innovators and a range of stakeholders so that social needs and demands have the opportunity to help shape innovation processes (Scot and Rip 1997). CTA is already being applied to nanoST through the work of Rip and colleagues in the Netherlands (see Rip 2008).

#### 10.4.3.2 Real-Time Technology Assessment

Real-time technology assessment (RTTA) has primarily been developed by Dave Guston and Daniel Sarawitz in the United States. The RTTA approach takes the same general starting point as CTA, namely a belief in the co-production of science and society and the need to engage directly in R&D processes. However, RTTA does not engage in experimentation and chooses to place more emphasis on situating the new technology within a historical context, as well as mapping how knowledge, values, and perceptions are changing through time (Guston and Sarewitz 2002). The RTTA approach is unique in the way that it brings together and integrates a range of established approaches, methods, and interests from across the social sciences. In doing so, it organizes itself around four types of activities that are said to be mutually supportive and simultaneously performed. The four interlinked components of RTTA are:

1. Analogical case studies (where relevant examples of past innovations are researched so as to understand patterns of societal response).
2. Research program mapping (where current R&D activities are mapped and monitored to understand who is doing what and where).
3. Communication and early warning (where studies are performed to understand media portrayals, public attitudes, and public responses to media portrayals, in this way studying how public knowledge and opinions change over time).
4. Technology assessment and choice (forecasting potential societal impacts, developing and deliberating various future scenarios, and evaluating the impact of RTTA activities on R&D).

Guston and Sarewitz are currently working with a range of collaborative partners in the United States to conduct RTTA style activities on nanoST (see CNS/ASU 2010).

### 10.4.4 Alternatives Assessment

The approaches mentioned so far go beyond risk and allow for, at least to some extent, the direct consideration of social aspects and discussions on different sociotechnical futures. They are, however, all focused on analyzing a particular technology rather than analyzing how the impacts and knowledge relating to this technology compare with available alternatives. Ideally, when considering whether to support the development of nanoST, we should be asking what the alternative routes are to achieving the same objectives that nanotechnologies aim to achieve and how all the alternatives compare on social, environmental, and ethical grounds. Multicriteria analysis is one decision-aiding tool that analyzes new technologies according to various criteria and does this in a way that allows for comparisons to be made across a range of options. Like technology assessment methodologies, the approach of multicriteria analysis has been practiced for many years; and like technology assessment, new forms have recently developed to respond to criticisms directed at older approaches. For multicriteria assessment, one of the significant new approaches is multicriteria mapping.

### 10.4.4.1 Multicriteria Mapping

Multicriteria mapping (MCM) is a method developed by Andy Stirling in the United Kingdom. It specifically aims to document stakeholder opinions on a range of different policy options in such a way that the various evaluative criteria, interests, and values that underpin their opinions are made clear. It therefore aims to open up assessment approaches by mapping a range of available alternative options and showing how these alternatives are evaluated by various forms of knowledge, framings, and values (Stirling 2009). The MCM approach shares the same four basic stages as multicriteria analysis:

1. Characterizing a range of alternative options available for achieving a particular aim.
2. Developing a set of relevant criteria for appraising the different options.
3. Evaluating each option by assigning numerical scores for its performance according to each criterion.
4. Assigning a weighting to each criterion in order to reflect its relative importance for the stakeholder involved.

In practical terms, MCM requires researchers to develop a list of alternative options available for achieving a particular objective and identifying a group of relevant and diverse stakeholders. Interviews with these stakeholders are then conducted to identify evaluative criteria, rank the different options according to these criteria, assign weightings to the criteria, and review and modify the results if required. This information is audiorecorded and entered in real time into a specifically developed software program (MC Mapper). In contrast to many multicriteria approaches, however, MCM allows the stakeholders involved (and not just the researchers) to add and redefine the various options presented to them and to define the criteria for evaluation themselves. Additionally, MCM is unique in its focus on also capturing uncertainties. This is achieved by having the stakeholders assign two performance scores for each criterion, such as how the option scores for a particular criterion under both optimistic and pessimistic assumptions. This allows for performance scores to be presented as a range of values, showing variance according to the deemed level of uncertainty. Unlike other approaches to multicriteria analysis, MCM is not focused on arriving at a singular best-performance option but acts as a tool that allows exploration of the maps of differences in option performance. In this way, MCM aims at opening up policy discussions through exploration of a range of different options according to varying values rather than simply closing down by trying to develop a consensus view on the best option (Stirling, Lobstein, and Millstone 2007).

### 10.4.8 Articulating Values

One of the important approaches to evaluating nanoST according to social—as opposed to simply technical—criteria has been the development of exercises in upstream engagement (Wilsdon 2005). However, many of the deliberative and socially inclusive public engagement exercises currently occurring around the advance of nanoST (such as consensus conferences, citizens' juries, focus groups, and science cafes) simply ask people to give their opinions on nanoST after considering the risks and benefits involved in hypothetical future scenarios (Delgado, Kjølberg, and Wickson 2010). In these approaches, the development and advance of nanoST is therefore essentially taken as a given, and discussions are aimed at understanding what is considered acceptable so as to avoid a public backlash similar to that against GMOs (Doubleday 2007). Holding public engagement exercises focused on nanoST in this way demands that those involved in the debate are well informed about technical details relating to what nanoST are and the potential risks that they may pose (either before attending the exercise or through information given early in the process). Not only does this create challenges in finding appropriate participants and in framing any provided information in an unbiased way, it also narrows the scope of the discussions dramatically in terms of how we would like our nanotechnology futures to be, rather than whether we want them at all. Another way to approach discussions, however, would be to begin by exploring wider avenues of enquiry, such as what participants value more generally, what they consider to be "the good life," how they understand progress, what kind of a future they would like to inhabit, and to then consider the potential desirability and role of nanoST within these visions and values. This approach would structure processes of public engagement and deliberation around discussions of fundamental values, rather than a specific technology per se. This approach would not imply that prior technical knowledge was required and would thereby allow all citizens to engage in discussions equally. It would also place social goals and ethical values rather than nanoST in the foreground of discussions.

## 10.5 Conclusion

In this chapter I have suggested that the discourse of human health and environmental risk dominates public discussions and political decision making on new technologies. By highlighting some of the broader social elements that are neglected in technical discussions of risk, I have aimed to show the limitations and narrow nature of this discourse. I then considered the case of nanotechnology and outlined not only some of the important risk issues involved but also the enormous uncertainties obstructing our ability to perform comprehensive scientific risk assessment in this field. I also suggested some of the social and ethical questions that are neglected by a sole focus on human health and environmental risk. Having criticized the narrow and technocratic nature of risk assessment as a decision-aiding tool (as well as our ability to apply it to nanoST), I then outlined a number of alternative tools available that are more specifically focused on integrating social and technical considerations, and in turn open up pathways toward potential futures. This included a description of different forms of knowledge assessment (for example, pedigree assessment), technology assessment (such as constructive technology assessment and real-time technology assessment), and alternatives assessment (like multicriteria mapping). I also considered what it might mean to think in more detail about the benefits involved in technological advance and its relationship to social and ethical values. In this way, I hope that this chapter has encouraged us all to move beyond simply discussing the risks associated with new technologies as they are thrust upon us and raise for discussion the multiple sociotechnical trajectories that we can possibly pursue to move toward sustainable futures.

## References

Adams, J. 1995. *Risk*. London: University College London Press.

Adams, J. and M. Thompson. 2002. *Taking Account of Societal Concerns about Risk: Framing the Problem*. London: Health and Safety Executive, 1–43.

Anonymous. 2007. Call for a moratorium on nanotechnology. *Sydney Morning Herald*, 17 March. Available at: http://www.smh.com.au/news/National/Call-for-moratorium-on-nanotechnology/2007/03/17/1174080202836.html (accessed 22 July 2009).

Balbus, J.M., A.D. Maynard, V.L. Colvin, et al. 2007. Meeting report: Hazard assessment for nanoparticles—Report from an interdisciplinary workshop. *Environmental Health Perspectives* 115(11): 1654–59.

Baun, A., S.N. Sørensen, R.F. Rasmussen, et al. 2008. Toxicity and bioaccumulation of xenobiotic organic compounds in the presence of aqueous suspensions of aggregates of nano-C60. *Aquatic Toxicology* 86:379–87.

Beck, U. 1986. *Risk Society: Towards a New Modernity*. London: Sage.

Bergamaschi, E. 2009. Occupational exposure to nanomaterials: present knowledge and future development. *Nanotoxicology*. 3(3): 194–201.

Bürgi, B.R. 2006. Societal implications of nanoscience and nanotechnology in developing countries. *Current Science* 90(5): 645–58.

Center for Nanotechnology in Society—Arizona State University (CNS-ASU). 2010. *Program: Research* http://cns.asu.edu/program/research.htm (accessed 26 January 2009).

Choi, J-W., G. Ramachandran, and M. Kandlikar. 2009. The impact of toxicity testing costs on nanomaterial regulation. *Environmental Science and Technology* 43(9): 3030–34.

Craye, M., S. Funtowicz, and J. van Der Sluijs. 2005. A reflexive approach to dealing with uncertainties in environmental health risk science and policy. *International Journal of Risk Assessment and Management* 5(2/3/4): 216–36.

de Melo-Martin, I. and Z. Meghani. 2008. Beyond risk. *EMBO Reports* 9(4): 302–6

Delgado, A., K. Kjølberg, and F. Wickson. 2010. Public engagement coming of age: From theory to practice in STS encounters with nanotechnology. *Public Understanding of Science* http://pus.sagepub.com/content/early/2010/04/13/09636625103630.54.full.pt.html (accessed July 7, 2011).

Donaldson, K., R. Aitken, L. Tran, et al. 2006. Carbon nanotubes: A review of their properties in relation to pulmonary toxicology and workplace safety, *Toxicological Sciences* 92(1): 5–22.

Doubleday, R. 2007. Risk, public engagement and reflexivity: Alternative framings of the public dimensions of nanotechnology. *Health, Risk and Society* 9(2): 211–27.

Douglas, M. and A. Wildavsky. 1982. *Risk and Culture: An Essay on the Selection of Technical and Environmental Dangers*. Berkeley: University of California Press.

Editor. 2008. The same old story. *Nature Nanotechnology* 3(12): 699–700.

EFSA. 2009. The potential risks arising from nanoscience and nanotechnologies on food and feed safety: Scientific opinion of the Scientific Committee. *EFSA Journal* 958: 1–39. Available at http://www.efsa.europa.eu/cs/BlobServer/Scientific_Opinion/sc_op_ej958_nano_en.pdf?ssbinary=true (accessed 21 July 2009).

ETC Group. 2009. *Nanotechnology*. Available at: http://www.etcgroup.org/en/issues/nanotechnology.html (accessed 22 July 2009)

European Commission. 2008. *Recommendation on a Code of Conduct for Responsible Nanosciences and Nanotechnologies Research*. Brussels: European Commission. Available at: http://ec.europa.eu/nanotechnology/pdf/nanocode-rec_pe0894c_en.pdf (accessed 23 July 2009).

Felt, U. and B. Wynne. 2007. *Taking European Knowledge Society Seriously*. Luxembourg: Office for Official Publications of the European Communities.

Friends of the Earth Australia. 2006. *Nanomaterials, Sunscreens and Cosmetics: Small Ingredients, Big Risks*. Available at: http://nano.foe.org.au/node/100 (accessed 22 July 2009).

Funtowicz, S.O. and J.R. Ravetz. 1990. *Uncertainty and Quality in Science for Policy*. Dordrecht: Kluwer Academic.

Funtowicz, S.O. and J.R. Ravetz. 1993. Science for the post-normal age. *Futures* 25(7): 739–55.

Funtowicz, S. and J.R. Ravetz. 1994. Uncertainty, complexity and post-normal science. *Environmental Toxicology and Chemistry* 13(12): 1881–85.

Grieger, K.D., S. Hansen, and A. Baun. 2009. The known unknowns of nanomaterials: Describing and characterising uncertainty within environmental, health and safety risks. *Nanotoxicology*. (advance access): 1–12. Available at: http://www.informaworld.com/smpp/content~db=all~content=a912823263 (accessed 21 July 2009).

Guston, D. and D. Sarewitz. 2002. Real-time technology assessment. *Technology in Society* 24:93–109.

Handy, R.D., and R. Owen. 2008. The ecotoxicology of nanoparticles and nanomaterials: current status, knowledge gaps, challenges and future needs. *Ecotoxicology* 17:315–25.

Holman, M.W. and D.I. Lackner. 2006. *The Nanotech Report*, 4th ed. New York: Lux Research.

Kandlikar, M., G. Ramachandran, A. Maynard, et al. 2007. Health risk assessment for nanoparticles: A case for using expert judgement. *Journal of Nanoparticle Research* 9:137–56.

Lam, C-W., J.T. James, R. McCluskey, et al. 2004. Pulmonary toxicity of single-wall carbon nanotubes in mice 7 and 90 days after intratracheal instillation. *Toxicological Sciences* 77:126–34.

Lockman, P.L., J.M. Koziara, R J. Mumper, et al. 2004. Nanoparticle surface charges alter blood-brain barrier integrity and permeability. *Journal of Drug Targeting* 12(9–10): 635–41.

Lucas, C. 2003. We must not be blinded by science. *Guardian*, 12 June. Available at: http://www.guardian.co.uk/politics/2003/jun/12/nanotechnology.science (accessed 22 July 2009).

Ma-Hock, L., S. Treumann, V. Strauss, et al. 2009. Inhalation toxicity of multi-wall carbon nanotubes in rats exposed for three months. *Toxicological Sciences* (advanced online access). Available at: http://toxsci.oxfordjournals.org/cgi/content/abstract/kfp146 (accessed 20 July 2009).

Maynard, A., P.A. Baron, M. Foley, et al. 2004. Exposure to carbon nanotube material: Aerosol release during the handling of unrefined single-walled carbon nanotube material. *Journal of Toxicology and Environmental Health* 67(1): 87–107.

Muller, J., F. Huaux, N. Moreau, et al. 2005. Respiratory toxicity of multi-wall carbon nanotubes. *Toxicology and Applied Pharmacology* 207:221–31.

Oberdörster, E. 2004. Manufactured nanomaterials (fullerenes, $C_{60}$) induce oxidative stress in the brain of juvenile largemouth bass. *Environmental Health Perspectives* 112(10): 1058–62.

Oberdörster, G., E. Oberdörster, and J. Oberdörster. 2005. Nanotoxicology: An emerging discipline evolving from studies of ultrafine particles. *Environmental Health Perspectives* 113(7): 824–39.

Otway, H. 1980. The perception of technological risks: A psychological perspective. In *Technological Risk: Its Perception and Handling in the European Community*, edited by M. Dierkes, S. Edward, and R. Coppock, 35–44. Cambridge: CEC Oelgeschlager.

Otway, H. 1987. Experts, risk communication and democracy. *Risk Analysis* 7(2): 125–29.

Owen, R. and R. Handy. 2007. Formulating the problems for environmental risk assessment of nanomaterials. *Environmental Science and Technology* 15:5582–88.

Poland, C.A., R. Duffin, I. Kinloch, et al. 2008. Carbon nanotubes introduced into the abdominal cavity of mice show asbestos-like pathogenicity in a pilot study. *Nature Nanotechnology* 3:423–28.

Rip, A. 2008. Nanoscience and nanotechnologies: Bridging gaps through constructive technology assessment. In *Handbook of Transdisciplinary Research*, edited by G. Hirsch Hadorn, H. Hoffmann-Riem, S. Biber-Klemm, W. Grossenbacher-Mansuy, D. Joye, C. Pohl, U. Wiesmann, and E. Zemp, 145–58. Dordrecht: Springer.

Robins, R. 2002. The realness of risk: Gene technology in Germany. *Social Studies of Science* 32(1): 7–35.

Royal Commission on Environmental Pollution. 2008. *Novel Materials in the Environment: The Case of Nanotechnology*. Norwich: Stationery Office.

Royal Society and the Royal Academy of Engineering. 2004. *Nanoscience and Nanotechnologies: Opportunities and Uncertainties.* London: Royal Society. Available at: http://www.nanotec.org.uk/finalReport.htm (accessed 21 July 2009).

SCENIHR. 2006. *The Appropriateness of Existing Methodologies to Assess the Potential Risks Associated with Engineered and Adventitious Products of Nanotechnologies.* European Commission, Health & Consumer Protection Directorate General. Available at: http://ec.europa.eu/health/ph_risk/committees/04_scenihr/docs/scenihr_o_003b.pdf (accessed 21 July 2009).

Schummer, J. 2007. The global institutionalisation of nanotechnology research: A bibliometric approach to the assessment of science policy. *Scientometrics* 70(3): 669–92.

Schwarz, M. and M. Thompson. 1990. *Divided We Stand: Redefining Politics, Technology and Social Choice.* Hemel Hempstead: Harvester Wheatsheaf.

Scot, J. and A. Rip. 1997. The past and future of constructive technology assessment. *Technological Forecasting and Social Change* 54(2/3): 251–68.

Shedova, A., V. Castranova, E. Kisin, et al. 2003. Exposure to carbon nanotube material: Assessment of nanotube cytotoxicity using human keratinocyte cells. *Journal of Toxicology and Environmental Health* 66(20): 1909–26.

Shedova, A.A., E.R. Kisin, R. Mercer, et al. 2005. Unusual inflammatory and fibrogenic pulmonary responses to single-walled carbon nanotubes in mice. *American Journal of Physiology—Lung Cellular and Molecular Physiology* 289:L698–L708.

Slovic, P. 1987. Perception of risk. *Science* 236:280–85.

Slovic, P. 1991. Beyond numbers: A broader perspective on risk perception and risk communication. In *Acceptable Evidence: Science and Values in Risk Management*, edited by D.G. Mayo and R.D. Hollander, 48–65. Oxford: Oxford University Press.

Slovic, P. 1998. Perceived risk, trust and democracy. In *Earthscan Reader in Risk and Modern Society*, edited by R.E. Lofstedt and L. Frewer, 181–92. London: Earthscan.

Slovic, P., B. Fischhoff, and S. Lichtenstein. 1982. Why study risk perception? *Risk Analysis* 2(2): 83–93.

Stirling, A. 1999. Risk at a turning point? *Journal of Environmental Medicine* 1:119–26.

Stirling, A. 2009. *Multicriteria Mapping.* Available at: http://www.multicriteriamapping.org/ (accessed 28 January 2009).

Stirling, A. and D. Gee. 2002. Science, precaution and practice. *Public Health Reports* 117:521–33.

Stirling, A., T. Lobstein, and E. Millstone. 2007. Methodology for obtaining stakeholder assessments of obesity policy options in the PorGrow project. *Obesity Reviews* 8(Suppl. 2): 17–2

Takeda, K., K. Suzuki, A. Ishihara, et al. 2009. Nanoparticles transferred from pregnant mice to their offspring can damage the genital and cranial nerve systems. *Journal of Health Science* 55(1): 95–102.

Thompson, M., R. Ellis, and A. Wildavsky. 1990. *Cultural Theory.* Boulder, Colo.: Westview Press.

van Der Sluijs, J.P., M. Craye, S. Funtowicz, et al. 2005. Combining quantitative and qualitative measures of uncertainty in model-based environmental assessment: The NUSAP system. *Risk Analysis* 25(2): 481–92.

Walker, N.J., and J.R. Bucher. 2009. A 21st century paradigm for evaluating the health hazards of nanoscale materials? *Toxicological Sciences* 110(2): 251–54.

Warheit, D.B., B.R. Laurence, K.L. Reed, et al. 2004. Comparative pulmonary toxicity assessment of single-wall carbon nanotubes in rats. *Toxicological Sciences* 77:117–25.
Wickson, F. 2009. Reliability rating and reflective questioning: A case study of extended review on Australia's risk assessment of *Bt* cotton. *Journal of Risk Research* 12(6): 749–70.
Wilsdon, J. 2005. Paddling upstream: New currents in European technology Assessment. In *The Future of Technology Assessment*, edited by M. Rodemeyer, D. Sarewitz, and J. Wilsdon, 22–29. Washington, D.C.: Woodrow Wilson International Center for Scholars.
Winner, L. 1986. *The Whale and the Reactor: A Search for Limits in an Age of High Technology*. Chicago: University of Chicago Press.
Wynne, B. 1992. Uncertainty and environmental learning: Reconceiving science and policy in the preventive paradigm. *Global Environmental Change* 2(2): 111–27.
Wynne, B. 2001. Creating public alienation: Expert cultures of risk and ethics on GMOs. *Science as Culture* 10(4): 445–81.
Zhu, L., D.W. Chang, L. Dai, et al. 2007. DNA damage induced by multiwalled carbon nanotubes in mouse embryonic stem cells. *Nano Letters* 7(12): 3592–97.

## Endnotes

1. The plural terms *nanosciences* and *nanotechnologies* are being distinguished here from the singular *nanotechnology*, as it is necessary to distinguish between different forms of nanotechnology as clearly as possible when delineating specific risks. The nanosciences are referred to here because there can also be risks associated with research practices themselves, as well as with products manufactured for nanoscience research.

2. The term *engineered nanoparticles* is used to indicate that the focus here is not on naturally occurring nanoscale particles but on those purposefully engineered and manufactured by humans. We can certainly learn from our experience with "natural" nanoscale particles (e.g., that they often have the potential to create harm) and from the methods developed for studying them (e.g., research on ultrafine particles), but it is also important to note that the purposeful generation of nanoparticles creates a range of novel materials and that the unique characteristics of these need to be researched to understand their (eco)toxicological potential.

3. Measures as to how this can be done are discussed in more depth in Section 10.3.3.

# 11

# Nanotechnology and State Regulation (India)

Nidhi Srivastava and Nupur Chowdhury

## CONTENTS

## 11.1 Introduction

As around the world, India has witnessed a remarkable surge of interest from academics and entrepreneurs in the development of nanotechnology. This has been further supported by the state, through the unveiling of the Nano Mission in 2007, through which it has hoped to streamline public investment into research and development (R&D). Infrastructure, science education, and entrepreneur-support programs in the field of nanotechnology are expected to bear fruit over the next decades, with the field anticipated to rival that of the information and communication technology (ICT) revolution in India. The institutional framework for nano R&D has implications in terms of determining the regulatory space available for undertaking product safety and regulation in the context of emerging technologies like nanotechnology. The Indian Ministry of Science and Technology is the nodal ministry for promotion of R&D in the area of nanotechnology. It administers its functions through three departments: the Department of Science and Technology (DST), the Department of Biotechnology (DBT), and the Department of Scientific and Industrial Research (DSIR). Through financial support, the DST has been the most instrumental agency within the government for encouraging nanotechnology development and application. In 2001, a Nano Science and Technology Initiative (NSTI) was launched and, as a follow-up, Nano Mission was set up in 2007. The department, since engaging with the agenda of promoting nanotechnology as a thrust area, has declared an investment of INR1000 crore[1] ($24 billion) for five years, commencing 2007, for basic and applied research promotion, infrastructure support, education, and international collaboration. The department provides the secretariat to the Nano Mission Council, which is the highest advisory policymaking body for nanotechnology in India. Besides the Council, the Nano Mission includes two other advisory groups: the Nano Applications and Technology Advisory Group and the Nano Science and Advisory Group.

Nano Mission is an umbrella program implemented by DST for capacity building toward the overall development of nanotechnology research in India. Of the total proposed outlay (INR9,300 crore) for DST under the XI five-year plan, INR1000 crore have been assigned for the Nano Mission. There are certain public–private initiatives in the form of industry-linked projects under the mission, half of which are with companies dealing with drugs and pharmaceuticals.[2] Other than the programs within the Ministry of Science and Technology, the Ministry of Information and Communication Technology and the Ministry of Defense have also shown interest in promoting nanotechnology R&D pertaining to their areas of activity.

On the ground, there have been a number of nanotechnology-based inventions that have been patented in India and will be marketed commercially over the short term. Several nanotechnology-based applications have also been introduced within the pharmaceutical and drug delivery sectors and have already hit the health market.

Given the still uncertain nature of the safety risks that underlie such applications in the health sector (as detailed in Chapters 3 and 10), obvious questions arise as to the capacity of the current regulatory framework for addressing these new challenges in an appropriate manner. This is fundamental to ensuring that the nanotechnology-based drugs or applications are safe for handling and human use. Since there are several sectors in which nanotechnology-based applications have been launched (including electronics, textiles, and food packaging), it is imperative that we focus on a specific sector so as to enable a critical discussion of stand-alone legislations. We focus on nanotechnology-based health applications for several reasons. First, this sector has witnessed the launch of several nanotechnology-enabled products and there are a several more in the pipeline. Second, the regulatory framework for drugs in India has undergone a process of revision and this is an opportune moment to discuss policy considerations that were addressed in this exercise. We have also analyzed the food regulatory regime, since a number of nano-applications in the food packaging sector are currently under development in India, and also because both the health and food regulatory regimes share common principles and norms and are among the most strictly regulated sectors in the country. A study of the regulatory regimes for health and food will provide a more holistic perspective on the manner in which nano-applications in food and health will be regulated in India.

In the following discussion we provide an overview of the kinds of nanotechnology-related drugs and applications that have been launched or expected to be launched in the Indian market. A number of these applications have been developed and manufactured by Indian pharmaceutical companies. This is unsurprising, given that many Indian pharmaceutical companies have both the financial strength and research capabilities required to develop nanotechnology-based drugs and drug delivery devices (henceforth nanomedicines). Such companies have been motivated by the desire to expand current product ranges—which at present are largely limited to generics. We aim this discussion toward understanding the product range and depth of nanotechnology applications in the pharmaceutical sector in India. Following this, we give an overview of the product safety and quality regulations that will govern the manufacture and marketing of nanomedicines in India and explore whether they equip India with ways to deal with the new challenges that are posed by nanotechnology in this sector as well as those applications in the food sector. Essentially, the aim is to identify critical points within the legal framework that would need to be reexamined in light of the changing characteristics of such new applications. Further, we also want to explore legislative and policy space available in the current regime to develop regulatory norms to address these new challenges.

In the following section we discuss the effectiveness of the institutional structure that oversees India's regulatory framework. Effectiveness of regulatory institutions is discussed in terms of their capacity (expertise) and operational mandate. In this section we also identify and analyze critical challenges that will influence the performance of the current institutional framework. We include some preliminary remarks that question what the ideal institutional framework for nanotechnology, more generally, should be. This debate is an important one in the context of India for two reasons: first, because previously there have been technology-specific horizontal regulatory arrangements (such as biotechnology) that are still largely untested in terms of their effectiveness in fulfilling regulatory objectives, and second, there has been a statement made by the head of the Nano Mission Council that seems to suggest plans are afoot for setting up a Nanotechnology Regulatory Board (Press Trust of India 2010). This implies that all nanotechnology products will be regulated by this body. In this context it is important we visit the debate between horizontal regulation and sector-specific regulations and what kind of permutations and combinations will be suitable. In the concluding section we reiterate specific comments on the current state of environmental health regulation in terms of preparedness to meet the challenges that will be posed by nanomedicine applications, and we particularly discuss certain policy options to evolve a robust and effective regulatory framework.

## 11.2  Product Range in the Health Sector

Of all the sectors in which nanotechnology is being applied, the health sector has attracted the most research interest, private sector involvement, and commercialization (Vivekanandan 2009). Moreover, several nanomedicines have already entered world markets or are ready to be commercialized. Within the health sector in India, it is pertinent to note that most of the current initiatives are focused toward the curative aspects of health research. Even the curative aspects are designed to treat lifestyle diseases rather than neglected and localized diseases, such as malaria, typhoid, and tuberculosis. However, recent research supported by the Indian Council of Medical Research (ICMR) is focusing on diseases of importance to India, like typhoid and tuberculosis. Table 11.1 gives an overview of health-related applications already launched or nearing launch in the Indian market.

All the nanotechnology-based drugs and pharmaceutical substances listed in Table 11.1 have been approved under the Indian drugs and cosmetics regulatory regime. Approval from the Drug Controller General of India (DCGI) is required to be obtained before they are launched in the market. In 2009 one of the drugs, Albupax of Natco Pharma, was ordered to be withdrawn from the market on the grounds of being substandard after the Central Drug Laboratory found it to contain high levels of endotoxins and chloroform (Dey 2009). The DCGI's decision has been under debate, and the Ministry of Health and Family Welfare has put a stay on the suspension order to the DCGI (Karnataka Drugs and Pharmaceuticals Manufacturers' Association 2010; Financial Express 2010). Natco alleged that the levels of toxins vary largely depending upon the test kit used. Standards, capacity, metrology, and associated risks present major challenges in regulating nanotechnology in India.

## 11.3  Regulatory Issues in Product Quality
##        and Safety Regulations

Despite several commentators underlining the need to develop international regulatory approaches to nanotechnology (Abbott et al. 2006), it is

**TABLE 11.1**

Health-Related Nanotechnology Applications Launched and Developed in India

| S. No. | Technology/ Process | Description | Owner/ Developer |
|--------|---------------------|-------------|------------------|
| 1 | Nanoxel | Paclitaxel-based drug delivery systems (DDS) for cancer drugs | Dabur (Fresenius Kabi Oncology Ltd.) |
| 2 | Albupax | Generic equivalent of Abraxane (nano drug-delivery system at tumor site that wraps the albumin around active drug) | Natco Pharma |
| 3 | Antimicrobial Gel | Silver/gold nanogel, bio-stabilized iron-palladium nanoparticles | Nanocet (Contract Research) |
| 4 | Verigene Platform | DNA functionalized gold gel for diagnostics. | Nanosphere |
| 5 | Estrasorb | DDS for estrogen therapy (drug loaded within the nanoparticle formulation) | Bharat Biotech (with Novavax) |
| 6 | MWCNT1–3 nano beads 1–2, nano fibers 1–4, fibrous carbon nanosize metals | Production of carbon nanomaterials for commercial purposes | Monad Nanotech |
| 7 | TB and typhoid diagnostic kit | Latex agglutination-based test | DRDE/IISc |
| 8 | iSens and silicon locket | Cardiac diagnosis product | IIT Bombay |
| 9 | Nano silver gel developed from nanoparticles of silver | Gel to cure burns | Virtuous Innovation |
| 10 | Drug delivery research | Drug delivery through use of mucoadhesive nanoparticles | Panacea Biotech |
| 11 | Drug delivery device for diseased coronary arteries | Introducing nanoparticles that release drugs to block cell proliferation in the narrowed diseased coronary arteries (through drug eluting stents) | Concept Medical Research Private Ltd. |
| 12 | Antimicrobial spray | Through use of silver nanoparticles | Bhaskar Center for Innovation and Scientific Research, Chennai |
| 13 | Abraxane | A formulation of paclitaxel for targeted drug delivery | Biocon |
| 14 | Hearing aids | The technology uses sensors that contain a giant magneto-resistance switch that uses electron spin rather than magnetic charge to sense signals and store information | Starkey India |
| 15 | Water-soluble carbon nanotube-based cancer drug delivery system | Water-soluble carbon nanotubes that have functional groups on the walls for conjugation with cancer drugs | Cromoz Inc./ IIT Kanpur |
| 16 | Drug scanner | Nanotechnology-based spurious drug detection scanner machine | Bilcare |

important to note that national regulation would seem to be the first necessary step within the regulatory paradigm. Effective national regulation is not only fundamental to ensuring responsible development of nanotechnology (see Chapter 4) but can also act as a foundational experience to developing international regulatory frameworks that could benefit from national regulatory experiments. From the outset, certain caveats need to be made. First, it is important to differentiate between regulation of technology and regulation of applications of that technology. Regulation of a technology per se is not only difficult but, as many regulatory experts have suggested, is also not desirable since technology developments often results in redundant regulatory categories that fail to fulfill regulatory objectives. Bowman and Hodge (see Chapter 12) comment on the impracticality of applying a moratorium on nanotechnology or even nanomaterials per se, given that they are largely descriptive terms that are not very well defined. It is pertinent, therefore, to discuss regulatory issues vis-à-vis specific nanotechnology applications. Nevertheless, in the context of emerging technology, one has witnessed the development of general regulatory principles (vis-à-vis uncertainty and risk regulation) that may have implications for the regulation of nanotechnologies. Thus, regulatory experience garnered in the biotechnology sector (faced within comparable risks) could have an influence on the architecture and design of regulatory structure and the choice of regulatory tools applied in the regulation of nanotechnologies. Third, given that there are a number of applications of nanotechnology, in this chapter we focus on the regulation of applications in health care—or what has generally been referred to as nanomedicines. However, in order to provide a more holistic picture, we also analyze food regulation (given that there is some overlap between substances characterized as food and those characterized as drugs) and environmental regulation (since that would cover all aspects of environmental impacts of nanomaterials used in manufacture and also those that might be discharged at the end of the product life cycle). Fourth, we comment generally on the philosophy of technological regulation in India and whether one can discern aspects of the underlying regulatory culture that might influence regulatory decision making within this context.

We now provide an overview of the scope of the Drugs and Cosmetics Act, 1940 (India), followed by an analysis of specific provisions that would have implications for nanomedicines. This analysis will primarily be a gap analysis, providing specific recommendations vis-à-vis the amendment or revision of provisions of this act. This analysis is followed by examination of the Food Safety and Standards Act 2006 and a broader consideration of whether the current regulatory framework for pharmaceutical regulation, food safety, and environmental protection is equipped to address the regulatory challenges stemming from developments within nanomedicines.

## 11.3.1 Drugs and Cosmetics Act 1940

The primary objective of the Drugs and Cosmetics Act 1940 is to regulate the "import, manufacture, distribution and sale of drugs and cosmetics" (Government of India 2003). The act clearly differentiates between the *ayurveda, siddha,* and *unani* systems of indigenous medicines and that of the modern allopathic medicine system, and sets up parallel systems of regulators and different sets of regulations for both. The act provides for a more liberalized regulatory regime in case of these indigenous systems. Thus traditional practitioners of such medicines—known as Vaidyas and Hakims—who manufacture such medicines for their own patients are exempt from any obligations related to manufacture and sale laid down under this act. This could raise legitimate questions about regulatory coverage in the event of nanomaterial usage within such formulations. For instance, there have already been news reports suggesting that the certain ayurvedic formulations like *bhasma* actually consist of nanoparticles (Bhasmas 2011).

The act also provides for the establishment of the Drugs Technical Advisory Board that would function as a general advisory body to the central and state governments on "technical matters arising out of the administration of this Act" (Section 5). The act also provides for the setting up of the Drugs Consultative Committee to advise the central and state governments and the Drugs Technical Advisory Board on aspects of legal uniformity and coherence within the national drug regulatory system (Section 7). The licensing authority for new drugs has been divided between the Central Drugs Standards Control Organisation (CDSCO) and the state drugs controllers. While the former has the power to issue licenses for the manufacture and imports of drugs listed in Schedules C and C1 (these include biological and special products), the latter oversees the manufacture and sale of other drugs and cosmetics.

The act provides for an expanded definition of the term *drug,* including medical devices for internal uses such as diagnostics and treatment (Section 3.b.iv and Section 3.b.iii). This is, therefore, sufficiently broad to include nano-related health applications. The act also empowers the central government to prohibit the import and manufacture of drugs and cosmetics in the public interest. Risk to human beings and animals is one of the circumstances under which the government can make such a prohibition (Sections 10A and 26A). The act also enables the government to specify the quality of its desired standard. Information disclosure forms an important part of the regulatory apparatus and is especially important in the case of entities that use, apply, or deal with nano-related health applications. The act provides for detailed penalties in the case of contaminated and spurious drug usage. The act prohibits the manufacture and sale of contaminated drugs and cosmetics, including cases where the "container is composed … of any poisonous or deleterious substance which *may* render the contents injurious to health or if it contains any harmful or toxic substance which *may* render it injurious to health" (Sections 17 A[c] and 17E[c] and [e]; our italicized emphasis).

If, for instance, carbon nanotubes (CNTs) for targeted drug delivery and gold nanoparticles used for diagnostics were found to be potentially hazardous, they could fall within this description. In this regard, it is useful to draw attention to the term *may* used in the above sections, as this considerably waters down the requirement for any substance (that is deemed to be toxic per se) proven to be harmful to health to be deemed adulterated. The threshold set refers to "possibility of" or "potential for" adverse health impacts and not the requirement of absolute proof.

Adequate support exists under the act to undertake measures ranging from labeling to prohibition of the manufacture and sale of nano-related health applications in India in cases where there is uncertainty about the safety of such drugs. Nevertheless, as we stated at the beginning of this discussion, the state drug controllers have the responsibility to oversee the regulation of drugs within India, and only seventeen of these have access to drug testing facilities within their state. This brings into sharp focus the capacity deficits faced by India's regulatory authorities. This problem is further exacerbated in the case of nanoparticles since limited studies have been conducted.

In 2003, the Ministry of Health and Family Welfare (MoHFW) set up the Mashelkar Committee[3] to undertake a comprehensive examination of drug regulatory issues, and specifically the problem of spurious drugs (Ministry of Health and Family Welfare, Government of India 2003). The committee's recommendations formed the basis of the Drugs and Cosmetics (Amendment) Act 2008 (Government of India 2008).

### 11.3.2 The Food Safety and Standards Act 2006

The Food Safety and Standards Act 2006 was passed by the Indian parliament to consolidate the laws relating to food safety (Government of India 2006). The primary objective of the act is to establish the Food Safety and Standards Authority (FSSA) with the mandate of standard setting for food items, as well as to regulate the manufacture, storage, distribution, sale, and importing of food—ensuring safe and available wholesome food. There are essentially four aspects of the legal provisions that could have an implication for nanoscale technologies that may be involved in the development of specific areas of the food industry (including new functional materials, food processing, and product development and storage) (Institute of Food Technologists 2006). These include definitions that specify the scope of the regulations, risk assessment by the FSSA, general principles of food administration, and the overriding effects of this legislation. It would be prudent to analyze these related provisions included in the act.

Section 3 of the act lays down the definitions. Section 3(k) defines food additives to mean "'any food additive' ... any substance not normally consumed as a food by itself or used as a typical ingredient of the food, whether or not it has nutritive value, the intentional addition of which to food for a technological (including organoleptic) purpose in the manufacture processing, preparation, treatment, packing, packaging, transport or holding of such food results, or may be reasonably expected to result (directly or indirectly), in it or its by-products becoming a component of or otherwise affecting the characteristics of such food but does not include 'contaminants' or substances added to food for maintaining or improving nutritional qualities."

This definition is broad enough to accommodate nanoparticles that may be used in packaging. Section 3(r) defines a "food safety audit" as "[a] systematic and functionally independent examination of food safety measures adopted by manufacturing units to determine whether such measures and related results meet with objectives of food safety and the claims made in that behalf."

Section 44 enables the FSSA to grant recognition to organizations for the purposes of conducting such food safety audits and ensuring compliance with food safety management systems as provided for under this act. This is an important mechanism available to the FSSA to provide adequate oversight in case of specific nanomaterials that may be potentially harmful and that may be used in food processing industries. Section 16(3) enumerates the duties and functions of the FSSA. It empowers the FSSA to undertake foresight activities in the case of emerging risks. The FSSA is therefore legally well grounded in investigating risks that might emanate from nanomaterials used within the food production process.

The General Principles of Food Safety are a distinctive and welcome addition within this act, given they provide much needed clarity and general coherence to the interpretation and implementation of the various provisions of this legislation (Section 18). One of the general principles relates to the adoption of a precautionary approach in specific circumstances. Section 18(1.c) states that "where in any specific circumstances, on the basis of assessment of available information, the possibility of harmful effects on health is identified but scientific uncertainty persists, provisional risk management measures necessary to ensure appropriate level of health protection may be adopted, pending further scientific information for a more comprehensive risk assessment."

Additionally, such kinds of measures have to conform to the principle of proportionality[4] and should be the least trade-restrictive option available. This is an important power in the hands of the FSSA, enabling it to act with necessary urgency in case of such potential risks—and one that can be very useful to regulate potentially harmful nanomaterials within food production. The other critical aspect relates to the provision of public information. Section 18(1.f) mandates the FSSA to provide for public information disclosure of risks to health that might emanate from potentially risky substances used in food production. Finally, Section 89 states that "the provisions of this Act shall have effect notwithstanding anything inconsistent therewith contained in any other law for the time being in force or in any instrument having effect by virtue of any law other than this Act."

This is essentially a supremacy clause that allows the provisions of this act to override the authority of any other piece of legislation. This is critical in cases of overlap between what are defined as food and as drugs, because the FSSA will have the authority to overrule. Given that the provisions of this act are broad and that the FSSA is specifically empowered to address uncertain risks that might emanate from potentially harmful nano-applications, the overriding authority essentially legitimizes the application of the highest level of risk regulation to nano-applications that could be classified both as drugs as well as food items.

### 11.3.3 Overall Comment on the Current Regulatory Framework

The above analysis presents a mixed bag of results vis-à-vis the preparedness of the current regulatory framework in the face of the regulatory challenges that emanate from the rapid proliferation of nanomaterials in the food and drugs sectors. Since the development of nano-applications in the food and drug sectors has already been reported across India, such preparedness is urgent. It would be pertinent here to mention that the above analysis has been restricted to pharmaceutical and food acts because these are the primary pieces of legislation regulating food and drugs in India. However, there are pieces of environmental legislation like the Environment Protection Act 1986 (especially the regulation of hazardous waste), the Water (Prevention and Control of Pollution) Act 1974, and the Air (Prevention and Control of Pollution) Act 1981 that could have a significant impact on aspects of environmental disposal of nanomaterials. Nevertheless, taking this approach (albeit narrowly) allows us to focus the analysis on specific normative principles and provisions that may come into play in the regulation of nanotechnology applications and products within these two sectors. Furthermore, it is important to underline that, at this stage, it would be very difficult to provide for an end-of-life product disposal regulatory perspective, given there is still uncertainty as to what characteristics of nanomaterials should form the basis for identification (what is referred to as a regulatory trigger within such discussions).

The primary question that the above analysis attempted to answer was whether the current regulatory framework is equipped to face the regulatory challenges posed by new applications of nanotechnology within the food and drugs sectors. Prima facie, the answer is that there is considerable textual flexibility within the current legislation to allow for the construction of specific regulatory guidelines and principles for enabling oversight of these new nanotechnology-based applications. This, of course, is only a partial answer; a more comprehensive analysis of the preparedness of the current regulatory framework will necessarily demand whether the regulatory authorities are even aware of the current reality and the regulatory challenge that it poses. The answer to that would have to be in the negative. There is little awareness among regulatory authorities of the challenges posed by newer applications of nanotechnology in these two sectors, and therefore, despite adequate avenues available within the current regulatory framework, there have been limited discussions and reflection on these issues.

## 11.4 Regulation in Action

The Ministry of Health is involved in the regulation of nanotechnology applications through its Directorate General of Health Services, under which the Central Drugs Standard Control Organisation (CDSCO) is situated. Health, being a state subject, is largely in the domain of state governments, but a lot of its direction is guided by the central government. Institutionally, the MoHFW is in charge of the prevention and control of health-related hazards, but the agenda of the MoHFW is already full with issues like providing basic health infrastructure, eradicating diseases like polio and *kala azar*, and checking counterfeit drugs. The CDSCO is responsible for the approval of various drugs and establishing standards, but the implementation takes place at the levels of states and union territories. There are thirty-five state drug controllers (SDCs) (Government of India 2010), which have the primary responsibility of overseeing the regulation, manufacture, sale, and distribution (including licensing) of drugs (Section 18). In their tasks, the SDCs are guided by the CDSCO and aided by government analysts and drug inspectors at state and local levels.

Another authority, although statutory, is administratively based under Director General of Health Services, operating under the MoHFW. It is the Food Safety and Standards Authority, which was recently established to lay down science-based standards for its manufacture, storage, distribution, sale, and importation to ensure the availability of safe and wholesome food for human consumption. The authority has the mandate of setting standards and guidelines and of developing accreditation systems. The FSSA is also entrusted to provide scientific advice and technical support to the central and state governments. In this regard, it is also responsible for collecting and collating data on incidence and prevalence of biological risk, contaminants in food, residues of various contaminants in food products, and identification of emerging risks. The FSSA is guided by a Central Advisory Committee and assisted by several scientific committee and scientific panels, consisting of experts from different fields of food production, processing, preservation, and contamination.

The Ministry of Environment and Forests (MoEF) deals with environmental impacts or hazards that may emanate from new applications of existing or new chemicals and substances. The Central Pollution Control Board (CPCB) discharges most of the functions relating to the prevention and control of pollution, including through hazardous materials. The State Pollution Control Boards (SPCBs) are the state-level authorities under the 1986 Environment Protection Act (EPA). The SPCBs do not look at nanotechnology applications or health applications, but any commercial establishment or manufacturing process will have to adhere to standards laid down by the EPA and Hazardous Materials Rules, thereby bringing them under supervision of SPCBs. The State Pollution Control Committees are responsible for granting authorization for the collection, reception, storage, treatment, and disposal of biomedical waste.

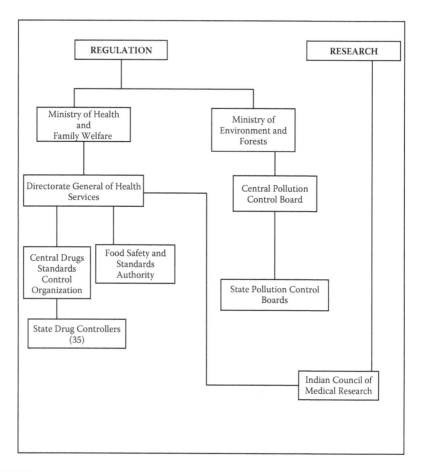

**FIGURE 11.1**
Overview of primary regulatory actors in India.

The range of authorities associated with the regulation of nanotechnologies in India is shown in Figure 11.1.

### 11.4.1 Research and Development

Set up as a body to fund and coordinate medical research in India, the Indian Council of Medical Research (ICMR) is a century-old agency. Today it is the primary agency under the aegis of the Ministry of Health and Family Welfare engaged with the development and implementation of biomedical research in India. The ICMR's priority areas concur with national health priorities aimed at enhancing the overall health and well-being of the nation. The council has a Scientific Advisory Board, committees, expert groups, and task forces. Although there are twenty-nine internal Research Institutes/

Centres, nanotechnology research is promoted by the ICMR primarily through extramural research support. It funded twenty-one research fellowships on nanotechnology-related projects and twelve extramural fellowships on nanotechnology and health research during 2007–2009.

Recently, the ICMR recognized the emerging importance of nanotechnology in the field of medicines and devices and set up a nanomedicine unit. However, the unit is yet to undertake any significant step in shaping the council's actions in promoting and governing nanotechnology applications in the health sector. With public health and national health priorities as its mandate, the ICMR can play an important role in the environmental health and safety of nano-applications in the health sector. The ICMR institutes— like the National Institute of Occupational Health—can take the lead in developing and promoting occupational health and safety protocols and standards for nanotechnology R&D.

Similar to the ICMR is the Council for Scientific and Industrial Research (CSIR), which directs its work toward scientific industrial R&D for economic, environmental, and social benefits. It has been supporting research in several areas including health, and as per CSIR reports, eleven of the fourteen new drugs developed since Indian independence are from the CSIR (Council of Scientific and Industrial Research, undated). Institutes and laboratories under the CSIR, such as the Central Drug Research Institute (CDRI) and the Indian Institute of Toxicology Research (IITR), are engaged in crucial, specialized research, fundamental for nanotechnology's effective regulation.

Other institutes, like the National Institute of Pharmaceutical Education and Research (NIPER) and Central Food Technology Research Institute (CFTRI), also have some expertise in toxicology and can help fill the information gap pertaining to nanotechnology-related risks in health applications. Over the past few years, several research institutes have started setting up nanotechnology units or centers of excellence. Currently, their focus is on imparting education and running degree programs on nanotechnology, with some research on nanosciences and nanotechnology. These centers—which are drawing financial support from the government—could set aside funds and infrastructure to research issues such as environmental and health impacts of nanomaterials and nanotechnology. This is specifically required as most of the basic and fundamental research in this field in India is dedicated toward industrial usage.

## 11.4.2 Institutional Challenges in the Regulation of Nanotechnology

### 11.4.2.1 *Regulatory Capacity*

Any regulation is only as efficient as those framing it and implementing it. One of the prerequisites of a competent institutional framework is a regulatory capacity for the formulation of rules, policies, and guidelines. Any regulatory intervention requires a great amount of technical expertise and foresight on the part of policymakers and regulators. The implementing agencies, which are usually at the state and local levels, need to be equipped to execute the rules and regulations that are already in place and are formulated from time to time. The agency responsible for regulation of drugs already faces challenges with respect to capacity, even in terms of responsibilities such as testing drugs. Furthermore, known capacity for testing nanoparticle toxicity exists at only a very few institutes, like the NIPER and IITR.

### 11.4.2.2 Flow of Information

Lack of capacity can also be linked to informational asymmetry. A smooth flow of information is necessary for building institutional capacity and taking regulatory measures. Since the main concerns around nanotechnology are the environmental, health, and safety risks, its regulation necessitates availability of information, both about the nature and extent of applications as well as the associated risks—and it is absolutely important that such information is readily available to regulators. With only a few institutions equipped to carry out risk and toxicity studies, information flow becomes even more important. Hence, the information among agencies with different mandates—such as the DST, DSIR, MoHFW, MoEF, and the research institutes—should be channeled in a way that each of these institutions performs its functions and furthers its mandate in an informed manner favorable to the well-being of the public at large.

### 11.4.2.3 Interagency Coordination

Governance structures in India are characterized by a multiplicity of ministries, local governments, and regulators at various levels (central, state), federal law-making institutional structures, and the existence of a sizeable number of public, private, and public–private sector entities. The multiplicity and diverse capabilities of actors pose challenges as well as offer opportunities for sharing roles and responsibilities at different levels.

The creation of various government agencies has resulted in the fracturing of regulatory jurisdiction between agencies. Environmental health is an important area of regulation, specifically in the context of the potentially adverse impacts of emerging technologies like nanotechnology and biotechnology. However, division of the regulatory mandate between the MoHFW and the MoEF has made it difficult to provide comprehensive and coherent regulatory cover on the issue of environmental health. In fact, environmental health—as a policy discipline—is underdeveloped in the Indian context (see World Bank 2001). Thus, regulatory fissures are further exacerbated when the state indirectly undermines regulatory overtures. It does so by privileging technology in its development agenda while also setting up individual state departments with the sole objective of technology promotion and facilitation. Even outside the health sector, nanotechnology applications may raise several health concerns that must somehow be addressed across a number of agencies. For example, occupational health is the prerogative of the Ministry of Labour, health is the mandate of the Ministry of Health and Family Welfare, and the environment is governed by the Ministry of Environment and Forests.

### 11.4.2.4 Regulatory Perception

Given the regulatory culture in India, the term *regulation* is implicitly assumed to mean government intervention and generally meddling with market mechanisms. Thus, industry organizations regularly lobby government, engaging quite vigorously in the rhetoric of technology development and its functionality in national development. Consequently, such groups consistently undermine efforts to strengthen the regulatory oversight of these technologies (see Damodaran 2005 for similar arguments in biotechnology). Interestingly, this skewed understanding of regulation and its implications is prevalent not only in industry but in government as well. For instance, most of the officials of DST with the mandate of technology promotion have been unwilling to engage with the issues of potential risks for nanotechnologies and the possible need for customized regulation. However, in February 2010, the chair of the Nano Mission Council announced at the International Conference on Nano Science and Technology (held at the Indian Institute of Technology in Mumbai), that a Nanotechnology Regulatory Board will be set up to regulate industrial nanotechnology products (Press Trust of India 2010). This appears to be a preemptive action to keep other ministries from deciding the course of the technology, especially given the recent *Bt* brinjal (transgenic brinjals created by inserting the bacterium *Bacillus thuringiensis* genome) controversy leading to a moratorium on the release of *Bt* brinjal in India—until such time that independent scientific studies can establish, to the satisfaction of both the public and industry professionals, the safety of the product from the point of view of its long-term impacts on human health and environment (Ministry of Environment and Forest, Government of India 2010).

## 11.5 Planning an Institutional Framework

Despite the existence of rules and regulations, one of the primary concerns with the institutional framework for either health (in the regulation of drugs) or environmental safety is poor implementation. This can be largely attributed to a capacity deficit in terms of the relevant institutions. The Ministry of Environment and Forest itself admits to this gap and cites the lack of capacity or resources to ensure compliance with various environmental regulations as the rationale for proposing the new National Environmental Protection Agency (NEPA) (Ministry of Environment and Forest, Government of India 2009).

Accepting that the present institutional framework is fraught with challenges—which are becoming increasingly multifarious in the case of nanotechnology—an obvious question arises: What might an ideal institutional framework for nanotechnology look like? This is not a question that is easy to answer and stipulates consideration of available options and models. Nanotechnology, especially in the discourse on regulation, is often compared to biotechnology, which, in India, is primarily governed by a Genetic Engineering Approval Committee (GEAC). Whether a GEAC model would work for nanotechnology or not could be best answered by looking at the performance of the GEAC, which has attracted considerable criticism on grounds of underperformance (see Ministry of Environment and Forest, Government of India 2010). Significantly, the issues in nanotechnology and biotechnology differ due to the different nature of both technologies; where the former focuses on materials, the latter is directed toward working with cells. Moreover, there is a difference in the scope of its current and possible applications. Although in biotechnology the majority of applications have been limited to the agricultural and pharmaceutical sectors, nanotechnology applications have already started entering markets across sectors and applications ranging from medicines and cosmetics to consumer goods and textiles.

The mandate for the newly proposed NEPA includes, among other things, chemical safety and biosafety. It proposes to take over the task of granting approvals for GMOs and products thereof. Even assuming that the NEPA, when established, can take nanotechnology regulation under its ambit as well, it is going to be an uphill task for the regulatory process, given the capacity deficit and near absence of regulatory toxicology data and risk management frameworks in place for nanotechnology in India.

It is important to understand how NEPA will interact with agencies such as the DST, CSIR institutes, the MoHFW, and Ministry of Food Processing Industry (MoFPI), which will be better positioned to generate and validate data on impacts as well as collate data on the impact and performance when released into the environment and markets.

Another problem with the NEPA's approach to regulating nanotechnology is that it would further centralize regulatory decision making, leaving out agencies at the subnational and substate levels for mere implementation. Given the nature of nanotechnology and its applications, nanotechnology regulation needs to be as multilevel-led and as multiscalar as possible (TERI 2009). This poses a challenge in India, as the government is structured federally and health falls under the umbrella of the state.

## 11.6 Conclusion

In summarizing, a few caveats need to be outlined. First, nano-based applications, both in the health and the food packaging sectors, have expanded rapidly over the past few years. Industry associations have confirmed reports of a further expansion of products, and more complex nanomedicines are expected to be launched in the short and medium term. This illustrates the urgent need to deliberate over regulatory options that will address the challenges emanating from such developments. Second, regulatory policymaking will have to be responsive to the rapidly evolving nature of the nanotechnology industry in these sectors. In practice this means that regulatory policy will have to be flexible and, therefore, that it is best to move forward incrementally and to design policies differentiating between the short, medium, and long term. Third, internationally there is a growing realization that regulators are negatively impacted by informational asymmetries that characterize such rapidly evolving technology domains. It is therefore critical that regulators realize that information disclosure by product manufacturers and marketers will have to be a prerequisite for developing any kind of regulatory guidance on such applications. Fourth, in such a scenario regulators will have to apply a mix of soft and hard regulatory instruments, enabling the design of mapping strategies that will allow critical, new scientific information to be fed into the regulatory mechanisms. For integrity's sake this information has to be peer-reviewed and supplemented by the government's own investment in toxicology and other EHS research being carried out by government laboratories.

In 2010, there were public discussions in India about establishing a Nanotechnology Regulatory Board—however, a year later little progress on this matter has been witnessed. The details are still unclear on what the mandate, scope, and dimensions of such a board would be. It is important, nevertheless, that a statement confirming such considerations was made by Dr. C. N. R. Rao, who heads the Nano Mission Council operating under the DST. As discussed, a potential mandate to regulate nanotechnology and, more specifically, nanotechnology-based applications (such as nanomedicines) could remain with various departments and ministries, like the MOEF, MoHFW, DST, and others. In such a situation, one can expect considerable competition between ministries seeking to establish their own regulatory mandates. However, the DST's foray into the debate about establishing a regulatory body for nanotechnology seems to suggest it is posturing in order to preempt other contenders from extending their mandate on this issue. Despite such reservations, the DST's statement does reflect an important development—an increasing realization within the Indian government that it is critical for policymakers to focus attention on developing regulatory measures to address the public health and safety challenges that underlie the nanotechnology-based applications already being launched in the Indian market.

Addressing challenges posed by new and emerging technologies like nanotechnology throws open the debate of whether there should be technology regulation per se or regulation of products using technology. Given the wide scope of nanotechnology applications, the lack of proper definitions, and various risk-related uncertainties, technology regulation may be difficult as well as undesirable. This leads to another question as to whether the regulatory regime for applications is equipped to attend to the concerns raised by nanotechnology applications. Our analysis of the regulatory instruments for the food and health sectors suggests that there is considerable flexibility, at least in the texts of these instruments, to allow for developing guidelines and protocols for regulatory oversight of nanotechnology. However, the existence of these flexibilities does not resolve the regulatory issues but rather puts forth further challenges in terms of institutional frameworks, uncertainties, and knowledge gaps. In order to make use of the existing flexibilities, the state has to be engaged with nanotechnology development in a holistic manner and move beyond the role of a technology promoter. The state needs to be aware, first, of the basic need for building regulatory capacity and enhanced preparedness, and second, of the nuanced and technical knowledge required to guide its decisions and actions, especially with respect to things such as standards, size, metrology, and exposure limits. To strengthen the institutional framework, issues such as regulatory capacity, informational asymmetries, and coordination need to be resolved urgently by the state and across multiple other levels, scales, and mandates.

## References

Abbott, K.W., Marchant G.E., et al. 2006. A framework convention for nanotechnology? *Environmental Law Reporter* 36:10931–942.

Abbott, K.W., Sylvester D.J., et al. 2010. Transnational regulation of nanotechnology: Reality or romanticism? In *International Handbook on Regulating Nanotechnologies*, edited by G.A. Hodge, D.M. Bowman, and A.D. Maynard, 525–39. Northampton, Mass.: Edward Elgar.

Bhasmas are nano-medicine of ancient times. 2011. *Times of India.* http://articles.timesofindia.indiatimes.com/2011–02–06/varanasi/28353173_1_bhasmas-herbo-mineral-formulations-rasa-shastra (accessed 8 April 2011).

Council of Scientific and Industrial Research, India. Undated. http://www.csir.res.in/External/Heads/csir_faqs.htm (accessed 4 February 2010).

Damodaran, A. 2005. Re-engineering biosafety regulations in India: Towards a critique of policy, law and prescriptions. *Law, Environment and Development Journal* 1/1. http://www.lead-journal.org/content/05001.pdf (accessed on 13 April 2011).

Dey, S. 2009. Govt orders recall as Natco cancer drug fails toxin test. *Economic Times*, 18 July 2009. http://economictimes.indiatimes.com/news/news-by-industry/healthcare/biotech/pharmaceuticals/Govt-orders-recall-as-Natco-cancer-drug-fails-toxin-test/articleshow/4791560.cms (accessed 4 February 2010)

Financial Express Bureau. 2010. Blow for Natco Pharma as order on nanotech drug licence stayed. *Financial Express*, 12 January 2010. http://www.financialexpress.com/news/blow-for-natco-pharma-as-order-on-nanotech-drug-licence-stayed/566243/0 (accessed 8 April 2011).

Government of India. 2003. The Drugs and Cosmetics Act 1940 (along with all its amendments) as corrected up to 30 April 2003. http://www.cdsco.nic.in/html/Copy%20of%201.%20D&CAct121.pdf (accessed 2 March 2011).

Government of India. 2006. The Food Safety and Standards Act, 2006. *Gazette of India* (extraordinary) Part-II, Section 1, Issue No. 40, Act No. 34 of 2006, dated 24 August 2006. http://supremecourtcaselaw.com/Ilbco's_Food_Safety_Standards_Act_(3rd_Edition_2009)_published_by_Rajan_Nijhawan.pdf (accessed 7 March 2010).

Government of India. 2008. The Drugs and Cosmetics Amendment Act 2008. *Gazette of India* (extraordinary) Part-II, Section 1, Issue No. 35. Act no. 26 of 2008, dated 5 December 2008.

Government of India. 2010. List of state drug controllers. List by central drugs standard control organization. *Ministry of Health and Family Welfare, Government of India.* http://cdsco.nic.in/html/STATE%20DRUGS1.htm (accessed 14 March 2010).

Institute of Food Technologists. 2006. Food nanotechnology. Scientific status summary on potential applications of nanotechnology in the food industry. http://members.ift.org/NR/rdonlyres/FA6CF1C9–5E70–4A0C-9300–6A8D450A7D06/0/ednanotech.pdf (accessed 7 March 2010)

Karnataka Drugs and Pharmaceuticals Manufacturers' Association. 2010. Indian Pharma News. *E-Sushruta*, February 2010 no 2. http://web.kdpma.in/web/html/E-Sushrutiya/E-Sushrutiya%20Feb%202010.pdf (accessed 4 February 2011).

Ministry of Environment and Forest, Government of India. 2009. *Proposal for a National Environment Protection Authority.* Discussion Paper. http://moef.nic.in/downloads/home/NEPA-Discussion-Paper.pdf (accessed 13 April 2011).

Ministry of Environment and Forest, Government of India. 2010. *Decision on Commercialisation of BT Brinjal.* http://moef.nic.in/downloads/public-information/minister_REPORT.pdf (accessed 20 February 2010).

Ministry of Health and Family Welfare, Government of India. 2003. *Report of the Expert Committee on a Comprehensive Examination of Drug Regulatory Issues, Including the Problem of Spurious Drugs.* November 2003.

Press Trust of India. 2010. India to have Nanotechnology Regulatory Board soon. *Business Standard*, Mumbai, February 18, 2010. http://www.business-standard.com/india/news/india-to-have-nanotechnology-regulatory-board-soon/86186/on (accessed 4 February 2010).

The Energy and Resources Institute (TERI). 2009. *Nanotechnology Development: Building Capability and Governing the Technology.* TERI Briefing Paper. http://www.teriin.org/div/ST_BriefingPap.pdf (accessed 14 April 2011).

Vivekanandan, J. 2009. Nano applications, mega challenges: The case of the health sector in India. *Studies in Ethics, Law, and Technology* **3(3)**: Article 3.

World Bank. 2001. *Environmental Health in India: Priorities in Andhra Pradesh: Environmental and Social Development Unit, South Asia Region, New Delhi.* http://siteresources.worldbank.org/INDIAEXTN/Resources/Reports-Publications/APHealth.pdf (accessed 14 March 2010).

## Endnotes

1. A crore is a unit in India equal to 10 million, or $10^7$.
2. Indian Institute of Technology, Madras, is working with Murugappa Chettiar and Orchid Pharma, University of Hyderabad with Dr. Reddy's Labs and NIPER, Chandigarh is also working with Pharma industry (Nano Mission, Government of India).
3. Previously also there have been several such expert committees set up by the government. These include the Hathi Committee 1975 and the Pharmaceutical Research and Development Committee (PRDC), 1999.
4. Principle of proportionality in law generally means that the measures taken by public authorities to effect certain objectives should be appropriate and in balance with the rights of the individual to whom it is directed.

# 12

## Nanotechnology and Global Regulation

Diana M. Bowman and Graeme A. Hodge

**CONTENTS**

## 12.1 Introduction

The world clearly cares about nanotechnologies. This is evident across industry, government, and civil society sectors. Aiming to protect their investments and intellectual property, industry has pursued an exponential growth in the number of nanotechnology-based patents granted by intellectual property offices (see, for example, Chen and Roco 2008; IPO 2009). Governments have poured large public-sector funding investments into research and development (R&D) of nanotechnologies and have also rolled out many public-sector initiatives over the past decade—from the launch of the United States' high-profile National Nanotechnology Initiative (NNI) in 2000 to the European Union's 2005 *Action Plan for Nanosciences and Nanotechnologies* (Lux Research 2005; European Commission 2005), as well as many others across a range of jurisdictions. Civil society, too, clearly cares about the impact of nanotechnology, with visible high-level activity evident in many nongovernmental organizations (NGOs). Some, such as Australia's Friends of the Earth (FoEA) and the ETC Group, have clearly been at the forefront of such efforts, and the diversity of groups, their core interests, and the range of agendas being pursued through these debates has been impressive (Miller and Scrinis 2010). The broad coalition of approximately seventy NGOs from around the world that developed and declared their commitment in 2007 to the *Principles for the Oversight of Nanotechnologies and Nanomaterials* is one such illustration (NanoAction 2007). The very action of such a large group of diverse NGOs joining up to make this statement of principles was, as Miller and Scrinis put it, "remarkable" (2010, 413). Such activities have helped move the phenomenon of nanotechnology out of the laboratory and into public consciousness. Many of these activities, too, have been part of a far wider set of ongoing tensions and debates around fundamental political philosophies and global inequalities, including a continued questioning of the merits of different varieties of capitalism and their winners and losers. There is also little doubt that international and national regulatory reforms—leading, for example, to the phasing out of chlorofluorocarbons (CFCs)—have led to real environmental improvements. So, is regulation also central to the nanotechnology futures?

The regulation of nanotechnologies has certainly been an area of increasing interest and debate. Although questions dealing with regulation can be traced back two decades to Forrest (1989) and Fiedler and Reynolds (1994), the most powerful landmark report was that of the Royal Society and Royal Academy of Engineering (RS-RAE) in 2004. Since then, there have been numerous regulatory publications. Contributions have included government-initiated reviews of the adequacy of national regulatory frameworks for nanotechnology-based products and processes (see Health and Safety Executive 2006; Chaudhry et al. 2006; Food and Drug Administration 2007; Ludlow, Bowman, and Hodge 2007; Food Safety Authority of Ireland 2008; European Commission 2008), as well as an increasing number of independent assessments (Taylor 2006, 2008; Fuhr et al. 2006; Davies 2006; Kimbrell 2006; Sadrieh and Espandiari 2006; Ludlow 2007; Marchant, Sylvester, and Abbott 2007; Gergely 2007; D'Silva and van Calster 2009) and parliamentary reports (Royal Commission on Environmental Pollution 2008; NSW LCSCSD 2008; House of Lords STC 2009). Other literature has considered, for example, the impact and regulation of nano-based weaponry or dual-use applications (Pinson 2004; Altmann 2010), current regulatory practices and impacts on the insurance sector (Munich Re Group 2002; Allianz 2005; Epprecht 2010), potential opportunities for regulatory convergence within specific sectors (Breggin et al. 2009), and the potential applicability of different regulatory approaches for nanotechnologies, including legally binding multilateral instruments and voluntary instruments (Abbott, Marchant, and Sylvester 2006; Abbott, Sylvester, and Marchant 2010; Pelley and Saner 2009; Bowman and Hodge 2009; Meili and Widmer 2010). Calls for a moratorium on the commercial release of certain types of nanomaterials have also made their way into the debates (Miller and Scrinis 2010), albeit with very limited traction (see, for example, NSW LCSCSD 2008). Governance debates continue.

The aim of our chapter is to build upon the evolving body of literature dealing with regulatory challenges and focus particularly on the role that international approaches may play in ensuring the responsible development of nanotechnologies. While recognizing that effective national regulation will continue to be fundamental in ensuring the safe development and commercialization of nanotechnologies, we acknowledge in this chapter the inherent limitations of such national approaches—especially when dealing with the all-pervasive nature of this platform technology. We also, however, recognize the complexities associated with negotiating multilateral agreements and issues of enforcement, for example, with the consequence that such mechanisms will not, by themselves, be any panacea. Rather, by employing multiple mechanisms operating at different levels and by drawing upon the strengths of many actors, international approaches will help provide the best foundation for ensuring that nanotechnologies are developed in a safe and responsible manner.

We commence this chapter with a brief exploration of regulatory concepts and methods. A more in-depth discussion of the key debates that have occurred to date is then presented, including major lessons from regulatory reviews conducted. We then explore those international regulatory mechanisms (such as framework conventions, self-regulation, and co-regulation) that best ensure that nanotechnologies are governed effectively, as well as the feasibility of such international approaches. Overall, we argue that both national and international regulatory regimes will, together, aim to ensure the safety of nanotechnologies and usher in sustainable futures. Neither of these regimes, however, is likely to fully satisfy the desires of policy critics who view the nanotechnology phenomenon as a useful rallying call to change broader existing social and economic power structures or inequalities across the globe.

## 12.2 Regulatory Purpose: How Governments and Other Actors Regulate and through What Mechanisms

### 12.2.1 Reconceptualizing Regulation

The term *regulation* has traditionally been viewed as an activity in which governments aim to manage or control risks through the application of law, including legislative instruments enacted by parliaments as well as developments in the common law (Brownsword 2010). But the fundamental notion of regulation and its place in society have both been reconceptualized over the past two decades. To begin with, the concept of regulation is firstly now seen to be much broader than simply law, within an ethos of command and control. Black's widely cited definition of regulation puts it eloquently when she outlined the contemporary view of regulation as "The sustained and focused attempt to alter the behaviour of others according to defined standards or purposes with the intention of producing a broadly identified outcome or outcomes" (2002, 19).

In other words, Black identified the underlying purpose of regulation as influencing behavior and therefore encompassing a wide variety of activities. As Roger Brownsword recently noted, these mechanisms could include "any instrument (legal or non-legal in its character, governmental or nongovernmental in its source, direct or indirect in its operation) ... that is designed to channel behaviour" (2010, 64). Additionally, the societal context of regulation has also been rethought. Black's definition clearly implies that regulation may be undertaken by a range of different actors across three sectors—government, civil society, and business—rather than just government alone. Governance scholars from economics and development speak of *regulatory governance* (Minogue and Carino 2006), those from political science talk of *regulatory capitalism* (Levi-Faur and Jordana 2005; Braithwaite 2008), and those from law and public policy now speak of living with a *regulatory state*, in its broadest sense (Glaeser and Shleifer 2003; Moran 2002; Sunstein 1990). The important point here is that, rather than viewing regulation in the old narrow sense of developing and enforcing rules or simply about managing risks, these scholars see regulation as crucial to a new order of governance and about reordering priorities and power. Regulation is now "a distinctive mode of policy-making" or an "alternative mode of public control" as Majone has stated (1999, 1). This reconceptualization of regulation and its place in society matter because they have major implications for how we all view regulation as leading to sustainable futures.

## 12.2.2 State-Based Regulation

As is outlined throughout this book, products and applications incorporating nanotechnologies have largely been regulated under existing national legislative instruments. These forms of command-and-control regulation have considerable legitimacy with the public. Their compulsory nature, the appearance of strong accountability, and a higher certainty of compliance are all characteristics that appeal to voters and societies more generally. Moreover, compared to no regulation at all, appropriate state-based regulation enables a level playing field to be achieved where firms compete on an equal basis. It also provides a degree of certainty that assists in securing capital and insurance (Ludlow, Bowman, and Kirk 2009, 615). In the context of regulating nanotechnologies, however, what is most appropriate is clearly contestable. It is open to our interpretation of past experience and our views on risks, benefits, innovation, and judgments on the importance of broader societal considerations. In short, what is seen as most appropriate depends on how effective we believe alternative regulatory instruments are, as well as our impressions of their unfavorable impacts.

State-based regulatory approaches, however, suffer from a number of criticisms: they are slow, cumbersome, and rigid, and often involve high transaction costs (Vogel 2006). Importantly, these criticisms may also be amplified in the context of rapidly evolving fields where requirements are more dynamic and information needed to craft frameworks is imperfect (Moran 1995; Sinclair 1997).

Given these limitations, it is unsurprising that some commentators have suggested that nano-specific state-based regulation may not be appropriate, at least in the short term. Renn and Roco (2006), for example, saw state-based regulatory instruments as taking substantially longer than alternative regulatory forms. The experience of BASF, one of the world's largest chemical companies, also supports this argument. Its 2004 in-house code of conduct predated by some five years the first piece of legislation specifically incorporating nanotechnology-specific provisions and introduced by a parliament at the national or supranational level (European Parliament 2009). In this instance, the recast of the European Union's Cosmetic Directive (finally adopted by the European Council in November 2009) took around two years to negotiate and was clearly a complex challenge. The regulation supersedes some fifty-five directives relating to cosmetics and is intended to streamline human safety requirements and increase transparency. As explained by Bowman, van Calster, and Friedrichs, "the adoption of the Regulation is significant, not least because it is the first piece of national or supranational legislation to incorporate rules relating specifically to the use of nanomaterials in any products" (2010, 92).

### 12.2.3 Civil Regulation

In light of our reframing of regulation and the limitations of command and control regimes, there has been increasing interest in the development of non-state-based regulatory mechanisms. Such soft law or civil regulatory mechanisms are part of a regulatory continuum and include various forms of self-regulation, voluntary regulation, and third-party regulation, as well hybrid arrangements. Importantly, "civil regulation extends regulatory authority 'sideways' beyond the state to civil society and to non-state actors," thereby removing the potential for legally binding standards (Vogel 2006, 3). Although under these mechanisms government no longer functions as the sole regulatory authority, the broad nature of this regulatory category nonetheless enables the state to be involved in the broader governance framework (Gunningham and Rees 1997). It is indeed modern regulatory governance at work.

The use of different types of civil regulation, such as voluntary codes of conduct, risk management frameworks, and industry codes, has steadily increased. One high-profile example is the U.S. chemical industry's Responsible Care Program (see, for example, Rees 1997; King and Lennox 2000). Less resource intensive to develop and administer, the industry can evolve and respond to changing needs quicker than via reactive, state-based regulation (Braithwaite 1982; Sinclair 1997), while breathing space for innovation remains. It is these strengths that have arguably inspired many organizations to develop and implement voluntary codes of conduct and risk management frameworks. These have included, for example, the ambitious principle-based Responsible NanoCode (RSIINIA, NKTN 2007) (see Chapter 10); the NanoSafe (Hull 2009) five-point risk management program, designed to minimize human and environmental exposure to nanomaterials; the NanoRisk Framework; and the Principles for the Oversight of Nanotechnology (NanoAction 2007). This latter document consists of eight overarching principles that the coalition of signatories believes "provide the foundation for adequate and effective oversight and assessment of the emerging field of nanotechnology" (NanoAction 2007, 1).

As noted by Marchant, Sylvester, and Abbott, many of these initiatives draw upon, and to varying degrees incorporate, traditional risk management principles, including "(a) acceptable risk, (b) cost–benefit analysis, and (c) feasibility (or best available technology)" (2008, 44). These programs operate within the shadow of formal regulatory obligations and do not seek to "roll back the state" or usurp the regulatory frameworks in which they operate. Investments in regulatory responses by industry are due—at least in part—to their need to be at the forefront of risk management in an increasingly technology-based economy (Gilligan and Bowman 2008, 241). Investments may also be attributed to industry's attempts to be at the forefront of development of any new rules (Webb 2004, 4).

Although self-regulation is hardly a new phenomenon, the increasing use of this and other soft law instruments has not been without controversy. Self-regulation, in particular, has been accused of serving the interests of industry above that of society, having variable standards of enforcement, and lacking the accountability and legitimacy of government regulation (Braithwaite 1993; Webb and Morrison 1996). In relation to self-regulation for nanotechnology-based foods and cosmetics, the International Risk Governance Council has suggested that "voluntary codes with no provisions to enforce action or compliance, in other words with 'no teeth,' would risk being branded as mere window-dressing for public relations purposes" (IRGC 2009, 35). Bowman and Hodge (2009) have therefore suggested that while voluntary initiatives will be increasingly employed by organizations to assist in the responsible development of nanotechnologies, they must only be part of a broader regulatory matrix.

### 12.2.4 Co-Regulation

Turning our minds back to the continuum of regulatory possibilities for nanotechnologies, we should acknowledge that state-based regulation and civil regulation sit at opposite ends of this line. Co-regulation—the "use of a panoply of tools and actors, formal and informal, governmental and nongovernmental, national and international" (NRC 2001, 200)—sits somewhere in the middle. This third category of regulation also encompasses a wide array of tools and may include a variety of actors across government, industry, and civil society. It also has the ability to draw upon the strengths of the other two approaches, while avoiding many of their weaknesses. Given this flexibility, governments have, not surprisingly, viewed it as a fertile area for regulatory innovation (Ayres and Braithwaite 1992; Utting 2005). It enables industry to shape regulatory outcomes while retaining government's general design oversight (Sinclair 1997, 544).

Evidence of using a co-regulatory approach to nanotechnologies has already emerged through the implementation of voluntary reporting schemes or data stewardship programs (DEFRA 2006; NICNAS 2007, 2008; EPA 2008; CDTSC 2009). The United Kingdom's voluntary scheme, for instance, was implemented in order to assist the government to "develop appropriate controls in respect of any risks to the environment and human health from free engineered nanoscale materials ... in the shortest time giving a predictable regulatory environment for all" (DEFRA 2006, 3). The development of the European Commission's voluntary Code of Conduct for Responsible Nanosciences and Nanotechnologies Research in 2007 provides a further example.

What is important here is to recognize that each regulatory approach has strengths and weaknesses, and that they should be seen as complementary rather than mutually exclusive. As Gunningham and Sinclair remind us, single-instrument approaches are misguided, because all instruments have strengths and weaknesses and because none are sufficiently flexible and resilient to be able to address all contextual problems (1999, 50). It is a question of an acceptable balance between the different approaches, depending on the available scientific knowledge, the legitimacy of alternatives, and the ways in which benefits and risks are perceived within a jurisdiction (Ludlow, Bowman, and Kirk 2009). Also important is the notion that regulatory activity is inherently a political activity, aiming to address scientific risks as well as broader societal values and other social preferences. Finding a workable and effective balance that has the necessary legitimacy with the public is an ongoing and evolving task as we seek to create sustainable futures.

## 12.3 Looking at the Debates

Despite the anticipated benefits of nanotechnologies (see Chapters 2, 5, and 6–9), there are clearly concerns over the potential risks posed to human as well as environmental health and safety (see Chapters 2, 4, 5, and 9). This has prompted widespread debate as to desirable regulatory arrangements. Interestingly, the maturation and increasing commercialization of nanotechnology has seen the debates moving from more open and blue-sky questions on whether nanotechnology-based products and process are regulated to questions primarily dealing with the *effectiveness* of our current regulatory regimes. Alongside this, we have also seen debates moving from calls to implement overarching nanotechnology-based regulatory frameworks to more focused calls for amendments in specific areas—such as in foods, cosmetics, industrial chemicals, and occupational health and safety. So, what are some of the key debates that have occurred here? It is to this question that we now turn.

The ETC Group was one of the earliest contributors to the global regulatory debate. Their 2003 report called for "a mandatory moratorium on the use of synthetic nanoparticles in the lab and in any new commercial products" (2003, 10). This appeal has since been echoed by a number of different organizations. The FoEA, for instance, has called upon governments to enact "a moratorium on the further commercial release of personal care products that contain engineered nanomaterials, and the withdrawal of such products currently on the market, until adequate, publicly available, peer-reviewed safety studies have been completed, and adequate regulations have been put in place to protect the general public, the workers manufacturing these products and the environmental systems in which waste products will be released" (2006, 3).

Drawing upon the precautionary principle, the FoEA has also called for governments to implement a moratorium on the commercial use or release of, for example, nanosilver, foods and food packaging containing nanomaterials, and carbon nanotubes until such times as appropriate laws have been implemented to ensure human and environmental health and safety (FoEA 2006a, 2006b, 2007; Miller and Senjen 2008). Despite these calls, Miller and Scrinis say that "decision makers have not been prepared to slow the rapid pace of nanotechnology commercialisation to address basic safety issues" (2010, 409).

These calls for a moratorium have seen considerable reflection but general dismissal by governments. In Australia, for example, the New South Wales (NSW) Standing Committee on State Development in its 2008 inquiry *Nanotechnology in NSW* argued that "it would be impractical to recommend or support a moratorium on nanotechnology or even nanomaterials, as both are broad descriptive terms rather than specific entities" (NSW LCSCSD 2008, xii). This statement highlights the complexities of simply defining the boundaries for any such moratorium, let alone enforcing a ban. In light of these difficulties, we see a general moratorium on the development or commercialization of nanotechnologies as unlikely. That said, we recognize fully that should the growing body of nanotoxicology literature show that certain nanoparticles pose an unacceptable health or environmental risk, then governments may rightly be forced to enact tighter controls in relation to such activities.

Against this backdrop, many commentators (see Warda 2003; RS-RAE 2004; Taylor 2006, 2008; Fuhr et al. 2006; Ludlow 2007) have highlighted the current limitations of existing state-based regulatory frameworks. In recognition of such limitations, commentators including Balbus et al. (2006) and Miller and Senjen (2008) have argued that governments should amend certain legislative instruments to take into account the additional and specific challenges posed by nanotechnologies. This increasing scrutiny over the adequacy of existing regulatory arrangements has covered not only statutory instruments but also their implementation and has encouraged further independent reviews. These have been undertaken in Australia, the European Union (EU), the United Kingdom, and the United States (see for example, Chaudhry et al. 2006; HSE 2003; EPA 2007; FDA 2007; Ludlow, Bowman, and Hodge 2007; FSAI 2008; FSA 2008; European Commission 2008). These reviews have varied in their scope and focus. One of the earliest reviews was undertaken for the United Kingdom and EU and broadly examined the "appropriateness of existing regulatory frameworks for environmental regulation" (Chaudhry et al. 2006, 10). In contrast, Ludlow, Bowman, and Hodge were required, pursuant to their contract, to focus more narrowly and "assess Australia's existing regulatory frameworks to determine if, and under what conditions, nanotechnology-based materials, products and applications, and their manufacture, use and handling, are covered by the existing regulatory frameworks" (Australian Government 2006, 5). Another example of a more limited assessment was that undertaken in the European Commission's in-house review that examined EU legislation most relevant to "nanomaterials currently in production and/or placed on the market" (2008, 2).

Despite the predictable jurisdictional flavors of these reviews, a number of common themes have been raised. For instance, in spite of concerns by groups such as the ETC Group and other commentators that nanotechnologies were "essentially unregulated," each of the reviews showed that nanotechnology-based products and processes fell within the ambit of the statutory instruments analyzed, and as such, are indeed regulated. This general finding was eloquently summarized by the European Commission in relation to their review: "existing Community regulatory frameworks cover in principle the potential health, safety and environmental risks related to nanomaterials. Without excluding regulatory change in the light of new information, the Commission stressed that the protection of health, safety and the environment needs to be enhanced mainly by improving the implementation of current legislation" (2009, 7–8).

In other words, the government reviews undertaken to date found that the instruments analyzed were, in principle, sufficiently broad in their application so as to capture nanotechnologies, without being technology specific. Notwithstanding this, authors such as Chaudhry et al. (2006) and Ludlow, Bowman, and Hodge (2007) pointed to some gaps in regulatory coverage. Chaudhry et al. identified a number of gaps in the broader regulatory framework, such as those relating to definition, thresholds, and exemptions, the effect or impacts of nanotechnologies in relation to their properties, and specific substances. In their view:

The regulatory gaps identified in this study derive either from exemptions (on tonnage basis) under legislative frameworks, or from the lack of information, or uncertainties over:

- Clear definition(s) encompassing the novel (or distinct) properties of nanotechnologies and [nanomaterials]; i.e. whether a [nanomaterial] should be considered a new or an existing material;
- Current scientific knowledge and understanding of hazards, and risks arising from exposure to [nanomaterials];
- Agreed dose units that can be used in hazard and exposure assessments;
- Reliable and validated methods for measurement and characterisation that can be used in monitoring potential exposure to [nanomaterials];
- Potential impacts of [nanomaterials] on human and environmental health (2006, 8).

Discussion of the significant scientific and technical issues posed by uncertainties, in relation to the statutory frameworks, were also common across reviews. They further highlighted the need for "risk-focused research" in order to address the knowledge gaps now challenging regulators and risk managers (Chaudhry et al. 2006, 267–69). Such fundamental research is needed by all jurisdictions in order to ensure that their regulatory frameworks are adequate. A scientifically driven, risk-orientated approach to addressing the potential risks posed by nanotechnology is not without its critics, however, as highlighted by Wickson in Chapter 10. In her view, a technical framing of the risks and uncertainties is too narrow, and a broader framework is required. While a broader framing may indeed be necessary, we believe it will still need to be underpinned by scientifically driven research.

Most of these reviews also highlighted the need for coordination not only between jurisdictions but with other key stakeholders (including industry, members of the research community, and nongovernmental organizations). The following section examines several of the present multilateral activities we believe are crucial to ensuring the responsible development of nanotechnologies.

## 12.4 Current State of Play: International Activities Aimed at Governing Nanotechnologies

International-level dialogue to address scientific, policy, and regulatory challenges posed by nanotechnologies has certainly gathered pace. Indeed, the Organisation for Economic Cooperation and Development (OECD), United Nations (UN), and World Health Organization (WHO) (in partnership with the Food and Agricultural Organization [FAO] of the UN) have all turned their attention to these issues. The International Organization for Standardization (ISO), a voluntary standards development body, has also played a crucial role through the development of definitions, common nomenclature, and standards for classification and testing of nanotechnology and nanomaterials (Miles 2007). As Miles notes, the development of these standards is not only important for research, commercialization, and trade but also to "support the development of appropriate national and international regulatory regimes" (2010, 87). Having said this, in this section we note that these organizations have not yet directly designed new regulatory instruments per se.

Looking first to the OECD, two working parties have been established: the Working Party on Manufactured Nanomaterials (WPMN) and the Working Party on Nanotechnology (WPN). The first of these, the WPMN, has been charged with a mandate "to promote international cooperation in human health and environmental safety related aspects of manufactured nanomaterials (MN), in order to assist in the development of rigorous safety evaluation of nanomaterials" (OECD 2008, 3). The focus of the WPMN is on finding globally oriented responses to the challenges posed by engineered nanomaterials. In contrast, the WPN was established to "advise on emerging policy-relevant issues in science, technology and innovation related to the responsible development of nanotechnology" (OECD 2007, 7). This is to be achieved through a range of projects, including public outreach, and by acting as a facilitator for international cooperation and collaboration on research activities (OECD 2007, 2008). While the objectives of the WPN are clearly ambitious, the proposed outputs should provide governments with well-informed insights into some of the potential impacts—both beneficial and detrimental—of nanotechnologies across a range of areas. This intergovernmental cooperation illustrates the fact that the challenges posed by nanotechnologies are clearly not bound by jurisdictions or by industry sectors. Moreover, it suggests that a coordinated and strategic approach to the most pressing research requirements will assist in addressing the current knowledge deficits in the shortest possible timeframe and inform all parties with the fundamental knowledge on which to build future risk management activities.

Similarly, committees and agencies of the UN have been active during the last five or six years. Attesting to this are the work agendas of UNESCO (2006, 2007) and the FAO (2008)—in partnership with the WHO and the UN Committee of Experts on the Transport of Dangerous Goods and on the Globally Harmonized System of Classification and Labelling of Chemicals (2009). The focus of efforts by UNESCO has been on "ethical reflection ... to address the potential benefits and harms of nanotechnologies but even more important is assessing and publicly discussing the goals for which these technologies will be used" (2007, 3). It has recognized both the current knowledge deficit as well as the scientific challenges facing nanotechnologies. As a consequence, in 2007 UNESCO advocated voluntary ethical guidelines. In its words, "The guidelines would represent a first attempt by

UNESCO to propose a harmonization of ethical principles related to nano-technologies and to recommend actions to be undertaken for research and applications in this field" (2007, 11).

Although purely aspirational and any impact of such guidelines is unlikely to be quantifiable, the approach by UNESCO highlights how it may be possible for longer-term regulatory instruments to focus not only on scientific risks but also matters of societal ethics as well (as per the discussions set out in Chapter 10).

Food and feed products processed with nanotechnologies or incorporating engineered nanomaterials have also been the subject of an expert meeting convened by the FAO and WHO in mid-2009. The aim was "to identify knowledge gaps including issues on food safety, review current risk assessment procedures, and consequently support further food safety research and develop global guidance on adequate and accurate methodologies to assess potential food safety risks that may arise from nanoparticles" (FAO/WHO 2008, 1). Issue identification and capacity building of national food safety regulators to meet the potential challenges posed were the major thrusts here.

In contrast, the work agenda of the United Nations Committee of Experts on the Transport of Dangerous Goods and on the Globally Harmonized System of Classification and Labelling of Chemicals in relation to nanomaterials appears to have been smaller. However, health and safety issues, along with international harmony, are both likely to be of increasing interest to this committee moving forward.

Against this backdrop of formal high-profile activities, in 2002, the ETC Group called for stronger regulatory action at the international level and specifically sought a legally binding international convention to cover both nanotechnologies as well as other emerging technologies. The proposed instrument, entitled *International Convention for the Evaluation of New Technologies*, sought to "consider the wider health, socioeconomic and environmental implications of nanoscale technologies" (ETC Group 2002, 10) and was to be overseen by an independent body whose role was to accept or reject emerging technologies according to benefits and risks (ETC Group 2003). Such a convention was clearly an ambitious idea and has gained little traction in public debates. This is hardly surprising given both the breadth of the phenomenon of nanotechnologies, as well as the multitude of challenges posed by transnational and international regulation. In our view, a more incremental approach to transnational regulation, such as that posed by Abbott, Sylvester, and Marchant (2010), would appear to be a more feasible option (see Section 12.5).

A number of informal transnational initiatives focused on the responsible development and governance of nanotechnologies—but not strictly regulation—have, however, been initiated. These have gained some, although arguably still limited, traction within the debates. These have included the establishment of bodies such as the International Council on Nanotechnology and the International Risk Governance Council's project on risk governance of nanotechnology (see, for example, IRGC 2006, 2009). They have drawn together representatives from different sectors and jurisdictions, and although high-level bodies, appear to operate effectively in relation to policy development for nanotechnologies. Unsurprisingly, these initiatives have not been lauded by all within the nanotech community with, for example, several NGOs opting not to participate in the programs (Powell 2004, 5).

## 12.5 Moving Forward: How and Which International Mechanisms May Be Best Employed for Regulating Nanotechnologies

The European Parliament and Council recently adopted nano-specific provisions as part of a push to regulate cosmetics. Marchant and Sylvester were therefore prescient when they argued that, despite nanotechnology's uncertainties, "we can have complete confidence in one aspect of nanotechnology's future—it will be subject to a host of regulations" (2006, 714). Whether these regulations will be developed at the national or international level, and what form they might take, however, remains unanswered.

Until recently, there had been some optimism about the possibility of creating and coordinating overarching international regulatory instruments. Nanotechnologies have been acknowledged as a global phenomenon, and we hoped that a new paradigm could be created—regulation that was proactive, global, and capable of adapting to rapidly changing conditions. Abbott, Sylvester, and Marchant (2010) proposed a two-stage approach for such action—a short-term informal approach in the form of expert groups—drawn from a range of fields—to engage in traditional dialogue and private codes of conduct, and in the medium term, the use of framework conventions that could be crafted to deal with specific issues as the technology evolved. Risks would therefore be dealt with under both soft and hard regulatory instruments and therefore harness the strengths of both.

Transatlantic cooperation and harmonization issues have also been considered by Breggin et al. (2009), although they, too, were not optimistic about the prospects of an overarching formal international regulatory approach for nanomaterials. They commented that "political energies that would need to be invested in such a project are better spent on strengthening existing forums for international coordination and adjusting domestic regulatory frameworks where needed" (2009, 94). This statement appears to reflect the political reality that fleshing out of any formal agreement would require more cost and time than governments are currently prepared to give, especially in the context of an uncertain payoff.

So, where does all of this leave us? The 2009 recast of the EU's Cosmetics Directive may have been the first national or supranational legislative instrument to be passed with nano-specific provisions, but it will certainly not be the last. It now appears that nano-specific amendments to the Registration, Evaluation, Authorisation and Restriction of Chemicals (REACH) Regulation (regulation [EC] no. 1907/2006) are inevitable. Accordingly, it appears that the window of opportunity for negotiating a legally binding convention or regulatory instrument at the international level is rapidly disappearing. Moreover, there is growing speculation that existing voluntary reporting or data collection activities may be hardened (see, for example, Monica and van Calster 2009; NICNAS 2009), and that national regulatory and policy approaches may be diverging as much as converging, with countries such as France proposing to adopt their own nano-specific regulations (Mayer Brown 2009). Such actions suggest that governments can put in place legislative instruments to specifically cover certain facets of the technology as they see fit. While this leaves regulators, governments, and industry with a huge challenge—to simply stay on top of the regulatory regimes—it is nonetheless in keeping with the dynamic commercial and political environment into which nanotechnologies are being born. The continuation and expansion of the current multiparty and multijurisdiction collaborations will, nonetheless, remain internationally beneficial, and perhaps it is through these channels that we can most sensibly identify and manage longer-term potential risks.

## 12.6 Conclusions

Together, both national and international regulatory regimes will be paramount in ensuring the safety of nanotechnologies into the future and help usher in sustainable alternatives. However, the existence of regulatory structures—albeit at the national or international level—are unlikely to fully satisfy the desires of policy critics who view the nanotechnology phenomenon as a useful rallying call to change broader existing social and economic power structures and inequalities across the globe. Uncertainties will no doubt continue to plague the nanotechnology debates. But uncertainty and the need for more scientifically sound data should not in itself give rise to paralysis in relation to governance frameworks. Although there is little consensus as to the nature and form hard regulatory frameworks for nanotechnologies should take, there have been many useful attempts to develop largely voluntary governance regimes. Such voluntary codes of conduct, risk management frameworks, and certification schemes may not be perfect, but they nevertheless provide a foundation upon which we can build as the body of scientific knowledge increases. Indeed, such innovative, consent-based governance regimes acknowledge inherent uncertainties and can be developed and implemented by institutions at the national and international levels with or without government will.

# References

Abbott, K.W., G.E. Marchant, and D.J. Sylvester. 2006. A framework convention for nanotechnology? *Environmental Law Reporter* 36:10931–942.

Abbott, K.W., D.J. Sylvester, and G.E. Marchant. 2010. Transnational regulation of nanotechnology: Reality or romanticism? In *International Handbook on Regulating Nanotechnologies*, edited by G.A. Hodge, D.M. Bowman, and A.D. Maynard, 525–44. Cheltenham: Edward Elgar.

Allianz Centre for Technology. 2005. *Small Sizes That Matter: Opportunities and Risks of Nanotechnologies*. Paris: OECD International Futures Programme.

Altmann, J. 2010. Military applications: Special conditions for regulation. In *International Handbook on Regulating Nanotechnologies*, edited by G.A. Hodge, D.M. Bowman, and A.D. Maynard, 372–87. Cheltenham: Edward Elgar.

Australian Government. 2006. *Request for Tender: Review of Possible Impacts of Nanotechnology on Australia's Regulatory Frameworks*. Canberra: DIITR.

Ayres, I. and J. Braithwaite. 1992. *Responsive Regulation: Transcending the Deregulation Debate*. New York: Oxford University Press.

Balbus, J.M. et al. 2006. Getting it right the first time—Developing nanotechnology while protecting workers, public health and the environment. *Annals of the New York Academy of Science* 1076:331–42.

Black, J. 2002. Critical reflections on regulation. *Australian Journal of Legal Philosophy* 27:1–36

Bowman, D.M. and G.A. Hodge. 2009. Counting on codes: An examination of transnational codes as a regulatory governance mechanism for nanotechnologies. *Regulation and Governance* 3(2): 145–64.

Bowman, D.M., G. van Calster, and S. Friedrichs. 2010. Nanomaterials and the regulation of cosmetics. *Nature Nanotechnology* 5(2): 92

Braithwaite, J. 1982. Enforced self-regulation: A new strategy for corporate crime control. *Michigan Law Review* 80(7): 1466–1507.

Braithwaite, J. 1993. Responsive regulation for Australia. In *Business Regulation and Australia's Future*, edited by P.N. Grabosky and J. Braithwaite, 81–96. Canberra: Australian Institute of Criminology.

Braithwaite, J. 2008. *Regulatory Capitalism—How It Works, Ideas for Making It Work Better*. Cheltenham: Edward Elgar.

Breggin, L. et al. 2009. *Securing the Promise of Nanotechnologies: Towards Transatlantic Regulatory Cooperation*. London: Royal Institute of International Affairs.

Brownsword, R. 2010. The age of regulatory governance and nanotechnologies. In *International Handbook on Regulating Nanotechnologies*, edited by G.A. Hodge, D.M. Bowman, and A.D. Maynard, 60–80. Cheltenham: Edward Elgar.

California Department of Toxic Substances Control (CDTSC). 2009. *Chemical Information Call-In: Carbon Nanotubes*, 22 January. Sacramento: DTSC.

Chaudhry, Q. et al. 2006. *Final Report: A Scoping Study to Identify Gaps in Environmental Regulation for the Products and Applications of Nanotechnologies*. London: DEFRA.

Chaudhry, Q. et al. 2007. Nanotechnology regulation: Developments in the United Kingdom. In *New Global Regulatory Frontiers in Regulation: The Age of Nanotechnology*, edited by G.A. Hodge, D.M. Bowman, and K. Ludlow, 212–38. Cheltenham: Edward Elgar.

Chen, H. and M.C. Roco. 2008. *Mapping Nanotechnology Innovations and Knowledge.* New York: Springer.

Davies, C. 2006. *Managing the Effects of Nanotechnology* Washington, D.C.: Project on Emerging Technologies.

Department of Environment Food and Rural Affairs (DEFRA). 2006. *UK Voluntary Reporting Scheme for Engineered Nanoscale Materials.* London: DEFRA.

D'Silva, J. and G. van Calster. 2009. Taking temperature: A review of European Union regulation in nanomedicine. *European Journal of Health Law* 16:249–69.

Environmental Protection Agency (EPA). 2007. *EPA Nanotechnology White Paper.* Washington, D.C.: Prepared for the US Environmental Protection Agency by members of the Nanotechnology Workgroup.

Environmental Protection Agency (EPA). 2008. Notice: Nanoscale Materials Stewardship Program. *Federal Register* 73(18): 4861–66.

Epprecht, T.K. 2010. Product safety or managing risks? How regulatory paradigms affect insurability. In *International Handbook on Regulating Nanotechnologies*, edited by G.A. Hodge, D.M. Bowman, and A.D. Maynard, 163–74. Cheltenham: Edward Elgar.

ETC Group. 2002. *No Small Matter! Nanotech Particles Penetrate Living Cells and Accumulate in Animal Organs.* Communiqué Issue #76, May. Ottawa: ETC Group.

ETC Group. 2003. *No Small Matter II: The Case for a Global Moratorium Size Matters!* Ottawa: ETC Group.

European Commission (EC). 2005. *Nanosciences and Nanotechnologies: An Action Plan for Europe 2005–2009.* Brussels: European Parliament.

European Commission (EC). 2008. *Regulatory Aspects of Nanomaterials* and *Regulatory Aspects of Nanomaterials: Summary of Legislation in Relation to Health, Safety and Environment Aspects of Nanomaterials, Regulatory Research Needs and Related Measures.* Brussels: European Commission.

European Commission (EC). 2009. *Nanosciences and Nanotechnologies: An Action Plan for Europe 2005–2009.* Second Implementation Report 2007–2009 (COM[2009]607 final). Brussels: European Commission.

European Parliament (EP). 2009. *MEPs Approve New Rules on Safer Cosmetics.* Press release, 24 March. Brussels: EP.

Fiedler, F.A. and G.H. Reynolds. 1994. Legal problems of nanotechnology: An overview. *Southern California Interdisciplinary Law Journal* 3:594–629.

Food and Agriculture Organization and World Health Organization (FAO/WHO). 2008. *Joint FAO/WHO Expert Meeting on the Application of Nanotechnologies in the Food and Agriculture Sectors: Potential Food Safety Implications.* Rome: FAO and WHO.

Food and Drug Administration (FDA). 2007. *Nanotechnology—A Report of the U.S. Food and Drug Administration Nanotechnology Task Force.* Washington, D.C.: FDA.

Food Safety Authority of Ireland (FSAI). 2008. *The Relevance for Food Safety of Applications of Nanotechnology in the Food and Feed Industries.* Dublin: FSAI.

Food Standards Agency. 2008. *A Review of the Potential Implications of Nanotechnologies for Regulations and Risk Assessment in Relation to Food.* London: FSA.

Forrest, D. 1989. *Regulating Nanotechnology Development.* Available at http://www.foresight.org/nano/Forrest1989.html (accessed 31 July 2011).

Friends of the Earth Australia (FoEA). 2006a. *An Analysis by Friends of the Earth of the National Nanotechnology Strategy Taskforce Report: "Options for a National Nanotechnology Strategy."* Melbourne: FoEA.

Friends of the Earth Australia (FoEA). 2006b. *Nanomaterials, Sunscreens and Cosmetics: Small Ingredients, Big Risks.* Melbourne: FoEA and FoEUS.

Friends of the Earth Australia (FoEA). 2007. *Nanosilver—A Threat to Soil, Water and Human Health?* Melbourne: FoEA.

Fuhr, M. et al. 2006. *Legal Appraisal of Nanotechnology: Existing Legal Frameworks, the Need for Regulation and Regulative Options at a European and National Level.* Darmstadt: Society for Institutional Analysis.

Gergely, A. 2007. Regulation of Nanotechnology—within REACH? *NanoNow* (February): 44–46

Gilligan, G. and D.M. Bowman. 2008. "Netting Nano": Regulatory challenges of the Internet and nanotechnologies. *International Review of Law Computer and Technology* 22(3): 231–46.

Glaeser, E.L. and A. Shleifer. 2003. *The Rise of the Regulatory State.* Boston: Harvard University

Gunningham, N. and J. Rees. 1997. Industry self-regulation: An institutional perspective. *Law and Policy* 19(4): 363–414.

Gunningham, N. and D. Sinclair. 1999. Regulatory pluralism: Designing policy mixes for environmental protection. *Law and Policy* 21(1): 49–76

Health and Safety Executive (HSE). 2003. *Health and Safety Regulations: A Short Guide.* London: HSE.

Health and Safety Executive (HSE). 2006. *Review of the Adequacy of Current Regulatory Regimes to Secure Effective Regulation of Nanoparticles Created by Nanotechnology: The Regulations Covered by HSE.* London: HSE

House of Lords Science and Technology Committee (STC). 2009. *Call for Evidence: Nanotechnologies and Food.* London: United Kingdom Parliament.

Hull, M. 2009. Nanotechnology environmental, health and safety: A guide for small business. In *Nanotechnology Risk Management*, edited by M. Hull and D.M Bowman, 243–97. London: Springer.

International Risk Governance Council (IRGC). 2006. *White Paper on Nanotechnology Risk Governance—Towards an Integrative Approach.* Geneva: IRGC.

International Risk Governance Council (IRGC). 2009. *Appropriate Risk Governance Strategies for Nanotechnology Applications in Food and Cosmetics.* Geneva: IRGC.

Intellectual Property Office. 2009. *UK Innovation Nanotechnology Patent Landscape Analysis.* London: IPO.

Kimbrell, G.A. 2006. Nanomaterial consumer products and FDA Regulation: Regulatory challenges and necessary amendments. *Nanotechnology Law and Business* 3(3): 329–38.

King, A.A. and M.J. Lenox. 2000. Industry self-regulation without sanctions: The chemical industry's Responsible Care Program. *Academy of Management Journal* 43(4): 698–716.

Levi-Faur, D. and J. Jordana. 2005. The rise of regulatory capitalism: The global diffusion of a new order. *Annals of the American Academy of Political and Social Sciences* 598(1): 12–32.

Ludlow, K. 2007. One size fits all? Australian regulation of nanoparticle exposure in the workplace. *Journal of Law and Medicine* 15:136–52.

Ludlow, K., D.M. Bowman, and G.A. Hodge. 2007. *A Review of Possible Impacts of Nanotechnology on Australia's Regulatory Framework*. Melbourne: Monash Centre for Regulatory Studies.

Ludlow, K., D.M. Bowman, and D. Kirk. 2009. Hitting the mark or falling short with nanotechnologies regulation? *Trends in Biotechnology* 27(11): 615–20.

Lux Research Inc. 2005. *Nanotechnology: Where Does the US Stand? Testimony before the Research Subcommittee of the House Committee on Science*. New York: Lux Research.

Majone, G., 1999. The regulatory state and its legitimacy problems. *West European Politics* 22(1): 1–24.

Marchant, G.E. and D.J. Sylvester. 2006. Transnational models for regulation of nano-technology. *Journal of Law, Medicine and Ethics* 34(4): 714–25.

Marchant, G.E., D.J. Sylvester, and K.W. Abbott. 2007. Nanotechnology regulation: The United States approach. In *New Global Regulatory Frontiers in Regulation: The Age of Nanotechnology*, edited by G.A. Hodge, D.M. Bowman, and K. Ludlow, 189–211. Cheltenham: Edward Elgar.

Marchant, G.E., D.J. Sylvester, and K.W. Abbott. 2008. Risk management principles for nanotechnology. *NanoEthics* 2(1): 43–60.

Mayer Brown. 2009. *EU Competition—Brussels Client Alert: France Might Take the Lead on Nanotechnology Regulation*. 5 March. Brussels: Mayer Brown.

Maynard, A.D. et al. 2006. Safe handling of nanotechnology. *Nature* 444:267–69.

Meili, C. and M. Widmer. 2010. Voluntary measures in nanotechnology risk gover-nance: The difficulty of holding the wolf by the ears. In *International Handbook on Regulating Nanotechnologies*, edited by G.A. Hodge, D.M. Bowman, and A.D. Maynard, 446–61. Cheltenham: Edward Elgar.

Miles, J. 2007. Metrology and standards for nanotechnology. In *New Global Regulatory Frontiers in Regulation: The Age of Nanotechnology*, edited by G.A. Hodge, D.M. Bowman, and K. Ludlow, 333–52. Cheltenham: Edward Elgar.

Miles, J. 2010. Nanotechnology captured. In *International Handbook on Regulating Nanotechnologies*, edited by G.A. Hodge, D.M. Bowman, and A.D. Maynard, 83–106. Cheltenham: Edward Elgar.

Miller, G. and G. Scrinis. 2010. The role of NGOs in governing nanotechnologies: Challenging the "benefits versus risks" framing of nanotech innovation. In *International Handbook on Regulating Nanotechnologies*, edited by G.A. Hodge, D.M. Bowman, and A.D. Maynard, 409–45. Cheltenham: Edward Elgar.

Miller, G. and R. Senjen. 2008. *Out of the Laboratory and on to Our Plates: Nanotechnology in Food and Agriculture*. Melbourne: FoEA, FoEEU, and FoEUS.

Minogue, M. and L. Carino. 2006. Introduction. In *Regulatory Governance in Developing Countries*, edited by M. Minogue and L. Carino, 3–16. CRC Series on Competition, Regulation and Development. Cheltenham: Edward Elgar.

Monica, J.C. and G. van Calster. 2010. A nanotechnology legal framework. In *Nanotechnology Risk Management*, edited by M. Hull and D.M. Bowman, 97–140. London: Springer.

Moran, A. 1995. Tools of environmental policy: Market instruments versus command-and-control. In *Markets, the State and the Environment: Towards Integration*, edited by R. Eckersley. South Melbourne: Macmillan Education.

Moran, M. 2002. Understanding the regulatory state. *British Journal of Political Science* 32:391–413.

Munich Re Group. 2002. *Nanotechnology—What Is in Store for Us?* Munich: Munchen Re.

NanoAction. 2007. *Principles for the Oversight of Nanotechnologies and Nanomaterials.* Washington, D.C.: International Center for Technology Assessment.

National Industrial Chemical Notification and Assessment Scheme (NICNAS). 2007. *Summary of Call for Information and the Use of Nanomaterials.* Canberra: Australian Government.

National Industrial Chemical Notification and Assessment Scheme (NICNAS). 2008. Industrial nanomaterials: Voluntary call for information 2008. *Australian Government Gazette* No. C 10 (7 October): 25–38.

National Industrial Chemical Notification and Assessment Scheme (NICNAS). 2009. *Proposal for Regulatory Reform of Industrial Nanomaterials: Public Discussion Paper—November 2009.* Canberra: NICNAS.

National Research Council (NRC). 2001. *Global Networks and Local Vales—A Comparative Look at Germany and the United States.* Washington, D.C.: National Academy Press.

New South Wales Legislative Council Standing Committee on State Development (NSW LCSCSD). 2008. *Nanotechnology in NSW.* Sydney: NSW Legislative Council.

Organisation for Economic Cooperation and Development (OECD). 2007. *OECD Working Party on Nanotechnology (WPN): Vision Statement.* Paris: OECD.

Organisation for Economic Cooperation and Development (OECD). 2008. *Nanotechnologies at the OECD.* Paris: OECD.

Pelley, J. and M. Saner. 2009. *International Approaches to the Regulatory Governance of Nanotechnology.* Ottawa: Carleton University.

Pinson, R. 2004. Is nanotechnology prohibited by the biological and chemical weapons conventions? *Berkeley Journal of International Law* 22(2): 279–309.

Powell, K. 2004. Green groups baulk at joining nanotechnology talks. *Nature* 432(7013): 5.

Rees, J. 1997. Development of communitarian regulation in the chemical industry. *Law and Policy* 19(4): 477–528.

Renn, O. and M.C. Roco. 2006. Nanotechnology and the need for risk governance. *Journal of Nanoparticle Research* 8:153–91.

Royal Commission on Environmental Pollution. 2008. *Novel Materials in the Environment: The Case of Nanotechnology.* London: Royal Commission.

Royal Society and Royal Academy of Engineering (RS-RAE). 2004. *Nanoscience and Nanotechnologies: Opportunities and Uncertainties.* London: RS-RAE.

Royal Society, Insight Investment, Nanotechnology Industries Association, Nanotechnology Knowledge Transfer Network (RSIINIA, NKTN). 2007. *Responsible Nanotechnologies Code: Consultation Draft—17 September 2007 (Version 5).* London: Responsible NanoCode Working Group.

Sadrieh, N. and P. Espandiari. 2006. Nanotechnology and the FDA: What are the scientific and regulatory considerations for products containing nanomaterials? *Nanotechnology Law and Business* 3(3): 339–49.

Sinclair, D. 1997. Self-regulation versus command and control? Beyond false dichotomies. *Law and Policy* 19(4): 529–59.

Sunstein, C. 1990. *After the Rights Revolution: Reconceiving the Regulatory State.* Boston: Harvard University Press.

Swiss Re. 2004. *Nanotechnology: Small Matter, Many Unknowns.* Geneva: Swiss Re.

Taylor, M.R. 2006. *Regulating the Products of Nanotechnology: Does FDA Have the Tools It Needs?* Washington, D.C.: Project on Emerging Technologies.

Taylor, M.R. 2008. *Assuring the Safety of Nanomaterials in Food Packaging: The Regulatory Process and Key Issues.* Washington, D.C.: Project on Emerging Nanotechnologies.

United Nations Committee of Experts on the Transport of Dangerous Goods and on the Globally Harmonized System of Classification and Labelling of Chemicals. 2009. *Ongoing Work on the Safety of Nanomaterials*. Geneva: UN.

United Nations Educational, Scientific and Cultural Organization (UNESCO). 2006. *The Ethics and Politics of Nanotechnology*. Paris: UNESCO.

United Nations Educational, Scientific and Cultural Organization (UNESCO). 2007. *Nanotechnology and Ethics: Policies and Actions—COMEST Policy Recommendations*. Paris: World Commission on the Ethics of Scientific Knowledge and Technology.

Utting, P. 2005. *Rethinking Business Regulation—From Self-Regulation to Social Control*. Geneva: United Nations Research Institute for Social Development.

Vogel, D. 2006. *The Private Regulation of Global Corporate Conduct*. Berkeley: University of California.

Wardark, A. 2003. *Nanotechnology and Regulation: A Case Study using the Toxic Substance Control Act (TSCA)*. Washington, D.C.: Project on Emerging Nanotechnologies.

Webb, K. 2004. Understanding the voluntary code phenomenon. In *Voluntary Codes: Private Governance, the Public Interest, and Innovation*, edited by K. Webb, 3–32. Ottawa: Carleton University.

Webb, K. and A. Morrison. 1996. *The Legal Aspects of Voluntary Codes*. Paper presented at the Exploring Voluntary Codes in the Marketplace Symposium, 12–13 September, Ottawa.

# 13

## Nanotechnology without Growth

Donnie Maclurcan and Natalia Radywyl

### CONTENTS

### 13.1 Introduction

Whatever one's views on the desirability of emerging technologies, it is worth pausing to consider the smallness of the scale on which this book has focused. Imagine splitting a human hair into 100,000 pieces—just one of those pieces would represent the realm on which nanotechnology operates. What happens in this realm is so integral to our existence that it has become the site for a worldwide exploration of some of the most fundamental challenges we collectively face. These are challenges that have already begun to appear with regularity in the headlines (such as climate change, food shortages, and restricted access to essential medicines), yet the full extent of their long-term impacts is still to be felt, both globally and locally. In this concluding chapter, we consider the way our authors addressed these challenges through their own studies and research into nanotechnology, and how, as a collective voice, their contributions addressed the four themes that make up this book: *limits*, *capacity*, *appropriateness*, and *governance*. These reviews lead us to consider nano-innovation in a broader context of the evolving relationships between science and society and to explore alternative paths for technological innovation, along with the new associations and challenges that may arise. Although somewhat speculative, we make an earnest attempt to explore the possibilities for innovation without growth, and conclude by placing nano-innovation within a larger context of futures beyond economic growth—a prospect toward which we believe we can work.

If to move away (for a moment) from nanotechnology-specific discussion and toward thinking about scientific innovation in more general terms, we can see that how we engage with technologies to respond to global challenges is increasingly significant, largely because of the sheer speed of innovation and its capacity for pervasive accumulation. Historians of technology increasingly work with the principle that technology is socially constructed; that "artifacts emerge as the expressions of social forces, personal needs, technical limits, markets, and political considerations" (Nye 2006, 49). As science and society become increasingly enmeshed, we believe that any study of innovation and sustainability cannot be divorced from examination of related issues such as political economy and sustainability. As well-known economist Joseph Schumpeter (1934) once noted: "innovation is the outstanding fact in the economic history of capitalist society" (86).

As one of the latest in a series of recent, controversial, and cross-sectoral technologies, we regard nanotechnology as a useful lens through which to explore the relationships between innovation, political economy, and sustainability, and how stakeholder agendas can be actively reshaped in the interest of creating positive and sustainable futures. In this book, we have therefore sought to open up debates about science and sustainability so as to move beyond a focus on issues of efficiency, productivity, and utility. By looking at the past, present, and possible futures, we have begun to assemble, in some very preliminary ways, what an alternative sustainability approach for science might look like beyond our current predominantly growth-driven paradigm.

When exploring the possibility of innovation without growth (given a limits-to-growth imperative), the issue of equity surfaces as critical, considering it is viewed as a precondition for a steady-state economy (Wilkinson and Pickett 2009). In this book, we thus sought a holistic consideration of equity in its environmental, power, needs-based, and participatory forms, given the role of each in ensuring technology is appropriate. We endeavored to look at equity as it relates to alternative starting points for innovation, infrastructure for innovation, approaches to technological design, and methods for overseeing innovation. To achieve this broad-ranging enquiry, we sought to draw together leading thinkers from a variety of backgrounds and interests: from social scientists, technical experts, and advocates who represent gender, disciplinary, and geographical diversity. This has allowed bodies of knowledge to uniquely coalesce between these pages, including environmental sociology, nanoscience, technology studies, anthropology, political economy, development theory, and regulatory studies. The diversity of interests and perspectives in this book led to original, valuable associations and a cross-fertilization of ideas, thereby traversing traditional barriers to create new assets for tackling some of our greatest challenges.

Although discussions of growth and engagement with political economy have not been the explicit focus of every chapter in this book, the contributions, when placed together and alongside external developments, shed introductory light on how we might innovate without growth. They essentially become an argument for a new approach to political economy.

## 13.2 Chapter and Themes Revisited

The first theme of the book, *Limits*, drew together authors whose perspectives on nanotechnology are acutely informed by an awareness of deeply rooted systemic constraints and foreshadowed approaches to using nanotechnology as an alternate form of thinking with which to surmount these barriers—either rising above and transcending them or working within them. We opened this book by authoring a chapter about the limits to growth as a fundamental scientific phenomenon, showing the need to rethink nanotechnology's implications for sustainability in a much broader and global manner.

In Chapter 2, "Nanotechnology and the Environment," David Hess and Anna Lamprou showed us how nanotechnology could play an important role in the shift to a steady-state economy if steered toward renewable technologies that facilitate the dematerialization of energy consumption. They argued that nanotechnology could play a particularly central role in making solar energy more competitive and to assist in efforts to go beyond grid parity. Clearly the technologies we might choose for innovation within limits are not immune from their own environmental and human health risks—a point recently acknowledged in an embodied energy assessment of fullerenes that documented their high energy requirements throughout their life cycle (Anctil et al. 2011). Hess and Lamprou therefore also warned of health risks posed by engineered nanoparticles within solar designs, as well as the dangers of ecological modernization without adequate foresight, reminding us that "even with a nanosolar revolution, aggregate levels of absolute withdrawals and deposits from the global ecosystem might continue to rise." Despite risks such as these, the authors advocated increased funding for nanosolar research, believing in its significant potential to assist us with dematerialization, as well as the ability to reduce specific risks associated with different types of nanosolar design, if appropriate attention is given to such concerns.

Discussion then moved from economic and risk-related considerations to those that cut through boundaries between bodies of knowledge. In Chapter 3, "Nanotechnology and Traditional Knowledge Systems," Ron Eglash detailed how nanotechnology can be tied to traditional forms of knowledge. Here he demonstrated that a more careful examination of the social histories of artifacts can shift our understanding from a purely epistemological perspective to understanding them as a more dynamic, vibrant set of practices, opening up new ground for sustainability science. The historical practices he mentioned have, in some cases, provided important points of dialogue with the development of Western science, and in other cases remain a resource for contemporary technological development and future applications. Presenting evidence grounded in sophisticated bodies of knowledge from the heritage cultures of high school students in the United States, he demonstrated that nanotechnology can be used to instill confidence in minority cultures about their contributions to science and, therein, sustainability. Yet, while drawing upon a variety of case studies that build a strong argument for the wisdom of working across both modern and traditional knowledge, Eglash also argued for sensitivity as to how we make indigenous innovations available in a just and responsible manner. His discussion clearly opens up a space for exploring the limitations of Western thinking, and in doing so highlights that innovation need not compulsively strive for the new. In fact, the recognition of limits can compel us to look inward, and enable us to re-discover, re-interpret, and re-innovate, and push a form of growth that moves from within, building upon existing wisdom.

Our book then moved to the second key theme—creating decentralized *Capacity*. This section collated contributions suggesting that, in terms of costs and feasibility, there is both an urgent need as well as some hope for nano-innovation to be decentralized, but that asset-based approaches are required in order to subvert embedded inequities. In Chapter 4, "Nanotechnology and Geopolitics: There's Plenty of Room at the Top," Stephanie Howard and Kathy Jo Wetter revealed the strong link between capacity for nano-innovation and early government funding and support. Here they detailed a contemporary regulatory environment that allocates a great deal of power to corporate entities, thereby limiting access to technologies. In addition, they argued that this drive for economic competitiveness and industrial growth has been at the expense of research into less risky and more environmentally sustainable systems and approaches. Investigation of nanotechnology's social application has similarly suffered. As noted by the authors, even despite a shift eastward in the gravitational center of innovation (to countries such as China, Russia, and India), there continues to be dramatic inequity in capacity.

Within this geopolitical picture, it is certainly difficult to envisage government leading the way to innovative futures determined by something other than growth imperatives. This is especially so, as explained by Howard and Wetter, because governments' economically framed arguments defending nanotechnology lack rigor, with evidence of there being little clarity around market projections, value-chain assessments, and the levels of funding required to ensure that nanotechnology delivers.

Hence, while the authors raised some of nanotechnology's potential benefits, they simultaneously showed how these benefits are restricted to a limited number of people. For example, the U.S. Patent and Trademark Office's resolution to confer "accelerated status" on technologies to combat climate change and foster job creation in the green technology sector are something of a double-edged sword, as they may very well exacerbate tensions between the Global South and North over energy-related intellectual property (IP). Furthermore, just because individual applications could prove to be technically viable, safe, superior, and accessible for the communities that typically exist outside the mainstream economy, the broader workings of the nano-economy could "undermine livelihoods or introduce new forms of contaminants that disturb resources upon which those same communities depend." In this light, the authors suggested that the lack of clarity around nanosafety makes claims of ecological sensitivity and sustainability highly premature. In response, they demand much greater scrutiny of developments like the nano clean-tech brand and suggest that avenues such as the International Forum on Chemical Safety might provide interim means for sustainability discussions internationally, until a permanent international forum for technology assessment and equitable distribution can be established.

In Chapter 5, "Nanotechnology, Agriculture, and Food," Kristen Lyons, Gyorgy Scrinis, and James Whelan examined the issue of capacity by first proposing that nanotechnology will soon have the capability to permeate the entire agri-food system. Although nanotechnology offers technical, Band-Aid approaches to social and systemic challenges, the authors argued that, overall, it is actually driving greater inequity in the agri-foods sector. They described how, by perpetuating and extending an inequitable system in which the capacity to provide food is controlled by fewer and fewer, social and ecological costs are increasingly borne by farmers, farmworkers, and citizens. Furthermore, much-touted environmental benefits (such as reduced waste in food packaging) may actually spawn new challenges (a subsequently reduced nutrient value within the foods, for example), given the high level of unknowns when it comes to nanotechnology's risks. The authors therefore positioned themselves as advocates of a precautionary approach with regards to mainstream technological innovation. Yet, in doing so, they mounted a very strong campaign for halting the technological treadmill, reducing our dependence on technologies driven by concentrations of power, embracing nonindustrial methods of agriculture and innovation, and therein supporting distributed capabilities, food sovereignty, biodiversity, and agro-ecology. Lyons, Scrinis, and Whelan also reminded us how consumers and voters can garner power through informed endorsement and suggested greater attention be given to building the capacity of nongovernment organizations across a range of movements, so as to ensure engagement with a breadth of nanotechnology debates.

In Chapter 6, "Poor Man's Nanotechnology—From the Bottom Up (Thailand)," Sunadan Baruah, Joydeep Dutta, and Louis Hornyak illustrated how, by taking an asset-based approach to nano-innovation, cost, resource, technical, and even environmental barriers can be overcome. The US$1,500 startup budget for their Centre of Excellence in Nanotechnology is nothing short of breathtaking when one considers their present outputs. Their self-proclaimed approach—"poor man's nanotechnology"—has been based upon international partnerships across varying levels of resource capacity, a culturally diverse mix of researchers, as well as hard work and an openness to innovation. Technically, they draw on the power of a bottom-up approach to science, including biomimicry (and, therein, the ability to build analytical instruments and experimental apparatus from scratch, or synthesize with inexpensive chemicals). Even with regard to top-down approaches, the authors argued that, through nanotechnology, fabrication procedures for items such as dye-sensitized solar cells demonstrate ways in which the production of energy-harvesting devices can be done more simply and in an environmentally friendly way. More broadly, the authors claimed that removing the cost barrier enables a greater focus to be placed on end-user, needs-based science (especially with respect to the needs of the poor). While Baruah, Dutta, and Hornyak openly shared the commercial aspirations for their laboratory's work, they envisaged their approach to nano-innovation as a means to circumvent traditional competition dynamics.

The third key theme of the book, *Appropriateness*, drew together contributions that explored contemporary mainstream and peripheral approaches to nanotechnology design, with particular consideration for levels of sensitivity to human needs, cultural norms, and environmental effects. In Chapter 7, "Nanotechnology and Global Health," Deb Bennett-Woods argued that only a contextually grounded approach to nano-innovation can be appropriate. She explained that traditional reductionist approaches to global health tend to focus upon fixing biomedical problems and therefore fail to mitigate the social foundations of ill health. For example, when turning to discussions about health care in the Global South, she contended that there needs to be a better match of technologies with southern capacities, as well as an improved prioritization of scarce resources through a more cohesive approach to funding allocation. And so, rather than rejecting the benefits of the medical model entirely (given its crucial importance to the direct management of disease burden), she advocated for engagement with nanotechnology using a strategic, systemic, and reprioritized approach.[1] This approach requires an initial appreciation for the broader context within which nanotechnology may be applied *before* assessment of its instrumental technological potential.

Bennett-Woods argued that the first step to understanding systemic contexts is facilitating a form of public engagement that is clearly supported by a research agenda. This agenda should clarify the critical systemic links between social determinants of health and the traditional biomedical paradigm that characterizes most current efforts to mitigate global health challenges. In this light, issues such as cost are central to matters of equity, and Bennett-Woods accordingly proposed that open licensing become more widespread—if not compulsory (as later taken up by Mushtaq and Pearce in Chapter 9). Bennett-Woods saw this as a part of advocating for a more collaborative, proactive, yet accountable approach toward dealing with global health issues. In this light, she reaffirmed the views of many of our contributors: that innovation should only occur at a pace relative to our level of understanding of the contexts in which it plays a role.

David Grimshaw tackled the idea of appropriateness in his contribution, *Toward Pro-Poor Nano-Innovation (Zimbabwe, Peru, Nepal)* (Chapter 8) by proposing that nanotechnology could help communities leapfrog to appropriate alternatives, given nanotechnology is not, as yet, entirely locked in to trajectories dictated by existing market forces. Grimshaw sought to challenge the "technology push" model, reminding us that, even though technology is largely adapted (rather than innovated), public input is critical in the (re) design of nanotechnologies seeking to address human needs. His arguments underscore the value of alternate systems of design, such as participatory design, that especially seek to include the views of those who are often silent (or silenced) in scientific research, including women, ethnic and racial minorities, indigenous peoples, peasants, and people with disabilities.[2]

In this sense, Grimshaw cautioned against automatically equating increased diffusion of technologies with increased equity. He contended that, by identifying the attributes of new technologies that need changing, we can take a first step toward the kinds of technology-related actions and policies that will ultimately help address human needs. In his eyes, a systems perspective means addressing power, price, promise, poverty, pervasiveness, promiscuous utility, and the surrounding paradigm. Drawing upon his work in Zimbabwe, Peru, and Nepal, Grimshaw revealed that local communities are willing to engage with nanotechnology, as long as consultation methods shift from tokenistic, upstream research to processes of participatory innovation that ensure community input into design and later prototype testing.[3] In light of new approaches to business that he mentioned (such as social enterprise), Grimshaw's "pro-poor innovation" model would seem to hold promise for futures beyond economic growth.

In "Open Source Appropriate Nanotechnology" (Chapter 9), Usman Mushtaq and Joshua Pearce claimed that the mainstream IP system presently governing technological innovation is designed to produce inequity and actually limits innovation, while also removing knowledge from the public domain of the Global South. The authors believe that nanotechnology is set to perpetuate these phenomena by restricting access to expertise and limiting diffusion of relevant artifacts. They therefore proposed that nanotechnology be developed in an open source model (referring to open information, software, hardware, and standards). Providing some encouraging examples of open source nanotechnology from the water, energy, and construction sectors, the authors suggested that such an approach would increase innovation through collaborative production, enhance access to appropriate technologies for the marginalized, and even more importantly, be "the people's way back in" for participatory, localized design and possibly for customized development. Self-determination could therefore be grounded in localized innovation, owing to, in their own words, "less dependence on a single supplier or information source, and a reduced risk of obsolescence."[4] In addition, open-sourcing nano-innovation could help to mitigate risks through greater public oversight, thereby emphasizing the positive influence of grassroots forms of governance.

Mushtaq and Pearce's work contributes to a rapidly strengthening trajectory of research and projects relating to Do-It-Yourself (DIY) and Do-It-With-Others (DIWO) innovation subcultures. As noted by Kera (2011), "Global and alternative innovation networks are developing around Do-It-Yourself (DIY) and Do-It-With-Others (DIWO) subcultures, such as Direct to Consumer (DTC) genomics, DIYbio labs, DIYgenomics, Clinical trials 2.0, Hackerspace hackathons, Maker fairs and FabLabs competitions" (49).[5] The development of these subcultures has significant and far-reaching potential, owing to the much-lowered barriers to access and participation in emerging innovation. For example, when the OS Nano initiative released its knowledge surrounding magnetite nanocrystals under a Creative Commons Attribution license, it allowed researchers access to a low-cost technique enabling them to "replace many of the expensive material requirements with everyday products such as soap, rust and vinegar while moving the crystallization process out of a lab and into a kitchen" (Thakur 2010, 339).

Similar DIY trajectories around appropriateness and decentralized capacity have been appearing in the open source hardware space, for example, "Home-scale machines, such as 3D printers, laser cutters, and programmable sewing machines, combined with the right electronic design blueprint, enable people to manufacture functioning products at home, on demand, at the press of a button" (Lipson and Kurman 2010, 4). For some, this is leading to an era of backyard, open source microfactories capable of disrupting capital intensive industrialism (Bauwens 2011; Jakubowski 2011).[6] Jakubowski (2011) calls this "Industry 2.0 … a scenario of distributive, local production via flexible fabrication, fueled by a global repository of open source design."

Not surprisingly, given the open source model runs on the notion of a gift economy, the prospect of open source nano-innovation is very good news for those hoping to innovate without growth, particularly as it offers new insights into radically different and possible forms of political economy. In this light, it is worth mentioning that, while open source *can* be profitable (as explored by Mushtaq and Pearce), this does not mean such profitability *must* fit within a for-profit business model. The Australian engineering firm engaging with nanotechnology and mentioned by Eglash in Chapter 3—the Myuma group—provides an interesting case in point. With an annual turnover of AUS$18 million and employing fifty staff members, this company is proving a successful business. However, it is a not-for-profit enterprise. That is, any financial surplus is reinvested back into the company, rather than paying dividends to distant shareholders.

Mushtaq and Pearce also envisaged a number of less-conventional links between nanotechnology and commerce in the future, such as smaller players engaging with open source nanotechnology having unprecedented access to a largely unexplored market, and nano-innovation funded through micro-investments managed through online gateways. However, overall they saw hybrid business models (mixing open source and proprietary approaches and the strategic use of patent pools) as the best possible situation for mainstream

nano-innovation in the short term, given the current predominantly proprietary status of existing mainstream knowledge.[7] Moreover, writing on a similar topic, Danuraj Thakur (2010) adds, "although seemingly counterintuitive,
any strategy for pushing open access nanotech in a developing country must
also be accompanied by adequate IPR [intellectual property rights] protection" (343).

While Mushtaq and Pearce described a longer-term picture comprising
user-developers, fully open knowledge and the associated move to low-foot-
print, service-based trade, they did caution that solid measures need to be in
place to avoid unethical commodification of open nano-innovation.

The fourth theme of this book, *Governance*, has explored approaches
to nanotechnology risk, crossing both national and international regulatory landscapes while seeking means for more participatory oversight. In
"Nanotechnology and Risk" (Chapter 10), Fern Wickson argued that the
dominant approaches to risk assessment for engineered nanomaterials are
too narrow in their focus to be considered truly comprehensive. This is not
to deny the serious gravity and uncertainties associated with present findings on the health and environmental risks of nanotechnology. Wickson
showed particular concern for the risks surrounding carbon nanotubes
and the need for more lab research, particularly into life-cycle assessments
(although she also highlighted that nanotechnology health and safety
research can routinely suffer shortfalls such as a lack of appropriate equipment, techniques, and funding). Rather, she astutely noted that framing
nanotechnology and its role in our futures in terms of scientific risks means
that physical harm to human and environmental health appear as the only
legitimate social concerns, relegating social and ethical issues to the margins of debate and policymaking.

In response, Wickson envisaged the need for a constructivist approach
that would incorporate both scientific research and broader forms of analysis
involving more direct integration of social science research and public values in research and development (R&D) processes. While Wickson saw some
hope in developments such as the European Commission's recent attempt
to actively steer nanotechnology in a socially responsible way through public consultation, she proposed benefits assessment, knowledge assessment,
pedigree assessment, technology assessment, and alternatives assessment as
more holistic methods for consideration. Wickson believes these strategies
will bring the bigger questions about nanotechnology to the fore by preemp-
tively exploring multiple sociotechnical trajectories, and, in doing so, allow us
to move beyond our solutions pathology in order to shape truly sustainable
futures. Looking outside Wickson's writing, it may be that such approaches
have begun to emerge of their own accord. For example, Kera (2011) reports
that Hackerspaces, FabLabs, Makerspaces, DIYbio labs, and citizen science
projects are showing the success of decentralized approaches to management
and policy, through participatory monitoring and crowdsourcing of data.
These alternative avenues for R&D "are becoming testbeds for new models

of public participation in Science and Technology but also new models for policy making in which political deliberation merges with design iteration and embraces citizen science paradigms of research" (Kera 2011, 49–50).

In Chapter 11, "Nanotechnology and State Regulation (India)," Nupur Chowdhury and Nidhi Srivastava engaged with the issue of governance by arguing that national regulation is foundational to developing international regulatory frameworks, and deficits in standards, capacity, and metrology present major challenges in regulating nanotechnology in India. The authors described how Indian nanotechnology products (such as in health care) have already hit the market. Yet, most Indian health-related nanotechnology research is focused toward curative aspects of health care, seeking to address lifestyle rather than neglected and localized diseases. Chowdhury and Srivastava detailed the very complex Indian regulatory system, illustrating how significant challenges arise from bodies that both overlap and undermine each other as they compete to establish their own regulatory mandates for nanotechnology. Compounding this situation is a lack of capacity to monitor regulatory compliance, little awareness among regulatory authorities of the challenges posed by newer applications of nanotechnology, and pro-innovation lobbyists driving a "skewed understanding of regulation." Furthermore, the authors described how there is a general aversion in India to open discussion of nanotechnology's potential risks and the possible need for customized regulation.

Turning to possible responses to this situation, they noted that a moratorium on nanotechnology or even nanomaterials per se would simply be impractical, and that it is pertinent to discuss regulatory issues vis-à-vis specific nanotechnology applications. In this light, the authors believe there is considerable flexibility in Indian legislation to allow the development of specific guidelines and principles for the regulatory oversight of nanotechnology-based applications. However, this flexibility does not presently appear to resolve regulatory issues, instead creating further challenges relating to institutional frameworks, uncertainties, and knowledge gaps. In response, the authors identified the need for a mix of soft and hard regulatory instruments, and for policy design to differentiate between the short, medium, and long term. Similarly to Mushtaq and Pearce in Chapter 9, they also advocated an open approach to nanotechnology knowledge, envisaging that it could reduce knowledge asymmetries and improve India's ability to more fully consider the health risks associated with nanotechnology.

In their contribution, Diana Bowman and Graeme Hodge examined and appraised the history of nanotechology's regulation. In "Nanotechnology and Global Regulation" (Chapter 12), the authors noted that, although regulatory questioning for nanotechnology goes back more than two decades, it is only since 2004 that more comprehensive reviews have been conducted. Importantly, recent government reviews in Australia, the United Kingdom, and the United States found that relevant legal instruments were, in principle, sufficiently broad in their application so as to encompass regulation for nanotechnologies. However, Bowman and Hodge highlighted what would seem to be some pretty fundamental gaps in these reviews to do with the novelty of nanomaterials and uncertainties around risks as well as testing capabilities. What they did record, however, is the noticeable shift in recent regulatory reviews from broader questions of whether nanotechnology-based products and process are regulated to more focused calls for amendments in specific areas—such as in foods, cosmetics, industrial chemicals, and occupational health and safety.

Similar to Chowdhury and Srivastava, Bowman and Hodge did not regard a general moratorium on the development or commercialization of nano-technologies a feasible approach, largely due to the complexities associated with both defining and enforcing such a moratorium. Yet, they also noted that, "should the growing body of nanotoxicology literature show that certain nanoparticles pose an unacceptable health or environmental risk, then governments may rightly be forced to enact tighter controls in relation to such activities." Bowman and Hodge therefore believe that regulatory flexibility is necessary to cope with the rapid advance of nanotechnology, proposing that this can be achieved through co-regulation that combines methods such as voluntary codes of conduct, risk management frameworks, industry codes, and certification schemes.

By employing multiple mechanisms operating at different levels, and by drawing upon the strengths of many contributors, international approaches will help provide the best foundation for ensuring that nanotechnologies are developed in a safe and responsible manner. However, given their belief that the window of opportunity for negotiating a legally binding convention or regulatory instrument at the international level has already disappeared, Bowman and Hodge proposed a two-stage process to developing effective co-regulation. In the short term, regulation may take an informal approach in the style of traditional dialogue between expert groups across a range of fields, and should seek to develop private codes of conduct. In the medium term, framework conventions could be crafted to deal with specific issues as the technology evolves. However, a further issue for consideration remains as to whether multiparty and multijurisdictional channels can most sensibly identify and manage longer-term potential risks.

## 13.3 Alternative Paths for Technological Innovation

We now step back from our survey of past chapters to place nano-innovation in a broader context of the evolving relationships between science and society. By using nanotechnology as a lens for investigating the relationships between scientific innovation and sustainability, in this book we have witnessed a broader potential for alternatives beyond the twin hype and over-zealous cautioning that can so often accompany emerging technology. This is a perspective that moves beyond solely socially or technologically determinist views, as it takes a more holistic approach. While recognizing the significant debates that each view offers, we seek to rise above polarity by pragmatically exploring substantial, integrated alternatives—such as notions like innovation without growth. This approach also seeks opportunities to democratize *through* science by considering the role of science in light of the heightened individualism—but also new forms of collectivity that are coming to define this current era. This is a phase of systemic change where highly individualized pursuits, mediated by technological development, are making it possible for forms of collectivity that transcend traditional temporal and spatial constraints (Beck 1999). In terms of technological change, this shift has more commonly been discussed in relation to media consumption and production (such as citizen journalism); however, we also see its influence in the R&D, production, and consumption of scientific innovation more broadly. As Denisa Kera (2011) states, "[the] future belongs to innovation that simultaneously and directly connects politics with design, community building with prototype testing, and offers an experimental setting for following the impact of emergent technologies on society. This trend is embodied by the global rise of alternative R&D places existing outside of the government funded universities or even corporate R&D labs. In them innovation is becoming an active expression of citizenship as much as it is a human pursuit to understand nature and create resilient and efficient tools" (49). Here, nanotechnology can be imagined as a protagonist rather than antagonist in the evolution of equitable futures beyond growth. And, as Mushtaq and Pearce argued in Chapter 9, if nanotechnology can address issues of equity, then its scientific potential may also be seen more favorably by its present critics.

## 13.4 New Associations and Further Research Avenues

In the meantime, the historically polarizing debates about nanotechnology have left many gaps within and between significant areas of discussion—and these are the gaps our authors have endeavored to address in this book. As has been demonstrated by our contributors, interdisciplinarity and alternative forms of scientific knowledge and production provide new vistas for grappling with sustainability more deeply *and* globally. This approach illustrates the immense value of seeking ideas and inspiration from the fringes and where boundaries cross. Therefore, upon reviewing our chapters, a number of new associations have arisen among the ideas our authors have raised. We now pose a number of subsequent questions we feel must be addressed, should nanotechnology seek to play a role in driving alternative paths for technological innovation. Understandably, the lines of enquiry arising from these new associations relate to the *process* of nano-innovation and lie largely at junctures where formal and informal (centralized and decentralized) systems coexist. Although we hope they may hold value in terms of driving diverse future research agendas, at this point we see their essence forming around the following question: How can a shift to decentralized nano-innovation evolve collaboratively from existing systems? As Howard and Wetter note in Chapter 4, the issue of governance is central to whether "a nano-economy will usher in the changes needed to restore a world of vastly compromised ecological health."

Several branches of questioning stem from this subject of nanotechnology and governance. For example: What are the implications of open source/ DIY nano-innovation for anticipatory governance? Could constructive technology assessment become the norm? In this sense, perhaps the methods Wickson raised (Chapter 10) could occur outside the lab (thereby using the multicriteria mapping for the kinds of processes outlined in Grimshaw's nanotechnology dialogues). Where, then, are the realistic limits to citizen input for nano-innovation? And if risk analysis were much more participatory, what might be the impacts on distributed innovation of a partial nanotechnology moratorium?

Perhaps, in following Grimshaw, Mushtaq, and Pearce's lead (Chapters 8 and 9), we can consider that pro-poor innovation might occur without public–private partnerships, through new funding and business models opened up by social enterprise (such as crowd-funded startup capital). Therefore, in terms of different business models, to what extent will open source nanotechnology licenses be legally enforceable across various jurisdictions? Where will be the tipping point for open versus closed licensing, and does creative commons licensing fundamentally change debates about ownership of discoveries relating to the biological and nonbiological fundamentals of nature? The issue of regulation also requires attention in this context, particularly in light of Chowdhury and Srivastava (Chapter 11) mentioning the parallel regulation of mainstream and indigenous innovation systems in India. In this case, what role might competition play in a future of nano-innovation without growth, and over what phases of innovation might the collaboration to which Grimshaw refers (Chapter 8) span? Is co-regulation a barrier or stepping-stone to alternative futures for innovation? Important too are the roles of advocacy groups in the picture, as outlined by Hess and Lamprou (Chapter 2). Therefore, overall, how do we accelerate a politicized nanotechnology at the margins without mainstreaming?

## 13.5 New Challenges

Clearly we have presented a substantial list of questions, none of which can be addressed briefly or with a simple response. A future of innovation without growth cannot be worked toward without surmounting significant challenges. We must consider, for example, the extent to which we *do* focus on efficiency gains and relative decoupling.[8] Approaches cognizant of limits to resource consumption should hardly deny the benefit of using resources more effectively, nor, as Wilkinson and Pickett (2009) note, should we expect a reduced pace of technological innovation. To the contrary; we need to embrace developments in spaces like industrial ecology, where significant rethinking has emerged around issues such as the value of waste (see McDonough and Braungart 2002). However, as raised in our introduction, we must remain aware of the rebound effect and the need for global economic contraction and convergence at steady-state levels. Efforts to improve technological efficiency can no longer be used as a means to displace labor if the primary goal is to produce economic gain, nor employed as a veneer to inhibit action on reducing overall resource use and a shift to greater frugality.

When it comes to matters of capacity, we also need to ask how long it is fair and sustainable for countries in the South to innovate *with* growth. In short, we believe it to be as long as the country's footprint remains commensurate with the limits proscribed by an equitable, global steady-state economy. However, as Rip and Laredo (2008) have argued, there are ample opportunities and reasons as to why the South need not imitate the system of countries in the North but can grow their own knowledge, research, and innovation systems. The evidence in this book provides further grounds to suggest that this *can* be the case.

## 13.6 The Bigger Picture: Nano-Innovation within a New Political Economy

Just as capitalism found full expression long after its earliest indicators emerged (Heilbroner 1976), so too may these developments (many outlined in this book) be driving post-growth futures that merely await articulation as a new form of political economy for a sustainable twenty-first century. Whatever the narrative, history suggests that the shifts, while possibly disruptive, will be evolutionary. That is, we will have to move from *within* capitalism so as to move *beyond* capitalism.[9] Indeed, the concept of innovation without growth is unlikely to be politically palatable in many nations, especially in those that have been long entrenched in growth-driven systems and been able to use these systems to assert their power and influence. However, despite this likely resistance—and as Eglash has reminded us in Chapter 3, as a human species—we have innovated without growth in the past. Relatively speaking, even up until recently, a great deal of scientific innovation occurred without a profit motive. As Daly and Farley (2010) note, "the heliocentric view of the universe, gravity, the periodic table of elements electromagnetic theory, and the laws of optics, mechanics, thermodynamics, and heredity were all discovered without the benefit of intellectual property rights and the profit motive" (375). The same can be said for more recent breakthroughs, such as Watson and Crick's discovery of DNA (Daly 2007).

In this light, and given the inextricable link between innovation and economy, we acknowledge the much greater challenge of exploring how innovation without growth might fit within a post-growth political economy. There is certainly a need for greater interrogation of how post-growth innovation could operate with respect to accumulation, labor, savings, pricing, taxation, innovation rents, geopolitics, and inflation. However, we must also recognize the new and significant trajectories in political economy that integrate recent forms of technological innovation, such as collaborative consumption, local exchange trading systems, peer-to-peer lending, ethical investment, fair trade, and various not-for-profit business models.

Far from providing a blueprint for innovation without growth, the work in this book has provided some insights into the kinds of factors that must align in order to usher in an alternative sustainability approach to science. Fulfillment of these factors will enable us to address human needs while also cultivating expressions of ingenuity. Just as a nongrowing society "does not mean a static society or a stagnating society" (Lowe 2009, 97), nongrowing innovation can still support a vibrant exploration of science and the profound truths that sustain our deepest enquiries. As Taylor (1976) daringly questions: Has innovation through growth actually prevented us from tackling the real changes required to develop as a species and planet? If so, and as confrontational as this might be, we believe that by removing the emphasis on growth, clearer paths can emerge—paths where technologies such as nanotechnology provide a means by which to reorient societies toward equitable and sustainable futures.

---

# References

Anctil, A., Babbitt, C.W., Raffaelle, R.P., and Landi, B.J. 2011. Material and energy intensity of fullerene production. *Environmental Science and Technology* 45(6): 2353–59.

Barrañón, A. 2010. Women in Mexican nanotechnology. In *Recent Advances in Applied Mathematics—Proceedings of the American Conference on Applied Mathematics (American-Math '10)*, edited by S. Lagakos, L. Perlovsky, M. Jha, B. Covaci, A. Zaharim, and N. Mastorakis. Stevens Point, Wis.: World Scientific and Engineering Academy and Society (WSEAS).

Bauwens, M. 2011. Why is open hardware inherently sustainable?' *P2P Foundation.* Available at: http://blog.p2pfoundation.net/why-is-open-hardware-inherently-sustainable/2011/05/24 (accessed 25 May 2011).

Beck, U. 1999. *World Risk Society.* Cambridge: Polity Press.

Cozzens, S. and Wetmore, J. 2010. Introduction. In *Nanotechnology and the Challenges of Equity, Equality and Development*, edited by S. Cozzens and J. Wetmore, ix–xx. New York: Springer.

Daly, H. 2007. *Ecological Economics and Sustainable Development, Selected Essays of Herman Daly*. Cheltenham: Edward Elgar.

Daly, H. and Farley, J. 2010. *Ecological Economics: Principles and Applications*. 2nd ed. Washington, D.C.: Island Press.

Heilbroner, R.L. 1976. *The Worldly Philosophers*. 6th ed. London: Penguin Books.

Hongladarom, S. 2009. Nanotechnology, development and Buddhist values. *Nanoethics* 3(1): 97–107.

Hope, J. 2008. *Biobazaar: The Open Source Revolution and Biotechnology*. Cambridge, Mass.: Harvard University Press.

Jackson, T. 2010. *Prosperity without Growth: Economics for a Finite Planet*. London: Earthscan.

Jakubowski, M. 2011. Open source micro-factory. *Factor E Farm Blog*. Available at: http://blog.opensourceecology.org/2011/03/open-source-micro-factory/ (accessed 21 March 2011).

Kera, D. 2011. Grassroots R&D, prototype cultures and DIY innovation: Global flows of data, kits and protocols. In *Pervasive Adaptation: The Next Generation Pervasive Computing Research Agenda*, edited by A. Ferscha and Th. Sc. Community. Linz: Institute for Pervasive Computing, Johannes Kepler University.

Kugler, H. 2011. Sun-harvesting textiles power remote villages. *SciDev.Net*. Available at: http://www.scidev.net/en/climate-change-and-energy/sun-harvesting-textiles-power-remote-villages-1.html (accessed 26 April 2011).

Latouche, S. 2010. *Farewell to Growth*. Cambridge: Polity Press.

Lipson, H. and Kurman, M. 2010. *Factory @ Home: The Emerging Economy of Personal Fabrication*. Occasional Papers in Science and Technology Policy, a report commissioned by the U.S. Office of Science and Technology Policy. Available at: www.mae.cornell.edu/lipson/factoryathome.pdf (accessed 31 July 2011).

Lowe, I. 2009. *A Big Fix*. 2nd ed. Melbourne: Black Inc.

Maclurcan, D. 2011. *Nanotechnology and Global Equality*. Singapore: Pan Stanford Publishing (World Scientific).

McDonough, W. and Braungart, M. 2002. *Cradle to Cradle: Remaking the Way We Make Things*. New York: North Point Press.

Meng, Y. and Shapira, P. 2010. Women and patenting in nanotechnology: Scale, scope and equity. In *Nanotechnology and the Challenges of Equity, Equality and Development*, edited by S. Cozzens and J. Wetmore, 23–46. New York: Springer.

Michelson, E.S. 2010. Nanotech ethics and the policymaking process: Lessons learned for advancing equity and equality in emerging nanotechnologies. In *Nanotechnology and the Challenges of Equity, Equality and Development*, edited by S. Cozzens and J. Wetmore, 423–32. New York: Springer.

Nieusma, D. 2010. Materializing nano equity: Lessons from design. In *Nanotechnology and the Challenges of Equity, Equality and Development*, edited by S. Cozzens and J. Wetmore, 209–30. New York: Springer.

Nye, D.E. 2006. *Technology Matters: Questions to Live With*. Cambridge, Mass.: MIT Press.

Pieterse, J.N. 1998. My paradigm or yours? Alternative development, post-development, reflexive development. *Development and Change* 29(2): 343–73.

Rip, A. and Laredo, P. 2008. *Knowledge, Research and Innovation Systems and Developing Countries*. Paper presented at the sixth Globelics Conference, September 22–24 2008, Mexico City, Available at: http://smartech.gatech.edu/bitstream/handle/1853/38861/Arie_Rip_Knowledge.pdf?sequence=1 (accessed 10 May 2011).

Schumpeter, J.A. 1934. *Business Cycles: A Theoretical, Historical, and Statistical Analysis of the Capitalist Process*. New York: McGraw-Hill.

Slade, C.P. 2010. Exploring societal impact of nanomedicine using public value mapping. In *Nanotechnology and the Challenges of Equity, Equality and Development*, edited by S. Cozzens and J. Wetmore, 69–88. New York: Springer.

Taylor, W. 1976. Innovation without growth. *Educational Management Administration Leadership* 4(2): 1–13.

Thakur, D. 2010. Open access nanotechnology for developing countries: Lessons from open source software. In *Nanotechnology and the Challenges of Equity, Equality and Development*, edited by S. Cozzens and J. Wetmore, 331–48. New York: Springer.

The Economist. 2009. How to live in a bubble. *Free Exchange*. Available at: http://www.economist.com/blogs/freeexchange/2009/11/how_to_live_with_bubbles (accessed 12 January 2011).

Wilkinson, R. and Pickett, K. 2009. *The Spirit Level: Why More Equal Societies Almost Always Do Better*. London: Penguin Books.

Wolbring, G. 2007. *Scoping Document on Nanotechnology and Disabled People for the Center for Nanotechnology in Society at Arizona State University*. Available at: http://cns.asu.edu/cns-library/documents/wolbring-scopingCDfinaledit.pdf (accessed 25 May 2011).

WPCCC. 2010. *Rights of Mother Earth: Proposal Universal Declaration of the Rights of Mother Earth*. Available at: http://pwccc.wordpress.com/2010/04/24/peoples-agreement/ (accessed 26 April 2010).

Youtie, J. and Shapira, P. 2010. Metropolitan development of nanotechnology: Concentration or dispersion. In *Nanotechnology and the Challenges of Equity, Equality and Development*, edited by S. Cozzens and J. Wetmore, 165–80. New York: Springer.

## Endnotes

1. Here it may be particularly worthwhile to consider Wolbring's (2007) writing on using the *Bias Free* framework—a practical tool for identifying and eliminating social biases in health research.

2. For more on these important debates as they relate to nanotechnology, see Wolbring (2007); Hongladarom (2009); Barrañón (2010); Meng and Shapira (2011); Nieusma (2010); and WPCCC (2010).

3. While Michelson (2010) notes the lack of prototyping beyond water purification devices in one recent example, pilot projects are underway in Brazil, Kenya, Haiti, Nicaragua, Madagascar, Mexico, and South Africa to trial cost-efficient, portable, nano solar-powered lights that can be inserted into shirts and other woven items using indigenous textile production (see Kugler 2011).

4. According to Bauwens (2011), open source hardware can bypass the vested interest of for-profit companies in creating nonsustainable products.

5. Kera (2011) continues, "Maker and hacker communities around the world prototype future gadgets and tools with open hardware platforms and feed the needs of various grassroots open labs for affordable equipment that offer opportunities for entrepreneurship" (49).

6. As Marcin Jakubowski (2011) explains, "the Open Source Micro-Factory is a robust, closed-loop manufacturing system for many kinds of mechanical and electronic devices. It includes the ability to provide its own fuel, electricity, and mechanical power. The designs are scalable in output. Break-through economics are included—such as building a $50k-value tractor at about $3k in parts, or 50 hp hydraulic motors at about $50 in parts via open source induction furnace, casting, and precision machining." With his team at Open Source Ecology, Jakubowski plans to "open-source ~12 of the most important, high-performance machines of industrial production and automation, provide plans for all these machines, and provide plans for certain key products that can be built with these machines starting from scrap metal as a feedstock" (2011).

7. This is a view reinforced by Janet Hope (2008) in her book *Biobazaar: The Open Source Revolution and Biotechnology*.

8. "Relative decoupling" is the ability to use fewer materials to produce more, in contrast to "absolute decoupling" in which resource use or environmental impact falls, absolutely (see Jackson 2010). To absolutely decouple would require accelerating returns from technological innovation to ensure the world's impact, per dollar of output, reduces relative to each doubling of economic output.

9. As Kera (2011) notes, many of the low-tech approaches to alternative and community R&D she sees springing up are "paradoxically inspired by both EU alternative squat cultures and the American spirit of entrepreneurship" (49).

# Index